AF411468

Algorithms and Methods for Designing and Scheduling Smart Manufacturing Systems

Algorithms and Methods for Designing and Scheduling Smart Manufacturing Systems

Editors

Vladimir Modrak
Zuzana Soltysova

MDPI • Basel • Beijing • Wuhan • Barcelona • Belgrade • Manchester • Tokyo • Cluj • Tianjin

Editors
Vladimir Modrak
Technical University of Košice
Slovakia

Zuzana Soltysova
Technical University of Košice
Slovakia

Editorial Office
MDPI
St. Alban-Anlage 66
4052 Basel, Switzerland

This is a reprint of articles from the Special Issue published online in the open access journal *Applied Sciences* (ISSN 2076-3417) (available at: https://www.mdpi.com/journal/applsci/special_issues/Algorithms_Methods_Smart_Manufacturing_Systems).

For citation purposes, cite each article independently as indicated on the article page online and as indicated below:

LastName, A.A.; LastName, B.B.; LastName, C.C. Article Title. *Journal Name* **Year**, *Volume Number*, Page Range.

ISBN 978-3-0365-4509-7 (Hbk)
ISBN 978-3-0365-4510-3 (PDF)

Contents

About the Editors

Vladimir Modrak

Prof. Dr. Vladimir Modrak is Full Professor of Manufacturing Technology at Faculty of Manufacturing Technologies. His research interests include manufacturing logistics, cellular manufacturing, lean manufacturing, mass customization and other related disciplines. He was the leading editor and co-author of the books "Operations Management Research and Cellular Manufacturing Systems: Innovative Methods and Approaches" (2012), "Handbook of Research on Design and Management of Lean Production Systems" (2014), "Mass Customized Manufacturing: Theoretical Concepts and Practical Approaches" (2016), "Industry 4.0 for SMEs Challenges, Opportunities and Requirements" (2020), and "Implementing Industry 4.0 in SMEs: Concepts, Examples and Applications" (2021). He is Fellow of the European Academy of Industrial Management (AIM).

Zuzana Soltysova

Dr. Zuzana Soltysova completed her bachelor, master and PhD. study at Technical University of Kosice, Faculty of Manufacturing Technologies. Currently, she is professor assistant at Technical University of Kosice, Faculty of Manufacturing Technologies, Department of Industrial Engineering and Informatics. Her doctoral thesis was titled "Research of product and production complexity in terms of mass customization". Moreover, her research activities include complexity problems, axiomatic design, product and process modularity and related disciplines to Manufacturing management.

applied sciences

Editorial

Algorithms and Methods for Designing and Scheduling Smart Manufacturing Systems

Vladimir Modrak * and Zuzana Soltysova *

Faculty of Manufacturing Technologies, Technical University of Kosice, Bayerova 1, 080 01 Presov, Slovakia
* Correspondence: vladimir.modrak@tuke.sk (V.M.); zuzana.soltysova@tuke.sk (Z.S.)

Citation: Modrak, V.; Soltysova, Z. Algorithms and Methods for Designing and Scheduling Smart Manufacturing Systems. *Appl. Sci.* **2022**, *12*, 3011. https://doi.org/ 10.3390/app12063011

Received: 10 March 2022
Accepted: 14 March 2022
Published: 16 March 2022

Publisher's Note: MDPI stays neutral with regard to jurisdictional claims in published maps and institutional affiliations.

1. Introduction

This Special Issue is a collection of some of the latest advancements in designing and scheduling smart manufacturing systems. The smart manufacturing concept is undoubtedly considered a paradigm shift in manufacturing technology. This conception is part of the Industry 4.0 strategy, or equivalent national policies, and brings new challenges and opportunities for the companies that are facing tough global competition. Industry 4.0 should not only be perceived as one of many possible strategies for manufacturing companies, but also as an important practice within organizations. The main focus of Industry 4.0 implementation is to combine production, information technology, and the internet [1]. Therefore, an introduction of smart manufacturing systems is primarily associated with the adaptation of the Internet of Things, cyber physical systems, artificial intelligence, advanced robotics, cloud technology, and so forth. In particular, web technologies act as enablers of smart manufacturing that can promote a disruptive innovation in small and medium enterprises. However, some recent studies (see, e.g., [2–4]) have shown that pre-existing managerial methods and philosophies such as lean manufacturing, reconfigurable manufacturing systems, or cellular manufacturing systems are of utter importance for the concept of smart manufacturing. In this context, manufacturing system design and scheduling methods also play a vital role in the era of the fourth industrial revolution. It is particularly evident from the link between Industry 4.0 and lean manufacturing that both domains are strongly interrelated [5]. Thus, we hope that this Special Issue may be helpful in solving the particular problems which are related to smart manufacturing systems and advanced manufacturing technologies.

2. Description of the Papers

The presented Special Issue consists of ten research papers presenting the latest works in the field. The papers include various topics, which can be divided into three categories— (i) designing and scheduling manufacturing systems (seven articles), (ii) machining process optimization (two articles), (iii) digital insurance platforms (one article). Most of the mentioned research problems are solved in these articles by using genetic algorithms, the harmony search algorithm, the hybrid bat algorithm, the combined whale optimization algorithm, and other optimization and decision-making methods.

The above-mentioned groups of articles are briefly described in this order in the rest of this editorial paper.

The paper written by J.S. Park, H.Y. Ng, T.J. Chua, Y.T. Ng, and J.W. Kim [6] presents a novel genetic algorithm approach that utilizes a multiple chromosome scheme to solve the flexible job-shop scheduling problem which involves two kinds of decisions—machine selection and operation sequencing. This novel genetic algorithm approach enables the application of an identical crossover strategy in the categorical and sequential parts. The authors used this unified approach for the extension of the existing candidate order-based genetic algorithm to the unified candidate order-based genetic algorithm to solve flexible job-shop scheduling problems.

The other paper titled "Calibration of GA parameters for layout design optimization problems using design of experiments" [7] focuses on finding out how the solutions of the cell formation problem are influenced by a set of probability parameters of genetic operators, namely crossover and mutation, including balanced weight factors. In view of this, the presented work attempts to employ the Taguchi approach to find an optimal combination of parameters that impact the efficiency of the genetic algorithm and to explore whether the optimal combination of the genetic operators for the given type of machine-cell formation problems can be influenced by the magnitude of the noise factors.

The next paper [8] introduces a robust cluster algorithm based on the concept of group technology. The paper is written by L.M. Li and its originality lies in its use of the algorithm to balance an assembly line by matching operators to workstations so that the line's workstations achieve the same targeted output rates. It enables managers to solve the problem of worker absences by assigning more than one operator with the required skillset to each workstation and rearranging them as needed. This algorithm has been applied to cellular manufacturing system problems. Four examples were presented to implement and validate this algorithm, where training time was reduced by matching operators' training and skills according to the workstations' skill requirements.

The paper written by L. Nagarajan, S.K. Mahalingam, S. Salunkhe, E.A. Nasr, J.P. Davim, and H. Hussein [9] proposes a novel methodology for simultaneous minimization of manufacturing objectives in tolerance allocation of complex assembly tasks. The methodology consists of a two-step process. For this purpose, a heuristic approach was applied to determine the best machine for each process. Subsequently, it applies a combined whale optimization algorithm with a univariate search method to allocate optimum tolerances with the best process selection for each sub-stage/operation. The proposed methodology was validated by solving two typical tolerance allocation problems of complex assemblies: a wheel mounting assembly and a knuckle joint assembly. Authors also compared their approach with existing ones. The comparison results showed that the proposed approach considerably reduces tolerance cost and machining time.

The next paper which is authored by S.K Mahalingam, L. Nagarajan, S. Salunkhe, E.A. Nasr, J.P. Davim, and H. Hussein aims to acquire the maximum number of non-linear assemblies with closer assembly tolerance specifications by mating the different bins' components [10]. Their proposed approach is based on using the combination of the univariate search method and the harmony search algorithm. The efficacy of the proposed method is demonstrated by showing 24.9% of cost savings while making overrunning clutch assembly in comparison with the existing method. Based on the obtained results in this work, the contribution of the proposed novel methodology is legitimate in solving selective assembly problems.

The originality of the following paper written by Zheng J. and Wang Y. [11] lies in an application of the hybrid bat optimization algorithm, which is based on the variable neighborhood structure, and two learning strategies to solve a three-stage distributed assembly permutation flow shop scheduling problem (DAPFSP) with the aim to minimize the makespan. The proposed algorithm is firstly designed to increase the population diversity by classifying the populations which solves the difficult trade-off between convergence and diversity of the bat algorithm. Secondly, a selection mechanism is used to update the bat's velocity and location, solving the difficulty of the algorithm trade-off of exploration and mining capacity. For this purpose, the Gaussian learning strategy and elite learning strategy assist the whole population in jumping out of the local optimal frontier. Based on the obtained simulation results, this algorithm can solve the DAPFSP problem well. In comparison with other metaheuristic algorithms, it provides better performance than the compared ones; thus, it is suitable to find the optimal solution(s).

The brief overview of the further paper titled "A multi-criteria assessment of manufacturing cell performance using the AHP method" [12] is as follows. It introduces the solution for how to find the optimal manufacturing cell design from alternative designs by using a multi-criteria evaluation. Alternative design solutions are mutually compared by using the selected performance criteria, namely operational complexity, production line balancing rate, and makespan. Then, multi-criteria decision analysis based on the analytic hierarchy process method is used to show that two more-cell solutions better satisfy the determined criteria of manufacturing cell design performance than three less-cell solutions. The main benefit of this approach lies in the exactly enumerated values of the selected criteria, according to which the points from the mentioned scale are assigned to the alternatives.

The first paper of the second group of articles is titled "Meta-heuristic technique-based parametric optimization for electrochemical machining of Monel 400 alloys to investigate the material removal rate and the sludge" [13]. The authors investigate in this work the predominant electrochemical machining process parameters, namely applied voltage, flow rate, and electrolyte concentration, and their effects on the performance measures, i.e., the material removal rate and the nickel presence. For this purpose, authors used a meta-heuristic algorithm named as the grey wolf optimizer for the multi-objective optimization of the process parameters for electrochemical machining. The obtained results were compared with the moth–flame optimization and particle swarm optimization algorithms. Based on the obtained results, it was observed that all the process variables significantly influenced the objectives. Then, it is confirmed that these metaheuristic algorithms—the moth–flame optimization algorithm and the grey wolf optimizer—are suitable for finding the optimum process parameters for machining Monel 400 alloys with electrochemical machining.

The second paper of this group is named "Optimization of process parameters for turning Hastelloy X under different machining environments using evolutionary algorithms: A comparative study" [14]. In their research, the authors investigated the machinability of turning Hastelloy X with a PVD Ti-Al-N coated insert tool in dry, wet, and cryogenic machining environments. The machinability indices, namely cutting force, surface roughness, and cutting temperature, are studied for the different set of input process parameters such as cutting speed, feed rate, and machining environment. They used the experiments conducted as per L27 orthogonal array. The authors proposed the moth–flame optimization to identify the optimal set of turning parameters through the multiple linear regression models, in view of minimizing the machinability indices. The effectiveness of the proposed algorithm is evaluated in comparison to the findings of genetic, Grass–Hooper, grey wolf, and particle swarm optimization algorithms. Based on the obtained results, this algorithm outperformed the others.

The last article in the given order is titled "Reflections on the customer decision-making process in the digital insurance platforms: An empirical study of the Baltic market" [15]. It aims to expand upon the existing scientific knowledge of end-user behavioral patterns and process frameworks in the Baltic insurance market, by including and examining a factor group of technological enablers. This paper is focused on research results in digitalization, personalization, and customization levels within the Baltic non-life insurance market in Estonia. There are also three major factor groups and process stages identified which influence insurance purchase decision-making in digital insurance platforms in the Baltic market.

3. Conclusions

It is believed that the collection of the ten papers in this Special Issue will be beneficial to readers who are interested in applying modern algorithms and methods for designing and scheduling smart manufacturing systems and related problems. Although this Special Issue has been closed, the current market challenges justify the need for an additional in-depth research in this domain.

Author Contributions: Conceptualization, V.M. and Z.S.; formal analysis, V.M. and Z.S.; writing—original draft preparation, V.M. and Z.S.; writing—review and editing, V.M. and Z.S. All authors have read and agreed to the published version of the manuscript.

Funding: This research was funded by the KEGA Project No. 025TUKE-4/2020 granted by the Ministry of Education of the Slovak Republic.

Informed Consent Statement: Not applicable.

Acknowledgments: Moreover, we would like to thank all the authors who participated in this Special Issue and all the reviewers, and give many thanks to the editorial team of the Applied Sciences journal.

Conflicts of Interest: The authors declare no conflict of interest.

References

1. Matt, D.T.; Modrak, V.; Zsifkovits, H. *Industry 4.0 for SMEs: Challenges, Opportunities and Requirements*; Springer Nature: Singapore, 2020; p. 412.
2. Yelles-Chaouche, A.R.; Gurevsky, E.; Brahimi, N.; Dolgui, A. Reconfigurable manufacturing systems from an optimisation perspective: A focused review of literature. *Int. J. Prod. Res.* **2021**, *59*, 6400–6418. [CrossRef]
3. Rauch, E.; Linder, C.; Dallasega, P. Anthropocentric perspective of production before and within Industry 4.0. *Comput. Ind. Eng.* **2020**, *139*, 105644. [CrossRef]
4. Modrak, V.; Soltysova, Z.; Poklemba, R. Mapping requirements and roadmap definition for introducing I 4.0 in SME environment. In *Advances in Manufacturing Engineering and Materials*; Springer: Cham, Switzerland, 2019; pp. 183–194.
5. Sanders, A.; Elangeswaran, C.; Wulfsberg, J.P. Industry 4.0 implies lean manufacturing: Research activities in industry 4.0 function as enablers for lean manufacturing. *J. Ind. Eng. Manag. (JIEM)* **2016**, *9*, 811–833. [CrossRef]
6. Park, J.S.; Ng, H.Y.; Chua, T.J.; Ng, Y.T.; Kim, J.W. Unified genetic algorithm approach for solving flexible job-shop scheduling problem. *Appl. Sci.* **2021**, *11*, 6454. [CrossRef]
7. Modrak, V.; Pandian, R.S.; Semanco, P. Calibration of GA parameters for layout design optimization problems using design of experiments. *Appl. Sci.* **2021**, *11*, 6940. [CrossRef]
8. Li, M.L. An Algorithm for Arranging Operators to Balance Assembly Lines and Reduce Operator Training Time. *Appl. Sci.* **2021**, *11*, 8544. [CrossRef]
9. Nagarajan, L.; Mahalingam, S.K.; Salunkhe, S.; Nasr, E.A.; Davim, J.P.; Hussein, H. A Novel Methodology for Simultaneous Minimization of Manufacturing Objectives in Tolerance Allocation of Complex Assembly. *Appl. Sci.* **2021**, *11*, 9164. [CrossRef]
10. Mahalingam, S.K.; Nagarajan, L.; Salunkhe, S.; Nasr, E.A.; Davim, J.P.; Hussein, H. Harmony Search Algorithm for Minimizing Assembly Variation in Non-linear Assembly. *Appl. Sci.* **2021**, *11*, 9213. [CrossRef]
11. Zheng, J.; Wang, Y. A Hybrid Bat Algorithm for Solving the Three-Stage Distributed Assembly Permutation Flowshop Scheduling Problem. *Appl. Sci.* **2021**, *11*, 10102. [CrossRef]
12. Soltysova, Z.; Modrak, V.; Nazarejova, J. A Multi-Criteria Assessment of Manufacturing Cell Performance Using the AHP Method. *Appl. Sci.* **2022**, *12*, 854. [CrossRef]
13. Nagarajan, N.; Solaiyappan, A.; Mahalingam, S.K.; Nagarajan, L.; Salunkhe, S.; Nasr, E.A.; Shanmugam, R.; Hussein, H.M.A.M. Meta-Heuristic Technique-Based Parametric Optimization for Electrochemical Machining of Monel 400 Alloys to Investigate the Material Removal Rate and the Sludge. *Appl. Sci.* **2022**, *12*, 2793. [CrossRef]
14. Sivalingam, V.; Sun, J.; Mahalingam, S.K.; Nagarajan, L.; Natarajan, Y.; Salunkhe, S.; Nasr, E.A.; Davim, J.P.; Hussein, H.M.A.M. Optimization of Process Parameters for Turning Hastelloy X under Different Machining Environments Using Evolutionary Algorithms: A Comparative Study. *Appl. Sci.* **2021**, *11*, 9725. [CrossRef]
15. Baranauskas, G.; Raišienė, A.G. Reflections on the Customer Decision-Making Process in the Digital Insurance Platforms: An Empirical Study of the Baltic Market. *Appl. Sci.* **2021**, *11*, 8524. [CrossRef]

Article

Unified Genetic Algorithm Approach for Solving Flexible Job-Shop Scheduling Problem

Jin-Sung Park [1], Huey-Yuen Ng [2], Tay-Jin Chua [2], Yen-Ting Ng [2] and Jun-Woo Kim [1,*]

[1] Department of Industrial and Management Systems Engineering, Dong-A University, Busan 49315, Korea; pjs0958@donga.ac.kr
[2] Singapore Institute of Manufacturing Technology, Singapore 138634, Singapore; nghy@simtech.a-star.edu.sg (H.-Y.N.); tjchua@simtech.a-star.edu.sg (T.-J.C.); ytng@simtech.a-star.edu.sg (Y.-T.N.)
* Correspondence: kjunwoo@dau.ac.kr; Tel.: +82-51-200-7687; Fax: +82-51-200-7697

Abstract: This paper proposes a novel genetic algorithm (GA) approach that utilizes a multichromosome to solve the flexible job-shop scheduling problem (FJSP), which involves two kinds of decisions: machine selection and operation sequencing. Typically, the former is represented by a string of categorical values, whereas the latter forms a sequence of operations. Consequently, the chromosome of conventional GAs for solving FJSP consists of a categorical part and a sequential part. Since these two parts are different from each other, different kinds of genetic operators are required to solve the FJSP using conventional GAs. In contrast, this paper proposes a unified GA approach that enables the application of an identical crossover strategy in both the categorical and sequential parts. In order to implement the unified approach, the sequential part is evolved by applying a candidate order-based GA (COGA), which can use traditional crossover strategies such as one-point or two-point crossovers. Such crossover strategies can also be used to evolve the categorical part. Thus, we can handle the categorical and sequential parts in an identical manner if identical crossover points are used for both. In this study, the unified approach was used to extend the existing COGA to a unified COGA (u-COGA), which can be used to solve FJSPs. Numerical experiments reveal that the u-COGA is useful for solving FJSPs with complex structures.

Keywords: flexible job-shop scheduling problem; combinatorial optimization; genetic algorithm; candidate order-based genetic algorithm; multichromosome

Citation: Park, J.-S.; Ng, H.-Y.; Chua, T.-J.; Ng, Y.-T.; Kim, J.-W. Unified Genetic Algorithm Approach for Solving Flexible Job-Shop Scheduling Problem. *Appl. Sci.* **2021**, *11*, 6454. https://doi.org/10.3390/app11146454

Academic Editors: Paolo Renna, Vladimir Modrak and Zuzana Soltysova

Received: 31 May 2021
Accepted: 9 July 2021
Published: 13 July 2021

Publisher's Note: MDPI stays neutral with regard to jurisdictional claims in published maps and institutional affiliations.

1. Introduction

Production scheduling is one of the most important decision-making procedures on manufacturing shopfloors, as it helps to utilize resources efficiently and maintain competitiveness in manufacturing companies [1–3]. Most production scheduling problems are NP-hard combinatorial optimization problems such that the optimal schedules are hard to find in polynomial time [4]. In this context, metaheuristic algorithms that can be used to find near optimal schedules in practical time are popular in production scheduling literature [5]. Examples of metaheuristic algorithms for solving production scheduling problems are the genetic algorithm (GA) [6], tabu search (TS) [7], simulated annealing (SA) [8], and particle swarm optimization (PSO) [9].

This paper aims to develop a novel GA for solving the flexible job-shop scheduling problem (FJSP). The classical job-shop scheduling problem (JSP) is one of the most well-known production scheduling problems. It consists of m machines and n jobs. A job contains a number of operations to be processed in a fixed order. In the JSP, each operation can be processed by one specific machine [10,11]. The FJSP is an extension of the classical JSP, which allows an operation to be processed by one of two or more machines. In other words, an operation is processed by one of alternative machines in the FJSP [12]. Since a machine for a specific operation is predetermined, the JSP can be solved by specifying the priorities of given operations such that a high-priority operation precedes others with lower priorities in the queue for given machines. Thus, the JSP can be regarded as a sequencing problem [13,14]. In comparison with the JSP, the FJSP requires an additional decision related to which machine is used to process a specific operation. Consequently, two kinds of decisions, i.e., operation sequencing and machine selection, are required to solve the FJSP [1].

GA has been a popular metaheuristic algorithm for solving the JSP and FJSP in recent decades [10–12,15]. In order to apply GA, a solution of the given problem must be encoded in the form of a string, known as a chromosome [6]. One issue is that a chromosome for the FJSP must consist of two subchromosomes to account for the operation sequence (OS) part and the machine selection (MS) part. In other words, existing GAs for the FJSP are based on a multichromosome. Typically, the MS part is represented by a string of categorical values, whereas the OS part forms a sequence of given operations. Since the MS and OS parts have quite different structures, it is difficult to apply an identical genetic operator to both of them. Consequently, the MS and OS parts are separately evolved by different kinds of genetic operators in existing GAs for the FJSP. This traditional approach has the following limitations: First, it makes the GA hard to implement since different kinds of genetic operators should be applied. In particular, crossover operators for sequencing problems are a lot more complicated than simple one-point or two-point crossovers [16]. Second, there is no correlation between the order and machine in a single operation. In other words, the evolution of the MS part occurs independently from that of the OS part. Assume that an operation has the optimal order and is assigned to the optimal machine in a chromosome. Then, it is desirable that this combination of order and machine should also be maintained in the chromosome's offspring, which is hard to achieve using the traditional approach. Consequently, the traditional approach can cause a loss of good combinations in chromosomes and unnecessary diversity in the population. To this end, this paper poses two research questions: how can we reduce the number of genetic operators in the GA for the FJSP, which will have a significant impact on the complexity of the GA? Can the GA with a reduced number of genetic operators, designed to maintain good combinations of OS and MS genes, deal with the FJSP effectively?

In order to overcome the limitations of the traditional approach, this paper proposes a novel unified GA approach for the FJSP, which enables the application of an identical crossover operator to both the OS and MS parts. The main idea of the proposed approach is to apply a candidate order-based GA (COGA) based on an identical crossover point to both the OS and MS parts. The COGA is a type of GA developed for solving sequencing problems. The most distinguished feature of the COGA is that simple point-based crossover strategies, including one-point and two-point and uniform crossovers, can be implemented [14]. This enables the application of an identical point-based crossover strategy with an identical crossover point to the OS and MS parts, which may help to maintain the combinations of good order and good machine in parent solutions.

The remainder of this paper is organized as follows: In Section 2, a literature review concerning the GA for the FJSP and COGA is provided. Section 3 outlines the concept of the unified GA approach, and procedures and genetic operators of the unified COGA (u-COGA) are introduced. Section 4 presents the experimental results obtained by applying the u-COGA to various benchmark FJSPs. Finally, Section 5 concludes the paper.

2. Research Background

2.1. Genetic Algorithm for Solving the Flexible Job-Shop Problem

FJSP is a well-known extension of the JSP, a classical production scheduling problem. During the past decades, GAs have been widely used to solve production scheduling problems, such as the JSP and FJSP [15]. Typically, a GA chromosome for the JSP contains only OS-type genes, since a solution for the JSP can be generated using a sequence of given operations [10,11,13,17]. In contrast, a GA chromosome for the FJSP typically consists of two types of genes: MS genes and OS genes, which are manipulated by applying different kinds of genetic operators. Genetic operators adopted by previous research papers on the GA for the FJSP are summarized in Table 1.

Gao et al. [18] used extended order crossover (OX) and uniform crossover for the OS and MS parts, respectively. Furthermore, the authors proposed two kinds of mutation operators: random alternative for the MS part and immigration for the OS part.

In Pezzella et al. [19], the MS part is manipulated by assignment crossover and two mutation operators: random alternative mutation and greedy mutation. Assignment crossover is designed to exchange machine assignment information for certain operations. In other words, only some of the MS genes in a chromosome participate in a single assignment crossover. Random alternative mutation is used to assign an operation to a machine randomly chosen from the set of all alternative machines that process the operation. On the contrary, greedy mutation assigns an operation assigned to the machine with the maximum workload to the machine with the minimum workload. On the contrary, precedence preserving order-based crossover (POX) and precedence preserving shift (PPS) mutation were applied to the OS part. Pazzella et al. [19] used a GA with the aforementioned operators to minimize the makespan of FJSP. Moreover, Defersha and Chen [20] utilized a similar GA for the purpose of minimization of the makespan of the FJSP with sequence-dependent setup times.

Lei [21] applied the GA to solve the FJSP with fuzzy processing times. A two-point crossover operator was adopted for the MS part, whereas two kinds of sequencing crossover operators, i.e., generalized position crossover (GPX) and generalization of the precedence preservative crossover (GPPX), were used to recombine the genes in the OS part. Swap mutation was used to randomly modify the OS part; however, the GA in Lei [21] did not contain the mutation procedure for the MS part.

Wang et al. [22] applied multipoint preservative crossover (MPX) to the MS part and improved precedence operation crossover (IPOX) to the OS part. They used random alternative mutation and greedy mutation to randomly modify the MS part in the chromosome. The authors proposed a greedy mutation operator designed to assign an operation to a machine that provides the shortest processing time (SPT). For the OS part, conventional insertion mutation was adopted.

Zhang et al. [1] used classical two-point crossover and uniform crossover operators to recombine the genes in the MS part of parent solutions, while the genes in the OS part are recombined using preserving order-based crossover (POX). As in Wang et al. [22], the genes in the MS part are mutated by greedy mutation based on an SPT machine. In contrast, the mutation for the OS part is performed by applying swap mutation.

Jiang and Du [23] used two-point crossover for the MS part and POX for the OS part. The mutation operations for the MS part and OS part were random alternative and swap mutation, respectively.

Türkyılmaz and Bulkan [24] adopted three types of crossover operators for solving the FJSP. Uniform crossover was used to perform the crossover operation for the MS part, while two variations of POX, modified-1 POX (MPX1), and modified-2 POX (MPX2), were applied to the OS part. The mutation operators for the MS part and the OS part were random replace and swap mutation, respectively.

Zhang et al. [25] used two-point crossover for the MS part and POX for the OS part. In order to maintain diversity in the population, the authors applied greedy mutation based on an SPT machine to the MS part. Moreover, they introduced a local search-based mutation operator called the adaptive neighbor search method for the mutation of the genes in the OS part.

Table 1 shows that the minimization of the makespan is the most popular objective function in the FJSP literature. This paper also considers minimization of the makespan of the FJSP. In addition, all of the previous research papers utilized a multichromosome that contains the MS and OS parts, which is also the case in this paper.

Most of the relevant research papers applied different crossover operators to the OS and MS parts of the FJSP multichromosome. Various previous research papers applied a

GA with a single crossover operator to the FJSP [26–28]. However, they typically focused on one of the OS and MS parts, and details of the other part are missing. In such papers, the issue of genetic operators for the multichromosome was not discussed appropriately. In contrast, this paper applies a single crossover operator to the OS and MS parts in an evident manner.

Table 1. Existing GAs for solving the FJSP.

Researchers	Chromosome Structure	Crossover	Mutation	Objective
Gao et al. [18]	MS + OS	MS: uniform OS: extended OX	MS: random alternative OS: immigration	Min. makespan, Min. maximal machine workload, Min. total workload
Pezzella et al. [19]	MS + OS	MS: assignment OS: POX	MS: random alternative, greedy (min. workload) OS: PPS	Min. makespan
Defersha and Chen [20]	MS + OS	MS: assignment OS: POX	MS: random alternative, greedy (min. workload) OS: PPS	Min. makespan
Lei [21]	MS + OS	MS: two-point OS: GPX, GPPX	MS: N/A OS: swap	Min. Maximum fuzzy completion time
Wang et al. [22]	MS + OS	MS: MPX OS: IPOX	MS: random alternative, greedy (SPT machine) OS: insertion	Min. makespan, Min. total workload of machines, Min. critical machine workload
Zhang et al. [1]	MS + OS	MS: two-point, uniform OS: POX	MS: random alternative, greedy (SPT machine) OS: swap	Min. makespan
Jiang and Du [23]	MS + OS	MS: two-point OS: POX	MS: random alternative OS: swap	Min. makespan
Türkyılmaz and Bulkan [24]	MS + OS	MS: uniform OS: MPX1, MPX2	MS: random alternative OS: swap	Min. total tardiness
Chang et al. [29]	MS + OS	MS, OS: two-point, uniform	MS: random alternative OS: neighborhood search	Min. makespan
Chen et al. [30]	MS + OS + TC	MS, TC: two-point OS: POX	MS: random alternative TC: swap OS: shift	Min. average earliness and tardiness
Driss et al. [31]	MS + OS	MS: uniform OS: POX	MS, OS: values mutation	Min. makesapn
Ishikawa et al. [32]	MS + OS	MS: uniform OS: POX	MS: random alternative OS: inversion	Min. makespan
Wang et al. [33]	MS + OS	MS: two-point OS: POX	MS: shift OS: swap	Min. makespan
Zhang et al. [25]	MS + OS	MS: two-point OS: POX	MS: greedy (SPT machine) OS: adaptive neighborhood search	Min. makespan, Min. total setup time, Min. total transportation time
This paper	MS + OS	MS, OS: COGO (one-point, two-point, uniform)	MS: random alternative OS: COGO	Min. makespan

Certain research papers developed GA hybrid solution methods and other algorithms for solving the FJSP, which are not considered in Table 1. Similar to the GAs in Table 1, almost all of the hybrid methods also adopted different genetic operators to handle the MS and OS parts.

The most important GA feature proposed in this paper is that crossover operations for both the MS and OS parts are performed by applying an identical operator: the candidate order-based genetic operator (COGO) provided by COGA. This can help to maintain a combination of a good MS gene and a good OS gene in a single operation during the search procedure. Moreover, we can see that the mutation for the OS part is also performed using the COGO. Note that the original version of the COGO is an integrated genetic operator, which can be used for the purposes of both crossover and mutation [14]. Consequently, the u-COGA for the FJSP can be developed by using only two kinds of genetic operators: the COGO and a mutation operator for the MS part. In other words, the structure of the u-COGO is simpler than that of previous GAs for the FJSP.

2.2. Candidate Order-Based Genetic Algorithm

The COGA is a type of GA that is based on the candidate order approach (COA). The main idea of the COA is to generate a feasible sequence of given items by appending one item at a time. Moreover, which item to append is chosen from a set of candidates; an item is a candidate if it is not appended to the sequence under construction and no constraints are violated after its appending [34]. COA can be used to develop metaheuristic algorithms for solving the sequencing problem.

The first example of a COA application is the COGA for solving the classical JSP (Kim, 2014). As a result of its flexibility, the COGA is quite effective for solving sequencing problems with additional constraints. Kim [14] applied the COGA to solve the job sequencing problem and the traveling salesman problem with precedence constraints. Kim and Kim [35] developed COGA for solving a variant of shortest path problem. Another important benefit of COA is that it can be integrated with various metaheuristic algorithms. For instance, Kim [34] developed a candidate order-based tabu search (COTS) for solving the job sequencing problem with precedence constraints. Kim and Kim [36] introduced the concept of the latest order constraint and applied COTS to solve the latest order constrained sequencing problem.

Typically, metaheuristic algorithms for solving sequencing problems have complex structures. However, COA-based metaheuristic algorithms are easier to implement than conventional metaheuristic algorithms. In particular, the COGA provides a distinguishing genetic operator called COGO, which has two important features. First, the COGO enables the application of simple point-based crossover strategies to sequencing problems, including one-point, two-point, and uniform strategies. Thus, both crossover operations for the MS and OS parts in a GA chromosome for the FJSP can be performed using the COGO. Second, the COGO also can be used to mutate a GA chromosome for the sequencing problem. The COGO generates a feasible sequence of given items in a constructive manner, which means a sequence is obtained by iteratively appending one item at a time. The item to be appended is chosen from a set of candidates. The objective of crossover, which is to create offspring similar to the parents, can be achieved by choosing an item with an earlier reference position. Note that the reference position of an item is its order in the parent solution. In contrast, the COGO performs the mutation by choosing an item with a later reference position, since the objective of the mutation is to create offspring dissimilar to the parents [14].

The main contributions of this paper are as follows. Firstly, we can implement the GA for solving the FJSP with only two genetic operators: the COGO and an additional operator for the mutation of the MS part. In other words, the u-COGA is easier to implement than previous GAs for the FJSP. Secondly, various kinds of crossover strategies, such as one-point, two-point, and uniform strategies, can be applied to the u-COGA. Thirdly, numerical experiments reveal that uniform crossover is the best crossover strategy for the u-COGA for the FJSP.

3. Unified Candidate Order-Based Genetic Algorithm

3.1. Existing GAs for the FJSP

Table 2 provides an example of the FJSP with three jobs (J_1, J_2, and J_3) and three machines (M_1, M_2, and M_3). Each job consists of two operations that should be processed in sequence, and jth operation of J_i (i = 1, 2, 3) is denoted by o_{ij}. An operation is processed by one of its alternative machines. For example, o_{11} has three alternative machines (M_1, M_2, and M_3). Different alternative machines can have different processing times for a single operation. In Table 2, M_2 is the SPT machine for o_{11} such that its processing time is shorter than other machines. In contrast, o_{12} has only two alternative machines (M_2 and M_3), since M_1 is not available for this operation.

Table 2. An example of the FJSP.

Job	Operation	Processing Time		
		M_1	M_2	M_3
J_1	o_{11}	4	3	5
	o_{12}	N/A	6	8
J_2	o_{21}	10	N/A	8
	o_{22}	5	6	4
J_3	o_{31}	7	10	N/A
	o_{32}	3	4	5

In order to create a schedule for the FJSP, two kinds of decisions are needed. First, each operation should be assigned to one of its alternative machines. Second, operations should be sequenced. Typically, GAs for the FJSP utilize a multichromosome that consists of the OS part and the MS part for representing the operation sequence and machine assignment, respectively. An example of a multichromosome for the FJSP is shown in the upper part of Figure 1. Let $MS(o_{ij})$ denote the machine to be used to process o_{ij}. It is clear that $MS(o_{ij})$ must be an element of a set of alternative machines for o_{ij}.

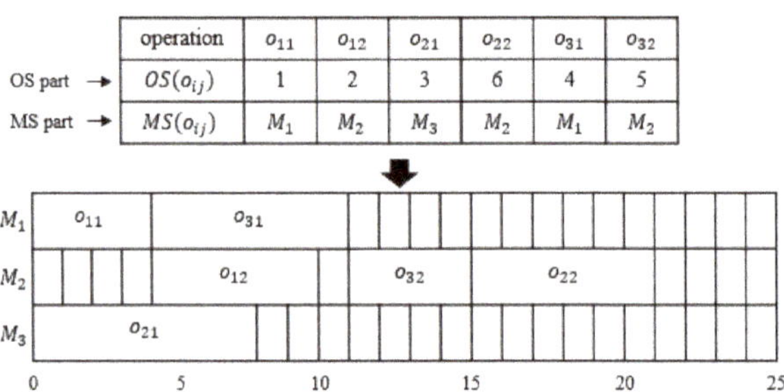

Figure 1. Decoding the multichromosome representation for the FJSP.

In this paper, a sequence of operations indicates the assignment orders. Let $OS(o_{ij})$ denote the order of o_{ij} in a sequence of given operations. Note that the encoding scheme for the OS part is the position listing representation, in which each gene specifies an order of the associated item. Thus, the OS part in Figure 1 can be converted into a sequence of given operations, $\langle os_{11}, os_{12}, os_{21}, os_{31}, os_{32}, os_{22} \rangle$. An operation o_{ab} is scheduled prior to o_{cd} if $OS(o_{ab}) < OS(o_{cd})$. For instance, o_{11} is the first operation to be scheduled in the Gantt chart in the lower part of Figure 1, since $OS(o_{11}) = 1$. Moreover, we can start o_{11} at time $t = 0$.

The order of given operations, $OS(o_{ij})$s, must satisfy the precedence relationships among operations within a single job. In other words, $OS(o_{ij})$ must be smaller than $OS(o_{ik})$ if $j < k$. In Figure 1, o_{12} is the second operation ($OS(o_{12}) = 2$) of a job and its start time is 4, because we can start o_{12} after its predecessor, o_{11}, is completed. Valid $MS(o_{ij})$ and $OS(o_{ij})$ values yield a feasible schedule for the FJSP, which can be represented in the form of a Gantt chart, as shown in Figure 1.

The crossover operation is the "backbone" of the GA, and crossover operators are designed to generate two offspring solutions by recombining the genes of two parent solutions. The offspring solutions inherit the features of their parents. In other words, crossover operators should generate offspring solutions similar to their parents. However, this is difficult to achieve when solving the FJSP by applying a GA based on a multichromosome.

An example of a crossover operation of the existing GA for the FJSP is illustrated in Figure 2. Consider two parent solutions, P_1 and P_2, in panel (a) in Figure 2. Typically, existing GAs for the FJSP apply different kinds of crossover operators to the OS and MS parts of parent solutions. In panel (b) in Figure 2, POX and classical one-point crossover are applied to the OS and MS part, respectively. POX is a well-known crossover operator for sequencing problems. In panel (b), the OS parts of the solutions are represented in the form of a sequence of operations, in order to apply POX. Job set 1 indicates that an operation o_{ij} has an identical position $OS(o_{ij})$ in both P_x and C_x ($x = 1, 2$) if it belongs to J_3, where C_x denotes an offspring solution. In contrast, job set 2 means C_1 (C_2) inherits precedence relationships among operations of J_1 and J_2 from P_2 (P_1).

$$P_1$$

OPERATION	o_{11}	o_{12}	o_{21}	o_{22}	o_{31}	o_{32}
$OS(o_{ij})$	2	5	4	6	1	3
$MS(o_{ij})$	M_1	M_3	M_2	M_3	M_2	M_1

$$P_2$$

OPERATION	o_{11}	o_{12}	o_{21}	o_{22}	o_{31}	o_{32}
$OS(o_{ij})$	1	2	3	4	5	6
$MS(o_{ij})$	M_2	M_1	M_1	M_3	M_1	M_2

(a) 2 parent solutions

POX for OS part   1-point crossover for MS part

Job set 1 = { J_3 } Job set 2 = { J_1, J_2 }

	OS part (POX)	MS part (1-point crossover, crossover point between columns 3 and 4)						
P_1	$< o_{31}, o_{11}, o_{32}, o_{21}, o_{12}, o_{22} >$	M_1	M_3	M_2	M_3	M_2	M_1	
P_2	$< o_{11}, o_{12}, o_{21}, o_{22}, o_{31}, o_{32} >$	M_2	M_1	M_1	M_3	M_1	M_2	
C_1	$< o_{31}, o_{11}, o_{32}, o_{12}, o_{21}, o_{22} >$	M_1	M_3	M_1	M_3	M_1	M_2	
C_2	$< o_{11}, o_{21}, o_{12}, o_{22}, o_{31}, o_{32} >$	M_2	M_1	M_2	M_3	M_2	M_1	

(b) Different crossover operators for OS and MS parts

$$C_1$$

OPERATION	o_{11}	o_{12}	o_{21}	o_{22}	o_{31}	o_{32}
$OS(o_{ij})$	2	4	5	6	1	3
$MS(o_{ij})$	M_1	M_3	M_1	M_3	M_1	M_2

$$C_2$$

OPERATION	o_{11}	o_{12}	o_{21}	o_{22}	o_{31}	o_{32}
$OS(o_{ij})$	1	3	2	4	5	6
$MS(o_{ij})$	M_2	M_1	M_2	M_3	M_2	M_1

(c) 2 offspring solutions

Figure 2. An example of crossover for an existing GA for solving the FJSP.

The one-point crossover operator divides a chromosome into two subparts, i.e., the earlier part and later part, and generates C_1 (C_2) by combining the earlier part of P_1 (P_2) and the later part of P_2 (P_1). Panel (c) in Figure 2 shows a multichromosome representation of C_1 and C_2, obtained by applying POX and one-point crossover. Similarly, most existing GAs for solving the FJSP use different crossover operators to recombine the multichromosome, which consists of the OS and MS parts.

In Figure 2, we can see that the OS part and MS part evolve separately. That is, the crossover procedure of the OS part is not correlated with that of the MS part. In this context, the traditional approach based on two different crossover operators is called the separate evolution strategy in this paper. This strategy has a critical limitation, i.e., a combination of the OS gene and MS gene for an identical operation can be easily broken. For instance, o_{31} has two associated genes: $OS(o_{31})$ and $MS(o_{31})$. For P_1 in panel (a) in Figure 2, $OS(o_{31}) = 1$ and $MS(o_{31}) = M_2$. If these two genes indicate the optimal assignment order and machine for o_{31}, the combination of $OS(o_{31}) = 1$ and $MS(o_{31}) = M_2$ should be preserved during the crossover procedure. However, this combination is not found in offspring solutions C_1 and C_2 in panel (c) in Figure 2. In more detail, $OS(o_{31})$ of P_1 is inherited to C_1, whereas $MS(o_{31})$ of P_1 is inherited to C_2. The u-COGA proposed in this paper is designed to overcome the limitation of the traditional separate evolution strategy for the FJSP.

3.2. Crossover of Unified COGA

This section outlines crossover procedure of the u-COGA. The COGO enables the application of simple crossover strategies to sequencing problems, such as one-point, two-point, and uniform crossovers. Thus, a solution of the FJSP is converted into a sequence of operations in order to apply the COGO, as shown in Figure 3. Note that an operation in the FJSP has an additional attribute, $MS(o_{ij})$, as shown in panel (b) in Figure 3.

P_1

OPERATION	o_{11}	o_{12}	o_{21}	o_{22}	o_{31}	o_{32}
$OS(o_{ij})$	2	5	4	6	1	3
$MS(o_{ij})$	M_1	M_3	M_2	M_3	M_2	M_1

P_2

OPERATION	o_{11}	o_{12}	o_{21}	o_{22}	o_{31}	o_{32}
$OS(o_{ij})$	1	2	3	4	5	6
$MS(o_{ij})$	M_2	M_1	M_1	M_3	M_1	M_2

(a) 2 parent solutions

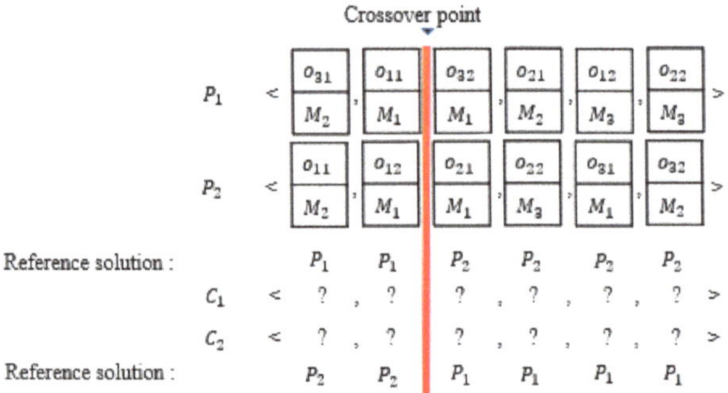

(b) Reference set assignment based on 1-point crossover strategy

Figure 3. Reference set assignment for parent solutions in the form of a sequence.

The COGO generates a solution in a constructive manner, which means that one operation is appended to a new solution at a time. An operation to be appended to a specific position is chosen with consideration of the reference solution, one of the parent solutions. Thus, the COGO determines the reference solution for each position in a sequence before generating a new solution. An important feature of the COGO is that simple crossover strategies can be used to determine reference solutions. In panel (b) in Figure 3, each sequence is divided into two parts, i.e., the earlier part with the first two operations and later part with the other four operations, by adopting the conventional one-point crossover strategy. We can see that P_1 (P_2) and P_2 (P_2) are used as reference solutions for the earlier part and later part of C_1 (C_2), respectively.

An operation to be used as the pth position of a new solution is determined after the $p - 1$th operation is appended. An operation to be the pth operation is chosen from a set of candidates; an operation is a candidate if and only if it has not been appended to the new solution under construction and no constraints are violated by appending the operation to the pth position. The main idea of the COGO is that the objectives of crossover and mutation can be achieved by choosing the appropriate operation from the set of candidates.

Figure 4 illustrates a procedure for generating the earlier part of C_1, an offspring of P_1 and P_2 in Figure 3. Let $RP(o_{ij}, X)$ denote the reference position of an operation o_{ij} in a solution X. In other words, o_{ij} is the $RP(o_{ij}, X)$th operation in X. The main idea of the COGO is that the objectives of crossover and mutation can be achieved by choosing candidates with the smallest and the largest reference position values, respectively. For the first position in C_1, all operations are candidates, since no operation is appended to C_1, yet. Among the candidates, an operation with the smallest reference position regarding P_1, o_{31} is chosen as the first operation. Similarly, o_{11} is appended to the second position in C_1. Note that $MS(o_{ij})$s of the offspring are inherited from the associated reference solution. That is, $MS(o_{31})$ and $MS(o_{11})$ of C_1 are the same as those of P_1.

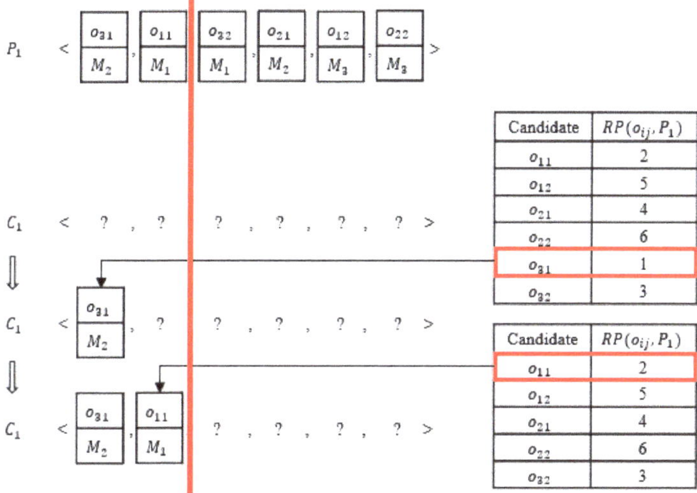

Figure 4. Generating the earlier part of the offspring solution.

Figure 5 shows the crossover procedure for the later part of C_1, where P_2 is used as the reference solution. We can see that the positions of $o_{12}, o_{21}, o_{22},$ and o_{32} in the later part of C_1 are different from those of P_2. Nevertheless, the precedence relationships among them in P_2 are also maintained in C_1. Moreover, $MS(o_{ij})$s of the operations in the later part are inherited from P_2. Consequently, C_1 inherits from both P_1 and P_2. Since the $MS(o_{ij})$ value of the reference solution is maintained in the offspring solution, the crossover procedure described in Figures 3–5 helps to preserve a combination of the OS and MS genes.

In contrast, the mutation operation for the pth position of a new solution is performed by choosing a candidate with the largest reference position in the COGO procedure. That is, crossover and mutation are integrated in the COGO. Assume that o_{ij} is the candidate with the largest reference position for the pth position of a new solution chosen by the COGO in mutation mode. Then, $OS(o_{ij})$ of the new solution is likely to be dissimilar to that of the reference solution. However, $MS(o_{ij})$ is not affected by the mutation procedure provided by the COGO. Hence, we need an additional mutation operator for $MS(o_{ij})$, and this paper applies random alternative mutation to the MS part of the multichromosome for the FJSP.

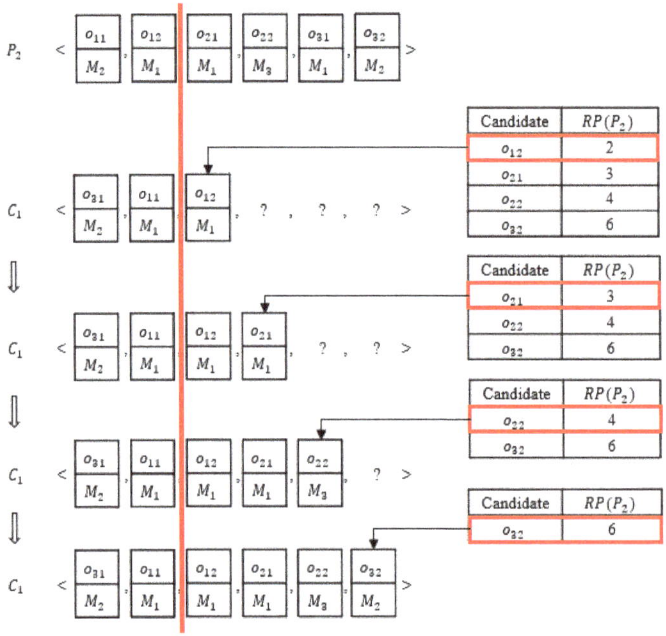

Figure 5. Generating the later part of the offspring solution.

3.3. Initialization, Fitness Function, Selection, and Elitism

Population initialization is the first step in the GA. Population can be defined as a set of chromosomes. There are several ways to initialize the population in the GA; however, the solutions in the initial population are randomly generated in this paper.

The fitness function of the GA is an objective function that needs to be maximized. In order to minimize the makespan of the FJSP, we use the reciprocal of the makespan as the fitness function of the u-COGA.

Selection is the procedure of creating a mating pool by copying the solutions in the current generation. The goal of the selection procedure is to choose the solutions with higher fitness values with larger probabilities. In this paper, we use conventional roulette wheel selection, in which a solution in the current population is chosen with a probability proportional to its fitness value.

Furthermore, the elitism strategy is applied to the u-COGA proposed in this paper. This strategy ensures that the best solutions in the current generation are also included in the mating pool. In other words, the elitism strategy helps to prevent desirable features of solutions being lost during the search procedure. Consequently, some solutions in the mating pool are determined by the elitism strategy, while the others are chosen by the roulette wheel selection procedure.

4. Numerical Experiments

In this paper, the u-COGA was written in Java language and was tested in a Windows 10 (64-bit) environment with an AMD Ryzen 7 2700X eight-Core processor (3.7 GHz) and 32 GB memory. The performance of the u-COGA was evaluated using the benchmark problems for the FJSP provided by the well-known Hurink library [37]. Depending on the average number of alternative machines for each operation, benchmark problems in the Hurink library are divided into three groups: eData, rData, and yData, as listed in Table 3.

Table 3. Groups of benchmark problems in the Hurink library.

Data Group	Description
eData	Few operations can be assigned to more than one machine (Low Complexity)
rData	Most of the work can be assigned to some machines (Moderate Complexity)
yData	Each operation can be assigned to several machines (High Complexity)

Benchmark problems from each data group are summarized in Table 4, where n is the number of operations, m is the number of machines, and M_{ij} is a set of alternative machines for an operation o_{ij}. Moreover, $Avg|M_{ij}|$ and $Max|M_{ij}|$ indicate the average and the maximum size of M_{ij}, respectively. Each data group contains three instances, and the instances of the yData group have the most complex structures, in that they generally have $Avg|M_{ij}|$ and $Max|M_{ij}|$ larger than the other groups. On the contrary, eData contains the simplest instances.

Table 4. Benchmark problem instances.

Data Group	Instance	n	m	$Avg\,|M_{ij}|$	$Max\,|M_{ij}|$
eData	mt06	6	6		$2\,(m \leq 6)$
	mt10	10	10	1.15	$3\,(m \geq 10)$
	mt20	20	5		
rData	mt06	6	6		
	mt10	10	10	2	3
	mt20	20	5		
vData	mt06	6	6		
	mt10	10	10	$\frac{1}{2}m$	$\frac{4}{5}m$
	mt20	20	5		

As shown in Table 5, we applied three types of u-COGA, SS, TT, and UU to the benchmark problems in Table 4. The symbols 'S', 'T', and 'U' indicate one-point, two-point, and uniform crossover, respectively. In addition, the first and second symbols in 'type' indicate the crossover strategies for the OS part and MS part, respectively. For example, the first type of u-COGA, i.e., SS, applies the one-point crossover strategy to both the OS and MS parts.

Table 5. Types of crossover operators applied to the existing GA and u-COGA.

	u-COGA			Existing GA	
Type	OS	MS	**Type**	OS	MS
SS	One-point	One-point	**PS**	POX	One-point
TT	Two-point	Two-point	**PT**	POX	Two-point
UU	Uniform	Uniform	**PU**	POX	Uniform

For comparison, the existing GA was also applied to the benchmark problems. The existing GA uses POX to recombine the OS parts of the chromosomes, while crossover for the MS part is performed in a one-point, two-point, and uniform manner. Thus, we have three types of existing GAs: PS, PT, and PU, as shown in Table 5.

The experiment results obtained by applying the u-COGA and the existing GA are summarized in Tables 6–8, where 20 experimental repetitions were performed for each case. $\overline{C_{max}}$ and $SD(\overline{C_{max}})$ are the average makespan and standard deviation of the makespan, respectively. The six types of GAs applied to a single benchmark problem instance are ranked in ascending order of $\overline{C_{max}}$. For instance, TT and PS are the best and worst algorithms for the mt06 instance of eData in Table 6, respectively. All experimental results were obtained with a population size = 50, a crossover rate = 0.8, and a mutation rate = 0.01. In addition, the elitism policy was used to preserve the five best solutions in the previous generation. A single experiment was terminated when its iteration number reached the maximum iteration limit = 300.

Table 6. Experiment results for eData.

		eData 6 × 6 (mt06)				
Algorithm	Existing GA			u-COGA		
	PS	PT	PU	SS	TT	UU
$\overline{C_{max}}$	49.95	50.3	50.85	49.05	48.9	47.9
SD($\overline{C_{max}}$)	1.43	1.72	1.93	1.67	1.48	1.33
Rank	4	5	6	3	2	1
		eData 10 × 10 (mt10)				
Algorithm	Existing GA			u-COGA		
	PS	PT	PU	SS	TT	UU
$\overline{C_{max}}$	1094.7	1090.1	1093.9	1078.8	1085	1050.85
SD$\overline{C_{max}}$	24.67	27.93	22.59	27.67	28.19	22.64
Rank	6	4	5	2	3	1
		eData 20 × 5 (mt20)				
Algorithm	Existing GA			u-COGA		
	PS	PT	PU	SS	TT	UU
$\overline{C_{max}}$	1134.15	1135.15	1134.75	1074.2	1101.9	1097.4
SD($\overline{C_{max}}$)	42.50	33.31	44.20	62.89	33.86	41.03
Rank	4	6	5	1	3	2

Table 7. Experiment results for rData.

		rData 6 × 6 (mt06)				
Algorithm	Existing GA			u-COGA		
	PS	PT	PU	SS	TT	UU
$\overline{C_{max}}$	49.45	48	47.9	48.25	47.85	47.25
SD($\overline{C_{max}}$)	1.79	1.56	1.51	1.74	1.53	1.74
Rank	6	4	3	5	2	1
		rData 10 × 10 (mt10)				
Algorithm	Existing GA			u-COGA		
	PS	PT	PU	SS	TT	UU
$\overline{C_{max}}$	997.95	1008.15	1002.05	996.35	997.8	962.2
SD($\overline{C_{max}}$)	28.14	23.04	19.33	29.50	25.35	26.02
Rank	4	6	5	2	3	1
		rData 20 × 5 (mt20)				
Algorithm	Existing GA			u-COGA		
	PS	PT	PU	SS	TT	UU
$\overline{C_{max}}$	1090.1	1091.1	1083.35	1090.3	1074.9	1056.55
SD($\overline{C_{max}}$)	41.07	41.74	37.27	34.45	36.89	31.46
Rank	4	6	3	5	2	1

Table 8. Experiment results for vData.

	vData 6 × 6 (mt06)					
Algorithm	**Existing GA**			**u-COGA**		
	PS	PT	PU	SS	TT	UU
$\overline{C_{max}}$	48	47.8	48.1	48.45	47.45	46.9
SD($\overline{C_{max}}$)	2.10	1.40	1.29	1.82	1.67	1.62
Rank	4	3	5	6	2	1
	vData 10 × 10 (mt10)					
Algorithm	**Existing GA**			**u-COGA**		
	PS	PT	PU	SS	TT	UU
$\overline{C_{max}}$	945.35	938.45	933	934.15	940.6	920.25
SD($\overline{C_{max}}$)	21.61	19.22	27.08	26.31	23.86	22.73
Rank	6	4	2	3	5	1
	vData 20 × 5 (mt20)					
Algorithm	**Existing GA**			**u-COGA**		
	PS	PT	PU	SS	TT	UU
$\overline{C_{max}}$	1074.8	1068.85	1072.5	1055.05	1053.2	1054.55
SD($\overline{C_{max}}$)	28.81	31.23	31.61	30.82	33.71	27.59
Rank	6	4	5	3	1	2

Tables 6–8 provide the following observations. First, u-COGAs generally produce shorter makespan values than existing GAs, which indicates that the u-COGA is a promising approach for solving the FJSP. Second, for seven out of nine instances in Tables 6–8, two instances (mt06 and mt10) of eData, three instances (mt06, mt10, and mt20) of rData, and two instances (mt06 and mt10) of vData, the shortest makespan values were found by the UU-type u-COGA. In other words, uniform crossover was the best crossover strategy for minimizing the makespan of the FJSP. The performances of SS- and TT-type u-COGAs were not as good as that of the UU-type. Third, excepting one benchmark problem instance, i.e., mt06 of vData in Table 8, the worst makespan value for each instance was obtained by one of the existing GAs, which reveals the limitation of the separate evolution strategy for solving the FJSP. In particular, PS- and PU-type GAs produced the worst makespan values for four instances and three instances, respectively. Finally, the u-COGA and existing GA did not demonstrate significant differences in SD($\overline{C_{max}}$) values.

Table 9 shows the average rank of the algorithms listed in Table 5. Again, we can see that the UU-type u-COGA has the smallest average rank, which means that it generally produces a shorter makespan than the other algorithms. In contrast, the average ranks of the individual existing GAs are larger than those of the u-COGAs. Consequently, the average rank of all u-COGAs (2.33) is noticeably smaller than that of all existing GAs (4.63), and we can conclude that the unified evolution strategy of the u-COGA helps to find better solutions for the FJSP.

Table 9. Average ranks of the algorithms.

Algorithm	**Existing GA**			**u-COGA**		
	PS	**PT**	**PU**	**SS**	**TT**	**UU**
$\overline{Rank}$	4.89	4.67	4.33	3.33	2.56	1.11
		4.63			2.33	

Table 10 compares the best makespan of the existing GAs and u-COGAs. For instance, the best makespan for the existing GAs for mt20 of vData is 1068.85, while the best makespan for a u-COGA for the same instance is 1053.20, as shown in Table 8. Moreover, the reduction ratio in Table 10 was calculated as follows: (the best makespan of the existing Gas—the best makespan of u-COGAs)/(the best makespan of the existing GAs). This index was used to measure how much improvement was achieved by the u-COGA.

Table 10. Comparison of the best makespan obtained by the existing GAs and u-COGAs.

	eData			rData			vData		
ALGORITHM	Existing GA	u-COGA	ReductionRatio (%)	Existing GA	u-COGA	ReductionRatio (%)	Existing GA	u-COGA	ReductionRatio (%)
MT06	49.95	47.9	4.10	47.9	47.25	1.36	47.8	46.9	1.88
MT10	1090.1	1050.85	3.60	997.95	962.2	3.58	933	920.25	1.37
MT20	1134.15	1074.2	5.29	1083.35	1056.55	2.47	1068.85	1053.2	1.46

In Table 10, we can see that the u-COGA always produces better solutions for the FJSP than the existing GA. The reduction ratio values range from 1.36% to 5.29% (average reduction ratio = 2.79%). That is, the existing GA sometimes demonstrates a similar performance to the u-COGA; however, the u-COGA generally produces superior results for the FJSP.

5. Conclusions and Further Remarks

In order to apply the GA to solve a combinatorial optimization problem, a solution of the problem must be represented in the form of a chromosome comprising a number of genes. A solution for the classical JSP can be encoded using only OS-type genes. In contrast, a solution for the FJSP is typically represented in the form of a multichromosome, which consists of the OS and MS parts. Since the structures of the OS and MS parts are quite different, existing GAs generally apply different kinds of genetic operators to them. Consequently, the evolution procedures of the two parts are independent from each other, and a combination of OS and MS genes for an identical operation is easily broken during the search procedure. Moreover, the application of different kinds of crossover operators can make the GA complex and hard to implement and maintain. In order to overcome these problems, this paper proposes the u-COGA, which can be used to solve the FJSP.

The u-COGA utilizes an integrated genetic operator called the COGO to handle the multichromosome for the FJSP. The COGO enables the application of conventional crossover strategies, such as one-point, two-point, and uniform crossovers to the OS part. The u-COGA is designed to apply an identical crossover strategy to both the OS and MS parts. Moreover, the COGO generates a new sequence of given operations by choosing one operation at a time, which is suitable for the purposes of crossover and mutation. In order to perform the crossover operation, the COGO chooses a candidate with the smallest reference position. On the contrary, the mutation operation is performed by choosing a candidate with the largest reference position. In other words, the COGO of the u-COGA is used to perform three operations: crossover for the OS part, crossover for the MS part, and mutation for the OS part. Mutation for the MS part should be performed in a different manner, and random alternative mutation is used in this paper. Consequently, crossover and mutation are performed using two genetic operators: the COGO and mutation operator for the MS part in the u-COGA. Note that four different kinds of operators, i.e., crossover for the OS part, crossover for the MS part, mutation for the OS part, and mutation for the MS part, are required to implement the existing GAs for the FJSP. In other words, the COGO enables one to solve the FJSP using a GA with a simpler structure.

In order to evaluate performance of the u-COGA, numerical experiments were performed using an FJSP benchmark problem library. The experiment results reveal that the performance of the u-COGA is generally superior to that of the existing GAs in terms of the makespan. In more detail, the u-COGA improves the makespan values for the FJSP by about 2.79%. Notably, the best solution for a benchmark problem instance was generally found using the u-COGA based on the uniform crossover strategy. In contrast, the worst solution was generally found by one of the existing GAs. This means that the u-COGA can explore the search space of the FJSP more effectively, and we can conclude that the GA for the FJSP should contain procedures for maintaining good combinations of good OS and MS genes; thus, the u-COGA is a promising approach for solving complex combinatorial optimization problems.

The authors propose three directions for future research on the u-COGA. First, this paper only considered one objective function, which is the minimization of the makespan. Thus, u-COGA performance regarding other objective functions should be investigated. Second, the primary advantage of the original version of the COGA is that it can flexibly handle additional constraints, such as precedence or position-related constraints. In this regard, in future research, we plan to apply the u-COGA to the FJSP with additional constraints. Third, it is expected that the u-COGA can help to solve other combinatorial optimization problems such that solutions have to be represented in the form of a multi-chromosome. Therefore, the authors will attempt to apply the u-COGA to combinatorial optimization problems in various other fields.

Author Contributions: Conceptualization, H.-Y.N. and J.-W.K.; methodology, J.-S.P. and J.-W.K.; software, J.-S.P.; validation, J.-S.P. and J.-W.K.; formal analysis, J.-S.P. and J.-W.K.; investigation, J.-S.P. and J.-W.K.; resource, H.-Y.N., T.-J.C. and Y.-T.N.; data curation, J.-S.P.; writing—original draft preparation, J.-S.P. and J.-W.K.; writing—review and editing, J.-S.P., H.-Y.N. and J.-W.K.; supervision, J.-W.K.; project administration, H.-Y.N., Y.-T.N. and J.-W.K.; funding acquisition, H.-Y.N., Y.-T.N. and J.-W.K. All authors have read and agreed to the published version of the manuscript.

Funding: This work was supported by the National Research Foundation of Korea (NRF) grant funded by the Korea government (MSIT) (NRF-2020R1F1A1049251) and the Korea Institute for Advancement of Technology (KIAT) grant funded by the Korea Government (MOTIE: Ministry of Trade Industry and Energy; Grant No. N0002429).

Institutional Review Board Statement: Not applicable.

Informed Consent Statement: Not applicable.

Conflicts of Interest: The authors declare no conflict of interest.

References

1. Zhang, G.; Gao, L.; Shi, Y. An effective genetic algorithm for the flexible job-shop scheduling problem. *Expert Syst. Appl.* **2011**, *38*, 3563–3573. [CrossRef]
2. Harjunkoski, I.; Maravelias, C.T.; Bongers, P.; Castro, P.M.; Engell, S.; Grossmann, I.E.; Hooker, J.; Mendez, C.; Sand, G.; Wassick, J. Scope for industrial applications of production scheduling models and solution methods. *Comput. Chem. Eng.* **2014**, *62*, 161–193. [CrossRef]
3. Pinedo, M.L. *Scheduling: Theory, Algorithms, and Systems*, 5th ed.; Springer: Berlin/Heidelberg, Germany, 2016.
4. Parente, M.; Figueira, G.; Amorim, P.; Marques, A. Production scheduling in the context of Industry 4.0: Review and trends. *Int. J. Prod. Res.* **2020**, *58*, 5401–5431. [CrossRef]
5. Hussain, K.; Salleh, M.N.M.; Cheng, S.; Shi, Y. Metaheuristic research: A comprehensive survey. *Artif. Intell. Rev.* **2019**, *52*, 2191–2233. [CrossRef]
6. Holland, J.H. *Adaptation in Natural and Artificial Systems: An Introductory Analysis with Applications to Biology, Control, and Artificial Intelligence*; University of Michigan Press: Ann Arbor, MI, USA, 1975.
7. Glover, F. Tabu Search: A Tutorial. *Interfaces* **1990**, *20*, 74–94. [CrossRef]
8. Kirkpatrick, S.; Gelatt, C.D.; Vecchi, M.P. Optimization by Simulated Annealing. *Science* **1983**, *220*, 671–680. [CrossRef] [PubMed]
9. Kennedy, J.; Eberhart, R. Particle swarm optimization. In Proceedings of the ICNN'95 International Conference on Neural Networks, Perth, Australia, 27 November–1 December 1995; Volume 4, pp. 1942–1948.
10. Cheng, R.; Gen, M.; Tsujimura, Y. A Tutorial Survey of Job-Shop Scheduling Problems using Genetic Algorithms-I. represent. *Comput. Ind. Eng.* **1996**, *30*, 983–997. [CrossRef]
11. Cheng, R.; Gen, M.; Tsujimura, Y. A tutorial survey of job-shop scheduling problems using genetic algorithms, part II: Hybrid genetic search strategies. *Comput. Ind. Eng.* **1999**, *36*, 343–364. [CrossRef]
12. Chaudhry, I.A.; Khan, A.A. A research survey: Review of flexible job shop scheduling techniques. *Int. Trans. Oper. Res.* **2016**, *23*, 551–591. [CrossRef]
13. Kim, J.W. Developing a job shop scheduling system through integration of graphic user interface and genetic algorithm. *Mul-timed. Tools Appl.* **2015**, *74*, 3329–3343. [CrossRef]

14. Kim, J.W. Candidate Order based Genetic Algorithm (COGA) for Constrained Sequencing Problems. *Int. J. Ind. Eng. Theory Appl. Pract.* **2016**, *23*, 1–12.
15. Katoch, S.; Chauhan, S.S.; Kumar, V. A review on genetic algorithm: Past, present, and future. *Multimed. Tools Appl.* **2021**, *80*, 8091–8126. [CrossRef]
16. Poon, P.; Carter, J. Genetic algorithm crossover operators for ordering applications. *Comput. Oper. Res.* **1995**, *22*, 135–147. [CrossRef]
17. Kim, J.W. A Job Shop Scheduling Game with GA-based Evaluation. *Appl. Math. Inf. Sci.* **2014**, *8*, 2627–2634.
18. Gao, J.; Sun, L.; Gen, M. A hybrid genetic and variable neighborhood descent algorithm for flexible job shop scheduling problems. *Comput. Oper. Res.* **2008**, *35*, 2892–2907. [CrossRef]
19. Pezzella, F.; Morganti, G.; Ciaschetti, G. A genetic algorithm for the Flexible Job-shop Scheduling Problem. *Comput. Oper. Res.* **2008**, *35*, 3202–3212. [CrossRef]
20. Defersha, F.M.; Chen, M. A Parallel Genetic Algorithm for a Flexible Job-shop Scheduling Problem with Sequence Dependent Setups. *Int. J. Adv. Manuf. Technol.* **2010**, *49*, 263–279. [CrossRef]
21. Lei, D. A Genetic Algorithm for Flexible Job Shop Scheduling with Fuzzy Processing Time. *Int. J. Prod. Res.* **2010**, *48*, 2995–3013. [CrossRef]
22. Wang, X.; Gao, L.; Zhang, C.; Shao, X. A multi-objective genetic algorithm based on immune and entropy principle for flexible job-shop scheduling problem. *Int. J. Adv. Manuf. Technol.* **2010**, *51*, 757–767. [CrossRef]
23. Jiang, L.; Du, Z. An Improved Genetic Algorithm for Flexible Job Shop Scheduling Problem. In Proceedings of the 2015 2nd International Conference on Information Science and Control Engineering, Shanghai, China, 24–26 April 2015; pp. 127–131.
24. Türkyılmaz, A.; Bulkan, S. A Hybrid Algorithm for Total Tardiness Minimisation in Flexible Job Shop: Genetic Algo-rithm with Parallel VNS Execution. *Int. J. Prod. Res.* **2015**, *53*, 1832–1848. [CrossRef]
25. Zhang, G.; Hu, Y.; Sun, J.; Zhang, W. An improved genetic algorithm for the flexible job shop scheduling problem with multiple time constraints. *Swarm Evol. Comput.* **2020**, *54*, 100664. [CrossRef]
26. Parjapati, S.K.; Jain, A. Optimization of Flexible Job Shop Scheduling Problem with Sequence Dependent Setup Times Using Genetic Algorithm Approach. *Int. J. Math. Comput. Nat. Phys. Eng.* **2015**, *9*, 41–47.
27. Phanden, R.K.; Jain, A. Assessment of makespan performance for flexible process plans in job shop scheduling. *IFAC-Paper* **2015**, *48*, 1948–1953. [CrossRef]
28. Elgendy, A.; Mohammed, H.; Elhakeem, A. Optimizing Dynamic Flexible Job Shop Scheduling Problem Based on Genetic Al-gorithm. *Int. J. Curr. Eng. Technol.* **2017**, *7*, 368–373.
29. Chang, H.-C.; Chen, Y.-P.; Liu, T.-K.; Chou, J.-H. Solving the Flexible Job Shop Scheduling Problem with Makespan Optimization by Using a Hybrid Taguchi-Genetic Algorithm. *IEEE Access* **2015**, *3*, 1740–1754. [CrossRef]
30. Chen, J.; Weng, W.; Rong, G.; Fujimura, S. Integrating Genetic Algorithm with Time Control for Just-In-Time Sched-uling Problems. *IFAC-Paper* **2015**, *48*, 893–897. [CrossRef]
31. Driss, I.; Mouss, K.N.; Laggoun, A. A New Genetic Algorithm for Flexible Job-Shop Scheduling Problems. *J. Mech. Sci. Technol.* **2015**, *29*, 1273–1281. [CrossRef]
32. Ishikawa, S.; Kubota, R.; Horio, K. Effective Hierarchical Optimization by a Hierarchical Multi-Space Competitive Genetic Al-gorithm for the Flexible Job-Shop Scheduling Problem. *Expert Syst. Appl.* **2015**, *42*, 9434–9440. [CrossRef]
33. Wang, L.; Luo, C.; Cai, J. A Variable Interval Rescheduling Strategy for Dynamic Flexible Job Shop Scheduling Problem by Improved Genetic Algorithm. *J. Adv. Transp.* **2017**, *2017*, 1–12. [CrossRef]
34. Kim, J.W. Performance Comparison of Neighborhood Structures of Tabu Search Algorithms for Sequencing Problems. *Adv. Sci. Lett.* **2017**, *23*, 10423–10426.
35. Kim, J.W.; Kim, S.K. Genetic Algorithms for Solving Shortest Path Problem in Maze-Type Network with Precedence Constraints. *Wirel. Pers. Commun.* **2019**, *105*, 427–442. [CrossRef]
36. Kim, J.W.; Kim, S.K. Interactive Job Sequencing System for Small Make-to-Order Manufacturers under Smart Manu-facturing Environment. *Peer-to-Peer Netw. Appl.* **2020**, *13*, 524–531. [CrossRef]
37. Hurink, J.; Jurisch, B. and Thole, M. Tabu Search for the Job-shop Scheduling Problem with Multi-purpose Machines. *Oper. Res. Spektrum* **1994**, *15*, 205–215. [CrossRef]

Article

Calibration of GA Parameters for Layout Design Optimization Problems Using Design of Experiments

Vladimir Modrak [1,*], **Ranjitharamasamy Sudhakara Pandian** [2] and **Pavol Semanco** [3]

[1] Department of Industrial Engineering and Informatics, Faculty of Manufacturing Technologies, Technical University of Kosice, 080 01 Presov, Slovakia
[2] Department of Manufacturing Engineering, Vellore Institute of Technology, School of Mechanical Engineering, Vellore 632014, India; sudhame@gmail.com
[3] Lear Corporation, Presov, Solivarska 1/A, 080 01 Presov, Slovakia; pavolsemanco@gmail.com
* Correspondence: vladimir.modrak@tuke.sk

Abstract: In manufacturing-cell-formation research, a major concern is to make groups of machines into machine cells and parts into part families. Extensive work has been carried out in this area using various models and techniques. Regarding these ideas, in this paper, experiments with varying parameters of the popular metaheuristic algorithm known as the genetic algorithm have been carried out with a bi-criteria objective function: the minimization of intercell moves and cell load variation. The probability of crossover (A), probability of mutation (B), and balance weight factor (C) are considered parameters for this study. The data sets used in this paper are taken from benchmarked literature in this field. The results are promising regarding determining the optimal combination of the genetic parameters for the machine-cell-formation problems considered in this study.

Keywords: facility layout; optimization; metaheuristic algorithm; cell formation; design of experiments

Citation: Modrak, V.; Pandian, R.S.; Semanco, P. Calibration of GA Parameters for Layout Design Optimization Problems Using Design of Experiments. *Appl. Sci.* **2021**, *11*, 6940. https://doi.org/10.3390/app11156940

Academic Editor: Paolo Renna

Received: 3 July 2021
Accepted: 22 July 2021
Published: 28 July 2021

Publisher's Note: MDPI stays neutral with regard to jurisdictional claims in published maps and institutional affiliations.

1. Introduction

In general, facility-layout optimization problems are nonlinear, nonconvex, and multimodal in their nature. Facility-layout problems (FLP) can be divided according to types of manufacturing systems into four basic categories, which are product layout, process layout, static layout, and cellular layout [1]. Taking this classification into account, the proposed study addresses the cellular manufacturing problem. In the past, the main objective of a facility-layout problem was to minimize the material handling cost of the manufacturing system [2]. Presently, the main goal of the FLP is to improve manufacturing efficiency. In line with this, a number of authors suggested different objective functions for facility-layout problems, e.g., to maximize the throughput rate and minimize the conveyance time per trip [3] or minimize cycle times in order to increase productivity [4], and so forth.

An important issue in the design of cellular manufacturing systems (CMS) is the manufacturing-cell-formation problem (MCFP), which is based on group technology principles. Several taxonomies of the MCFP are reviewed in the literature, e.g., by Selim et al. [5], Bidanda et al. [6] and Modrak and Pandian [7]. The core of MCFP procedures is that machines are grouped into machine cells and parts into part families. In practical applications, it is not easy to arrange all parts and machines into autonomic cells, and therefore some operations have to be performed on separate machines. The cost of duplicating machines is often high, and therefore, related managerial decisions are usually trade-offs between economic and technological criteria [8–10]. During previous decades, numerous heuristic and metaheuristic methods and their variations were developed, tested and compared for this problem. The metaheuristics include the genetic algorithm (GA), simulated annealing, ant colony optimization, tabu search, scatter search, particle swarm optimization, GRASP and hybridized metaheuristics. There are several survey papers, such as by Herroelen et al. [11], Kolisch and Hartmann [12] and others, that present and compare the methods from various perspectives.

In this work, the focus is on application of GAs, and especially on finding out how the solutions of the cell-formation problem are influenced by a set of probability parameters of genetic operators, namely crossover and mutation, including balanced weight factors. In view of this, the presented work attempts to employ the Taguchi approach to find an optimal combination of parameters that impact the efficiency of the genetic algorithm and to explore whether the optimal combination of the genetic operators for the given type of MCFP can be influenced by the magnitude of the noise factors, which is represented by matrix size in our case.

2. A Brief Literature Review

In this section, a short review of the selected research on manufacturing cell formation will be given. John Holland, inspired by population genetics, introduced the concept of the GA in the 1970s. In 1975, the book [13] is published by him and his colleagues. The first GA-based approach for the cell-formation research area was proposed by Venugopal and Narendran [14]. It was a proposed mathematical model with a solution procedure based on a GA that can be purposely implemented in a cellular manufacturing environment. Joines and Houck [15] in their work applied a non-stationary penalty to solve CFP with a genetic algorithm. Gupta et al. [16] used a GA approach to solve the layout design problem with a predetermined number of manufacturing cells. Alsultan and Fedjki [17] utilized a GA approach to solve the machine-cell-part clustering problem in order to minimize total intercell and intracell moves. Gravel et al. [18] developed a GA with a double-loop, able to solve large-scale capacitated cell-formation problems with multiple routings. Moon and Gen [19] proposed a genetic algorithm to solve an integer programming model with consideration of alternative process plans and machine-duplication consideration. An adaptive genetic approach to solve the manufacturing-cell-formation problem in order to enhance the performance of the genetic search process was proposed by Mak et al. [20]. Zhao and Wu [21] evaluated the solutions of a multi-objective GA that applies minimizing costs due to intercell and intracell part flows, minimizing the total within-cell load variation and minimizing exceptional elements. They also incorporated in their research the multiple routes of parts. Arzi et al. [22] proposed a genetic algorithm with grouping efficiency and capacity requirements as objectives for large-scale systems design. Zolfaghari and Liang [23] tested and evaluated a genetic algorithm against simulated annealing and tabu search using binary cell-formation problems. Yasuda et al. [24] in their research proposed a method to solve the multi-objective CFP, partially adopting Falkenauer's grouping genetic algorithm. It was also aimed at improving the efficiency of their algorithm with regards to initialization of the population, fitness valuation, and keeping the crossover operator from cloning. Other similar approaches were published with many innovative algorithms. For example, Farahani et al. [25] described ant colony optimization to solve machine–part cell-formation problems. Mohammad Mohammadi et al. [26] approached a layout problem in cellular manufacturing systems with alternative processing routings. Dmytryshyn et al. [27] suggested a novel modeling approach for solving the cell-formation problem. Kamalakannan et al. [28] developed a simulated annealing algorithm for solving CFP with ratio level data. Shashikumar et al. [29] determined the solution for CFP using a heuristics approach. Sharma et al. [30] had done research on an implementation model using AHP and ANP for CFP. Octavio et al. [31] developed metaheuristic algorithms to solve grouping problems. Mourtzis et al. [32] brought the idea of adaptive scheduling in cellular manufacturing systems. Firouzian et al. [33] developed an artificial immune system for part family clustering. Vitayasak et al. [34] proposed a GA-based layout design approach to solve robust machine layout design problems for systems subject to demand uncertainties and maintenance. Their innovative method includes an experimental design that was used to test the robust design approach with corrective, preventative, and combined maintenance regimes. Dolgui et al. [35] explored a reconfigurable manufacturing system that exhibited some crucial design and control characteristics for complex value-adding systems in highly dynamic scenarios. Such systems are designed at the outset for rapid change in the structure of machines, in order to quickly adjust production capacity and functionality within a part family in response to frequent market changes or intrinsic system changes. An interesting and very useful approach to the facility-layout design optimization was proposed by Bucki and Suchanek [36]. They proposed an effective tool for performance analysis of manufacturing systems from logistics viewpoint by using a mathematical simulation model.

It is worth mentioning other works that directly relate to the manufacturing cell-formation problem, such as the hybrid GA/branch and bound approach to solve the manufacturing-cell-formation problem using a graph partitioning formulation, which was proposed by Boulif and Atif [37]. Their effort has been made to take into account the natural constraints of real-life production systems. A typical CFP with the objective of minimizing the exceptional elements was explored by Mahdavi et al. [38]. Deljoo et al. [39] used a GA to solve the dynamic cell-formation problem. Based on their studies, they reconsidered the shortcomings related to machine relocation cost and machine purchasing cost and developed a model for dynamic CFP in order to resolve these two shortcomings. Arkat et al. [40] developed a multi-objective GA for CFP considering cellular layout and operations scheduling. Cell-formation problems related to scheduling problems with the objective to minimize makespan are available in references [41–44]. A new algorithm for CFP with alternative machines and multiple-operation-type machines was developed by Li [45]. Its purpose was to improve traditional group technology cell-formation methods by considering alternative machines and multiple-operation-type machines. Boulif [46] proposed a new graph-cut-based encoding representation in order to solve the CFP with the genetic algorithm. Obviously, there are other related works, since this domain attracts a large research interest.

3. Structure of Genetic Algorithm

In a GA, a high-quality candidate solution is represented by a collection of genes called chromosome. A chromosome's potential is given by its fitness function. A population consists of a set of selected chromosomes, and the population is subjected to generations (or iterations). Finally, crossover and mutation operators are performed with defined probabilities to improve the solutions. A GA has several advantages over the traditional optimization methods. It may quickly arrive at a good solution set. As worse cases are eliminated, they will never affect the generated solution. GAs are one of the best methods to solve a problem about which little is known. This is because a genetic algorithm works by its own rules. This is a very useful strategy for a GA to solve highly complex problems in nature. In a GA, it is necessary for the problem solver to choose the appropriate coding method. If the solutions are coded in different combinations, then the GA will start its searching operation using its operators known as selection, crossover, and mutation, respectively. As is known, a suitable type of crossover technique for a particular problem can improve the GA's performance [47]. All these methods are probabilistic in nature. The proper stopping criteria will be given as input for the GA to stop its searching process. This is done purely based on the experience of the problem-solver. Based on the stopping criteria, the GA will stop running and give the solution that it finds at that point of time. The solution of the problem has to be represented in the GA as a genome (or chromosome). The genetic algorithm then creates a population of solutions and applies genetic operators such as mutation, crossover, and selection to evolve the solutions in order to find the best one(s). The crucial aspects of using GAs are: the definition of the objective function, the definition and implementation of the genetic representation, the definition and implementation of the genetic operators [48].

GAs are frequently adopted to find out how to make the machine clusters form cells in accordance with the rules of production flow analysis. In cell formation, a GA works well and finds a good solution, as given in the abovementioned literature studies. It gives high-quality solutions even if the problem has a high level of complexity. There are few other traditional metaheuristics approaches that performing good searches in such a situation [49]. The following representation is used in a typical cell-formation problem solved using GA. This representation is popularly known as real coding (see Figure 1).

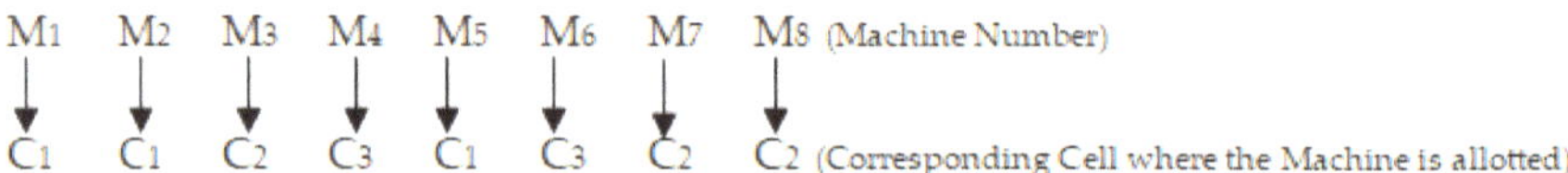

The Values are ENCODED for representing the Chromosome in GA as follows

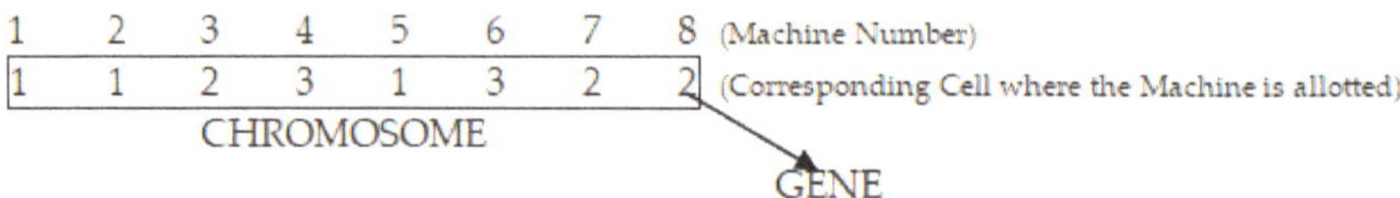

Figure 1. Representation of a chromosome using real coding.

Depending on the number of cells to be accommodated in the layout, the number of genes on a chromosome increases. In the above example there are three cells and hence there are only 1, 2, and 3 values present in the chromosome of 8 genes (8 machines).

4. Mathematical Model of the Cell-Formation Problem

According to the review of the literature, the minimization of intercell flows and the total cell load variation can be considered the essential objectives in the manufacturing cell-formation research. Thus, the bi-objective fitness function used to evaluate the solution incorporates intercell flows and cell load variation. The mathematical model is given as:

Intercell flow fitness function:

$$\text{Minimize} f_1 = \sum_i \sum_j \sum_k a_{ij} |X_{ik} - Y_{ik}| \tag{1}$$

Cell load variation fitness function:
Minimize

$$f_2 = \sum_i \sum_k X_{ik} \sum_j \left(\omega_{ij} - m_{ik} \right)^2 \tag{2}$$

Variable:

$$m_{ik} = \sum_i X_{ik} \times Y_{jk} \times \omega_{ij} / \sum_i X_{ik} \tag{3}$$

Decision variables:

$$\omega_{ij} = t_{ij}; \text{ if part } j \text{ is processed on machine } i \tag{4}$$

0; otherwise

$$X_{ik} = 1; \text{ if machine } i \text{ is in cell } k \tag{5}$$

0; otherwise

$$Y_{jk} = 1; \text{ if part } j \text{ is in cell } k \tag{6}$$

0; otherwise

$$a_{ij} = 1; \text{ if part } j \text{ needs to be processed on machine } i \tag{7}$$

0; otherwise
Constraints:

$$\sum_k^c X_{ik} = 1 \ \forall \in \{1, 2, \ldots, m\} \tag{8}$$

$$\sum_{k}^{c} Y_{jk} = 1 \; \forall \in \{1, 2, \ldots, p\} \tag{9}$$

$$\sum X_{ik} \geq 1 \; \forall k \in \{1, 2, \ldots, c\} \tag{10}$$

$$\sum Y_{jk} \geq 1 \; \forall k \in \{1, 2, \ldots, c\} \tag{11}$$

Bi-objective fitness function:

$$Z(t) = \alpha \cdot f_1 + (1 - \alpha) \cdot f_2 \tag{12}$$

The given model of cell-formation problem works with processing time. According to this model, we search to obtain the number of cells and the number of machines and parts within each cell. Equation (1) shows the calculation of the intercell flows, and Equation (2) shows the cell load variation. Equation (3) shows the average intercell processing time for the j-th part and k-th cell. Equation (4) shows the decision variables that assign the time t_{ij} of the i-th machine required to process the j-th part. Equations (5)–(7) are decision variables that state that x_{ik}, y_{jk} and a_{jj} are 0–1 binary numbers. Equations (8) and (9) ensure that each machine and part is attached to only one cell. Equations (10) and (11) ensures that, in each cell, there must be allocated at least one machine and one part, respectively. Equation (12) is the bi-objective fitness function for a non-binary cell-formation problem balanced by weight factor α.

A fitness function value is computed for each chromosome in the population, and the objective is to find a chromosome with the maximum fitness function value. Due to objective of minimizing both the total cell load variation and the exceptional elements, it is necessary to map it inversely and then maximize the result. Goldberg [50] suggested a mapping function given as:

$$F(t) = Z_{max} - Z(t) \tag{13}$$

The symbol $F(t)$ stands for the fitness function of the t-th chromosome, and Z_{max} is the $max[Z(t)]$ of all chromosomes (t). The advantage is that the worst chromosomes obtain a zero-fitness function value, so they are not going to be reproduced into the next generation.

The following procedure given by Zolfaghari and Liang [51] is used to assign parts into the machine cells:

$$P_{kj} = \left(\frac{f_{kj}}{f_k} \right) \cdot \left(\frac{f_{kj}}{f_j} \right) \cdot \left(\frac{T_{kj}}{T_j} \right) \tag{14}$$

where, P_{kj} is the membership index of the j-th part belonging to the k-th cell; f_{kj} is the number of machines in the k-th cell required by the j-th part; f_k is the total number of machines in the k-th cell; f_j is total number of machines required by the j-th part; T_{kj} is the processing time of the j-th part in the k-th cell; and T_j is the total processing time required by the j-th part.

5. Case Study on Layout Design Optimization

The procedure we decided to apply to analyzing the influence of genetic parameters on the final solution quality for reorganization of machines and parts into cells is an experimental design method developed by Genichi Taguchi. Here we established a P-diagram (see Figure 2) that identifies the inputs and outputs of the system together with control and noise factors.

The Taguchi experimental method consists of performing selected experiments to study the influence of several operating factors on output-parameter values. Taguchi separates the factors into two domains: control and noise factors. The difference between these two factor groups is that we cannot control them directly. The elimination of the noise factors is often impractical and impossible so the Taguchi method seeks to minimize the effect of noise, using optimal levels of important control factors based on the concept of robustness [52]. These three fundamental control factors are the probability of crossover

(A), probability of mutation (B), and balance weight factor (C). The levels of each set of control factors are shown in Table 1.

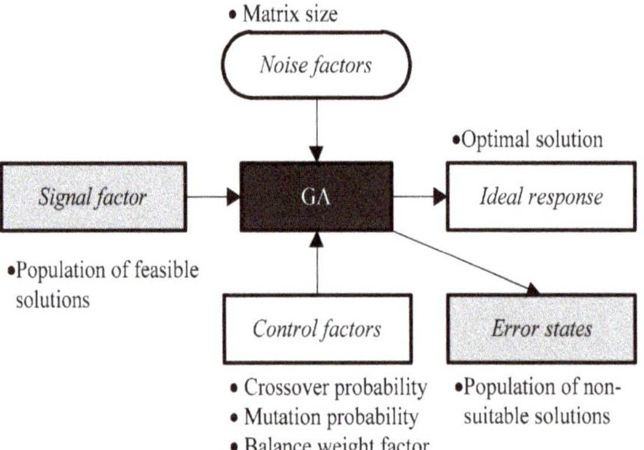

Figure 2. Parameter diagram for a genetic algorithm.

Table 1. Input signal factors and levels.

Control Factors	Description	Levels		
		−	0	+
A	Probability of crossover (P_c)	0.3	0.6	0.8
B	Probability of mutation (P_m)	0.01	0.05	0.1
C	Balance weight factor (α)	0.4	0.6	0.8

Based on the principles of the genetic algorithm, these control-factor values are decided taking a reference from the optimization of control parameters for a genetic algorithms using an image registration problem [53]. In the evaluation of a final feasible solution, there also exist the so-called noise factors that have an impact on the results. These factors cause deviation in the search space size and the noise factors, such as the size of the machine-part matrix (D). The offered factors are composed of individual levels, wherein each level is completely independent. This particular case is under consideration with three levels of control and two levels of noise factors.

It is proposed to apply Taguchi quality concept for the assessment of the final solution, and, in this context, the L9 orthogonal array has been chosen. The results of the experiments are subsequently transformed into a signal-to-noise ratio. For reducing the variability of solutions around a target, the smaller-is-better S/N ratio (SNR) calculation is applied [54]:

$$SNR = -10 \log[\frac{i}{n} \sum_{i=1}^{n} Y_i^2]$$ (15)

A signal-to-noise ratio is a measure of robustness that can be used to identify the control factor settings that minimize the effect of noise on the response. In the parameter design, there are two factors: the signal factor that can be controlled and the noise factor that is too expensive to control. In this research work, the signal factors are A, B, and C, wherein the noise factor is the exceptional element. Yi^2 is obtained from the mean sqare deviation (MSD) of signal factors and of exceptional elements. Here the 'n' represents the number of experiments that are performed using Taguchi's experiment design.

Higher values of SNR will control the noise factors that lead to the minimization of the objective function. Since, in this research work, the objective is to minimize the combination of two functions, it is recommended from Taguchi's SNR principle that the 'smaller is better' is suitable for this approach.

The stopping criterion used to test algorithms was set at a number of generations [55], fixed to min 120. In order to conduct the experiment, we used a set of four instances and implemented the algorithms in PHP script. As a common performance measure, the number of exceptional elements (EE) has been used. The total set of experiments that were performed was obtained by combining the L9 array of the control factors. In order to obtain more objective results, we decided on two groups of size problems; the first of them consisted of 4 size problems (24×16, 24×40, 30×16, 30×40) as shown in Table 2, and the second one consisted of 19 size problems (see Table 3).

Table 2. Combination of factors and the resulting values of the trials of experiments (Group #1).

#	Factors			D	24×16	24×40	30×16	30×40	Results	
	A	B	C	1	2	3	4		MSD	SNR
1	−	−	−	12	6	19	13		12.5	−22.49
2	−	0	0	15	5	21	5		11.5	−22.53
3	−	+	+	13	3	26	7		12.25	−23.54
4	0	−	0	11	0	12	4		6.75	−18.47
5	0	0	+	27	6	25	6		16	−25.52
6	0	+	−	11	0	9	4		6	−17.36
7	+	−	+	15	6	19	9		12.25	−22.45
8	+	0	−	12	3	17	7		9.75	−20.89
9	+	+	0	10	0	7	4		5.25	**−16.15**

Table 3. Combination of factors and resulting values of the trials of experiments (Group #2).

#	Factors		D	7×5	8×6	10×10	11×7	18×5	12×8	15×10	20×8	20×20	20×23	23×14	24×14	24×16	24×40	30×16	35×20	30×40	40×24	41×30	Results	
	A	B	C	1	2	3	4	5	6	7	8	9	10	11	12	13	14	15	16	17	18	19	MSD	SNR
1	−	−	−	2	2	1	3	5	1	0	5	20	8	0	0	12	6	19	0	13	2	2	65.84	−18.19
2	−	0	0	2	2	0	3	5	1	0	9	17	4	0	0	15	5	21	0	5	0	3	60.74	−17.83
3	−	+	+	2	2	0	3	5	1	0	5	19	13	0	0	13	3	26	0	7	0	3	79.47	−19.00
4	0	−	0	2	2	0	3	5	1	0	12	24	10	0	0	11	0	12	0	4	0	3	60.68	−17.83
5	0	0	+	2	2	0	3	5	1	0	9	17	12	0	0	27	6	25	0	6	0	3	104.84	−20.21
6	0	+	−	2	2	0	3	5	1	0	27	32	14	0	0	11	0	9	0	4	0	3	116.79	−20.67
7	+	−	+	2	2	0	3	5	1	0	27	17	22	0	0	15	6	19	0	9	0	3	118.79	−20.75
8	+	0	−	2	2	0	3	5	1	0	9	19	8	0	0	12	3	17	0	7	0	3	55.21	−17.42
9	+	+	0	2	2	0	3	5	1	0	9	15	10	0	0	10	0	7	0	4	0	3	32.79	**−15.16**

The total number of experiments for the first group of size problems is 36 and for the second group of size problems is 171. The objective in this factorial experiment is to search values of the signal-to-noise ratio for "smaller-is-better", which is computed for each set of experiments. After obtaining the results of Taguchi's experiment design, EEs are transformed into S/N ratios. The results of the experiment for each series of experiments are presented in Tables 2 and 3.

6. Conclusions

Based on Table 1 (input signal factors and levels), the crossover probability is considered in ascending order (lower to higher), which is mentioned in Tables 2 and 3. The change in value for A occurs in every three rows; for B, the change in value occurs in every row in ascending order, and for factor C (1,2,3 then 2,3,1 then 3,1,2) as per the L9 orthogonal array representation. It happens incidentally that the last value (9th row) is better compared to other values, as far as this research concerned. The optimal level of the factors is the level with the highest SNR. Tables 2 and 3 show that the optimal level has been obtained in the 9th row (refer to Tables 2 and 3) where the value of factor A is 0.8, the value of factor B is 0.1, and the value of factor C is 0.6, mentioned as (+, +, 0) as per Table 1. The general view is that whenever the probability of crossover and/or mutation is increased, the number of searches in the solution space will also increase; thereby the chance of obtaining a better solution is increased. The Taguchi's design of experiments based on this research work provides evidence to the above statement.

Building on this analysis the following conclusions and suggestions for further research have been made.

It can be stated that the Taguchi design of experiments method can be used as an effective tool to determine the optimal combination of the genetic operators for the given type of MCF problems, because two different experiments brought out that the optimal level of the factors A, B, and C are identical.

Due to the limited number of experimental groups, we can only anticipate that an optimal combination of the genetic operators for the given type of MCF problem is not influenced by the magnitude of the noise factors. Therefore, in our further research this dependence/independence relation will be investigated.

This work provides further evidence that the efficiency of a genetic algorithm is dependent on the mentioned control factors and their parameters. As a direction for future studies, it could be interesting to extend the parameters (for example, the probability of reproduction and the number of generations) and develop effective genetic algorithm incorporating advanced features. For more realistic models, it would be useful to consider several practical assumptions, such as machine-availability constraints or changeover times.

Author Contributions: Writing—original draft preparation was done by R.S.P. Conceptualization and methodology were performed by all the authors. Writing—review was done by V.M. Experimentation was carried out by P.S. Results of the investigations were analyzed and evaluated by all the authors. All authors have read and agreed to the published version of the manuscript.

Funding: The authors would like to thank the Ministry of Education, Science, Research and Sport of the Slovak Republic for the financial support of this work by grant KEGA, project No. 025TUKE-4/2020.

Data Availability Statement: The detailed input numerical data used in this research work are available from the corresponding author upon request. Due to the actual fact that data were collected from various literature, it is used for research purposes only.

Conflicts of Interest: The authors declare no conflict of interest.

References

1. Hosseini-Nasab, H.; Fereidouni, S.; Ghomi, S.M.T.F.; Fakhrzad, M.B. Classification of facility layout problems: A review study. *Int. J. Adv. Manuf. Technol.* **2018**, *94*, 957–977. [CrossRef]
2. Azadivar, F.; Wang, J. Facility layout optimization using simulation and genetic algorithms. *Int. J. Prod. Res.* **2000**, *38*, 4369–4383. [CrossRef]
3. Hamamoto, S.; Yie, Y.; Salvendy, G. Development and Validation of Genetic Algorithm-Based Facility Layout-A Case Study in the Pharmaceutical Industry. *Int. J. Prod. Res.* **1999**, *37*, 749–768. [CrossRef]
4. Li, X.; Fung, R.Y.K. Optimal K-unit cycle scheduling of two-cluster tools with residency constraints and general robot moving times. *J. Sched.* **2016**, *19*, 165–176. [CrossRef]
5. Selim, H.M.; Aal, R.M.A.; Mahdi, A.I. Formation of machine groups and part families: A modified SLC method and comparative study. *Integr. Manuf. Syst.* **2003**, *14*, 123–137. [CrossRef]
6. Bidanda, B.; Ariyawongrat, P.; Needy, K.L.; Norman, B.A.; Tharmmaphornphilas, W. Human related issues in manufacturing cell design, implementation, and operation: A review and survey. *Comput. Ind. Eng.* **2005**, *48*, 507–523. [CrossRef]
7. Modrak, V.; Pandian, R.S. *Operations Management Research and Cellular Manufacturing Systems: Innovative Methods and Approaches*; Business Science Reference: Hershey, PA, USA, 2012.
8. Hertz, A.; Jaumard, B.; Ribeiro, C.C.; Filho, W.P.F. A multi-criteria tabu search approach to cell formation problems in group technology with multiple objectives. *RAIRO-Oper. Res.* **1994**, *28*, 303–328. [CrossRef]
9. Süer, G.A.; Vazquez, R.; Cortes, M. A hybrid approach of genetic algorithms and local optimizers in cell loading. *Comput. Ind. Eng.* **2005**, *48*, 625–641. [CrossRef]

10. Semančo, P.; Modrák, V. Hybrid GA-Based Improvement Heuristic with Makespan Criterion for Flow-Shop Scheduling Problems. *Commun. Comput. Inf. Sci.* **2011**, *220*, 11–18. [CrossRef]
11. Herroelen, W.; De Reyck, B.; Demeulemeester, E. Resource-constrained project scheduling: A survey of recent developments. *Comput. Oper. Res.* **1998**, *25*, 279–302. [CrossRef]
12. Kolisch, R.; Hartmann, S. Experimental investigation of heuristics for resource-constrained project scheduling: An update. *Eur. J. Oper. Res.* **2006**, *174*, 23–37. [CrossRef]
13. Holland, J.H. *Adaptation in Natural and Artificial Systems*; The MIT Press: Cambridge, MA, USA, 1992.
14. Venugopal, V.; Narendran, T. A genetic algorithm approach to the machine-component grouping problem with multiple objectives. *Comput. Ind. Eng.* **1992**, *22*, 469–480. [CrossRef]
15. Joines, J.A.; Houck, C.R. On the use of non-stationary penalty functions to solve constrained optimization problems with genetic algorithms. In Proceedings of the IEEE International Symposium EC, Orlando, FL, USA, 27–29 June 1994; pp. 579–584.
16. Gupta, Y.P.; Gupta, M.C.; Kumar, A.; Sundaram, C.A. A genetic algorithm-based approach to cell composition and layout design problems. *Int. J. Prod. Res.* **1996**, *34*, 447–482. [CrossRef]
17. Alsultan, K.S.; Fedjki, C.A. A genetic algorithm for the part family formation problem. *Prod. Plan. Conotrol* **1997**, *8*, 788–796. [CrossRef]
18. Gravel, M.; Nsakanda, A.L.; Price, W. Efficient solutions to the cell-formation problem with multiple routings via a double-loop genetic algorithm. *Eur. J. Oper. Res.* **1998**, *109*, 286–298. [CrossRef]
19. Moon, C.; Gen, M. A genetic algorithm-based approach for design of independent manufacturing cells. *Int. J. Prod. Econ.* **1999**, *60/61*, 421–426. [CrossRef]
20. Mak, K.L.; Wong, Y.S.; Wang, X.X. An Adaptive Genetic Algorithm for Manufacturing Cell Formation. *Int. J. Adv. Manuf. Technol.* **2000**, *16*, 491–497. [CrossRef]
21. Zhao, C.W.; Wu, Z.M. A genetic algorithm for manufacturing cell formation with multiple routes and multiple objectives. *Int. J. Prod. Res.* **2000**, *38*, 385–395. [CrossRef]
22. Arzi, Y.; Bukchin, J.; Masin, M. An efficiency frontier approach for the design of cellular manufacturing systems in a lumpy demand environment. *Eur. J. Oper. Res.* **2001**, *134*, 346–364. [CrossRef]
23. Zolfaghari, S.; Liang, M. A new genetic algorithm for the machine/part grouping problem involving processing times and lot sizes. *Comput. Ind. Eng.* **2003**, *45*, 713–731. [CrossRef]
24. Hu, L.; Yin, Y. A grouping genetic algorithm for the multi-objective cell formation problem. *Int. J. Prod. Res.* **2005**, *43*, 829–853. [CrossRef]
25. Farahani, M.H.; Hosseini, L. An Ant ColonyOptimization Approach for the Machine–Part Cell Formation Problem. *Int. J. Comput. Intell. Syst.* **2011**, *4*, 486–496. [CrossRef]
26. Mohammadi, M.; Forghani, K. A novel approach for considering layout problem in cellular manufacturing systems with alternative processing routings and subcontracting approach. *Appl. Math. Model.* **2014**, *38*, 3624–3640. [CrossRef]
27. Dmytryshyn, T.; Ismail, M.; Rashwan, O. A novelmodeling approach for solving the cell formation problem. In Proceedings of the ASME International Mechanical Engineering Congress and Exposition, Proceedings (IMECE), Pittsburgh, PA, USA, 9–15 November 2018; Volume 2.
28. Kamalakannan, R.; Sudhakara Pandian, R.; Sivakumar, P. A simulated annealing for the cell formation problemwith ratio level data. *Int. J. Enterp. Netw. Manag.* **2019**, *10*, 78–90.
29. Shashikumar, S.; Raut, R.D.; Narwane, V.S.; Gardas, B.B.; Narkhede, B.E.; Awasthi, A. A novel approach to determine the cell formation using heuristics approach. *OPSEARCH* **2019**, *56*, 628–656. [CrossRef]
30. Sharma, V.; Gidwani, B.D.; Sharma, V.; Meena, M.L. Implementation model for cellular manufacturing system using AHP and ANP approach. *Benchmark. Int. J.* **2019**, *26*, 1605–1630. [CrossRef]
31. Ramos-Figueroa, O.; Quiroz-Castellanos, M.; Mezura-Montes, E.; Schütze, O. Metaheuristics to solve grouping problems: A review and a case study. *Swarm Evol. Comput.* **2020**, *53*, 100643. [CrossRef]
32. Mourtzis, D.; Siatras, V.; Synodinos, G.; Angelopoulos, J.; Panopoulos, N. A framework for adaptive scheduling incellular manufacturing systems. *Proced. CIRP* **2020**, *93*, 989–994. [CrossRef]
33. Firouzian, S.; Mahdavi, I.; Paydar, M.M.; Saadat, M. Simulated annealing and artificial immune system algorithms for cell formation with part family clustering. *J. Ind. Eng. Manag. Stud.* **2020**, *7*, 191–219.
34. Vitayasak, S.; Pongcharoen, P.; Hicks, C.; Hicks, C. Robust machine layout design under dynamic environment: Dynamic customer demand and machine maintenance. *Expert Syst. Appl. X* **2019**, *3*, 1–17. [CrossRef]
35. Dolgui, A.; Ivanov, D.; Sokolov, B. Reconfigurable supp.ly chain: The X-network. *Int. J. Prod. Res.* **2020**, *58*, 4138–4163. [CrossRef]
36. Bucki, R.; Suchánek, P. Comparative Simulation Analysis of the Performance of the Logistics Manufacturing System at the Operative Level. *Complexity* **2019**, *2019*, 7237585. [CrossRef]
37. Boulif, M.; Atif, K. A new branch-&-bound-enhanced genetic algorithm for the manufacturing cell formation problem. *Comput. Oper. Res.* **2006**, *33*, 2219–2245. [CrossRef]
38. Mahdavi, I.; Paydar, M.M.; Solimanpur, M.; Heidarzade, A. Genetic algorithm approach for solving a cell formation problem in cellular manufacturing. *Expert Syst. Appl.* **2009**, *36*, 6598–6604. [CrossRef]
39. Deljoo, V.; Al-E-Hashem, S.M.; Deljoo, F.; Aryanezhad, M. Using genetic algorithm to solve dynamic cell formation problem. *Appl. Math. Model.* **2010**, *34*, 1078–1092. [CrossRef]

40. Arkat, J.; Farahani, M.H.; Ahmadizar, F. Multi-objective genetic algorithm for cell formation problem considering cellular layout and operations scheduling. *Int. J. Comput. Integr. Manuf.* **2012**, *25*, 625–635. [CrossRef]
41. Modrák, V.; Pandian, R.S. Flow shop scheduling algorithm to minimize completion time for n-jobs m-machines problem. *Tehnički vjesnik* **2010**, *17*, 273–278.
42. Rafiei, H.; Rabbani, M.; Gholizadeh, H.; Dashti, H. A novel hybrid SA/GA algorithm for solving an integrated cell formation–job scheduling problem with sequence-dependent set-up times. *Int. J. Manag. Sci. Eng. Manag.* **2015**, *11*, 1–9. [CrossRef]
43. Semanco, P.; Modrak, V. A comparison of constructive heuristics with the objective of minimizing makespan in the flow-shop scheduling problem. *Acta Polytech. Hungarica* **2012**, *9*, 177–190.
44. Goli, A.; Tirkolaee, E.B.; Aydin, N.S. Fuzzy Integrated Cell Formation and Production Scheduling considering Automated Guided Vehicles and Human Factors. *IEEE Trans. Fuzzy Syst.* **2021**, 1. [CrossRef]
45. Li, M.-L. A novel algorithm of cell formation with alternative machines and multiple-operation-type machines. *Comput. Ind. Eng.* **2021**, *154*, 107172. [CrossRef]
46. Boulif, M. A New Cut-Based Genetic Algorithm for Graph Partitioning Applied to Cell Formation. In *Heuristics for Optimization and Learning*; Springer Science and Business Media LLC: Berlin, Germany, 2021; pp. 269–284.
47. Zhang, J.; Chung, H.; Lo, W.-L. Clustering-Based Adaptive Crossover and Mutation Probabilities for Genetic Algorithms. *IEEE Trans. Evol. Comput.* **2007**, *11*, 326–335. [CrossRef]
48. Man, K.F.; Tang, K.S.; Kwong, S. *Genetic Algorithms: Concepts and Designs*; Springer Science & Business Media: Berlin, Germany, 2001.
49. Fan, S.-K.S.; Jen, C.-H. An Enhanced Partial Search to Particle Swarm Optimization for Unconstrained Optimization. *Mathematics* **2019**, *7*, 357. [CrossRef]
50. Goldberg, D.E. *Genetic Algorithms in Search Optimization, and Machine Learning, Reading*; Addison-Wesley Press: Boston, MA, USA, 1989.
51. Zolfaghari, S.; Liang, M. Comparative study of simulated annealing, genetic algorithms and tabu search for solving binary and comprehensive machine-grouping problems. *Int. J. Prod. Res.* **2002**, *40*, 2141–2158. [CrossRef]
52. Phadke, M.S. *Quality Engineering Using Robust Design*; Prentice-Hall: Hoboken, NJ, USA, 1989.
53. Cao, Y.J.; Wu, Q.H. Optimization of control parameters in genetic algorithms: A stochastic approach. *Int. J. Syst. Sci.* **1999**, *30*, 551–559. [CrossRef]
54. Taguchi, G.; Chowdhury, S.; Wu, Y. *Taguchi's Quality Engineering Handbook*; Wiley: Hoboken, NJ, USA, 2005.
55. Zupan, H.; Herakovic, N.; Starbek, M. Hybrid Algorithm Based on Priority Rules for Simulation of Workshop Production. *Int. J. Simul. Model.* **2016**, *15*, 29–41. [CrossRef]

Article

An Algorithm for Arranging Operators to Balance Assembly Lines and Reduce Operator Training Time

Ming-Liang Li

Department of Industrial Management, Asia Eastern University of Science and Technology, Banqiao Dist., New Taipei City 22061, Taiwan; fg017@mail.aeust.edu.tw; Tel.: +886-(2)-7738-0145 (ext. 5113)

Abstract: Industry 4.0 is transforming how costs, including labor costs, are managed in manufacturing and remanufacturing systems. Managers must balance assembly lines and reduce the training time of workstation operators to achieve sustainable operations. This study's originality lies in its use of an algorithm to balance an assembly line by matching operators to workstations so that the line's workstations achieve the same targeted output rates. First, the maximum output rate of the assembly line is found, and then the number of operators needed at each workstation is determined. Training time is reduced by matching operators' training and skills to workstations' skill requirements. The study obtains a robust, cluster algorithm based on the concept of group technology, then forms operator skill cells and determines operator families. Four numerical examples are presented to demonstrate the algorithm's implementation. The proposed algorithm can solve the problem of arranging operators to balance assembly lines. Managers can also solve the problem of worker absences by assigning more than one operator with the required skillset to each workstation and rearranging them as needed.

Keywords: assembly line balancing; group technology; cluster algorithm; bottleneck station; output rate

Citation: Li, M.-L. An Algorithm for Arranging Operators to Balance Assembly Lines and Reduce Operator Training Time. *Appl. Sci.* **2021**, *11*, 8544. https://doi.org/10.3390/app11188544

Academic Editors: Vladimir Modrak and Zuzana Soltysova

Received: 26 July 2021
Accepted: 12 September 2021
Published: 14 September 2021

Publisher's Note: MDPI stays neutral with regard to jurisdictional claims in published maps and institutional affiliations.

1. Introduction

Netessine and Taylor [1] demonstrated that expensive production technology usually leads to low product prices and may simultaneously lead to high-quality products. However, the product price is usually proportional to the product cost, part of which is labor cost. Assembly line balancing (ALB) is a crucial part of production/operations management [2]. On most mass-production assembly lines, workers repetitively perform a set of tasks predefined by using assembly line balancing techniques [3], which assign a number of work elements to various workstations to maximize the assembly line's balancing efficiency [4]. A balancing of tasks across the workstations results in the optimization of a given objective function without violating preceding constraints [5]. An unbalanced assembly line can be dangerous. For example, an operator without adequate training or who is focused on personal problems can cause an accident, decrease the efficiency of the assembly line, and increase production costs. Assembly lines maximize assembly operations, and most manufacturing plants have one or more lines, making assembly line balancing one of the oldest challenges in industry [6]. Moreover, a balanced assembly line can reduce a factory's work-in-process (WIP) inventory. Kim et al. [7] suggested a mathematically mixed integer programming model that accounts for the WIP balance and setup time in a semiconductor fabrication line. Maximum productivity and minimum load smoothness are benefits of solving the assembly line balancing problem (ALBP) [8,9].

Many types of production lines exist. Kuroda [10] presented a robust design for a cellular-line production system with unreliable facilities. A cellular-line production system comprises multiple flow shops, which are individual sets composed of functionally different facilities. Kuroda grouped facilities that perform similar operations.

This study initially considers a traditional assembly line that has four workstations with outputs λ_A, λ_B, λ_C, and λ_D (Figure 1). Thus, the production rate, or output rate, of this assembly line is equal to $\lambda_{line} = min.\{\lambda_A, \lambda_B, \lambda_C, \lambda_D\}$.

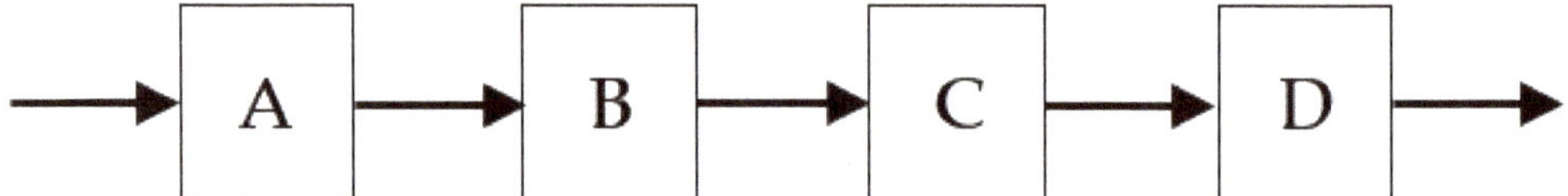

Figure 1. Traditional assembly line.

The cycle time (CT) refers to the longest period required for workstations to complete operations. Therefore, all station operation times are less than or equal to the CT. The aim of balancing the assembly line is to ensure that the output rate and operating time of each workstation are identical. Previous studies have examined assembly line balancing. According to Bartholdi and Eisenstein [11], the assembly line balances itself. They found that if workers are sequenced from slowest to fastest, a stable partition of work spontaneously emerges independent of the stations at which the workers begin. Moreover, the production rate converges to a value. Huang [12] determined that the CT and labor efficiency are negatively exponential to the number of assigned workers. Sawik [13] presented new mixed integer programming formulations for scheduling a flexible flow line with blocking. In this flexible flow line, each stage has one or more identical parallel machines. This study uses parallel operators instead of parallel machines. Rearranging operators may increase the production rate. Muth [14] examined whether a rearrangement in the order of servers affects the production rate. Muth speculated that the production rate remains invariant if the production line is reversed.

If the number of workers at a workstation is more than the number of machines, the machines are expected to limit the output rate and cause a bottleneck. However, Zavadlav et al. [15] showed that the workers, rather than the machines, limit the rate of output.

The manager of a factory always works under "cost" pressure; the company wants the manager to make continual improvements without increasing the cost. Managers typically try to increase the output by adding machines and workers, but managers must first consider whether assembly lines are balanced. When assembly lines are not balanced, productivity is low and workers complain. Workers might not only complain that they have to do more work at their station for the same pay, but also that their skills are not suited to their workstation. This work environment might lead workers to miss shifts on purpose or to resign.

Many ALBP methods have been examined in the study of manufacturing systems, including the heuristic algorithm [16–24], artificial bee colony algorithm [25–27], genetic algorithm [28–30], branch-and-bound algorithm [31–34], imperialist competitive algorithm [35], Pareto Greedy algorithm [36], memetic ant algorithm [37], whale optimization algorithm [38], recursive algorithm [39], fuzzy linear programming [40], swarm algorithm [41,42], simulated annealing algorithm [43], and data envelopment analysis [44]. The genetic algorithm [45,46] and artificial bee colony algorithm [47] have been used to solve the ALBP in remanufacturing systems. However, studies have seldom discussed how to arrange operators and group similarly skilled operators at the same workstation. This paper considers workers, not machines, in determining whether a workstation has the right equipment and machinery for operators to use. To achieve this goal, the study works out a method for rapidly balancing assembly lines by arranging operators among the workstations and uses a robust, cluster algorithm based on the concept of group technology (GT) to form operator skill cells and determine operator families.

2. Problem Statement

Traditionally, when arranging a limited number of operators on an assembly line, a manager begins with one operator per workstation, before adding an operator to the workstation with the lowest output. Step by step, all of the operators are arranged and the output rate can be found.

When a workstation has a lower output rate, the manager typically adds another operator. However, the lower output might be due to the "human" factor; an operator who is trained on many machines might be assigned to a workstation outside of their skill set. For example, if an operator knows how to work an auto-optical inspection machine but is assigned to a workstation with an in-circuit test machine, the station's output rate will drop because the operator does not have the required skills. Output can also fall when a worker is absent or resigns unexpectedly and the manager fills the vacancy with a random worker who has not been trained to operate that workstation.

Thus, the problem statement of this study is:

Ideal: The maximum output rate is found, all of the workstations have the same output rate, and the number of operators at each workstation can be found, with all operators assigned to a workstation that matches their skill set.

Reality: The arrangement algorithm is step by step, and the maximum output rate cannot be determined until all of the operators have been arranged. Operators can be assigned to unsuitable workstations.

Consequences: The maximum output rate is an unknown that will affect a boss's decisions, especially before setting up a new plant. An operator who works at an unsuitable workstation might complain, be absent, or quit unexpectedly, causing a lower output rate.

Proposal: An arrangement algorithm with a constant number of operators must consider the maximum output rate and balance the assembly line without any increase in cost. A sorting algorithm must consider the skills required at each workstation and the skill set of each operator.

3. Methodology

A manager adds a worker to a workstation that has a lower output than at other workstations. When the output of another workstation is lower, the manager adds another worker to that workstation. The use of this step-by-step method balances the assembly lines, especially when setting up a new plant.

3.1. Example of Step-by-Step Method

Suppose an assembly line has six workstations, as shown in Figure 2.

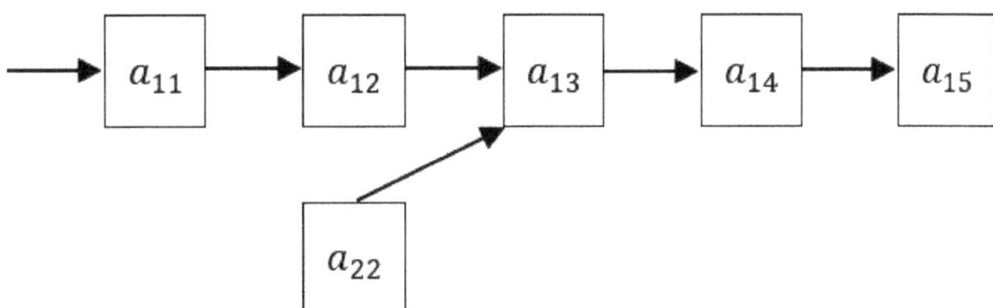

Figure 2. Assembly line with six workstations.

In addition, suppose that the terms t_{11}, t_{12}, t_{22}, t_{13}, t_{14}, and t_{15} are equal to 12, 15, 10, 16, 20, and 8 min, respectively. If one workday comprises 8 h, then the output per operator per day at each workstation, γ_{11}, γ_{12}, γ_{22}, γ_{13}, γ_{14}, and γ_{15}, are 40, 32, 48, 30, 24, and 60, respectively. Suppose that 500 workers are assigned to this assembly line. The workers can be assigned step by step, as follows (Table 1).

Table 1. Traditional ALB assignment.

	a_{11}	a_{12}	a_{22}	a_{13}	a_{14}	a_{15}	
p_{ij}	1	1	1	1	1	1	$\sum p_{ij} = 6$
Υ_{ij}	40	32	48	30	24	60	$\Upsilon_{line} = 24$

The term λ_{line} is limited in workstation a_{14}, which is a bottleneck workstation. Thus, this workstation should have one more operator to increase the output rate. The seventh worker is added to workstation a_{14}, and the following is obtained (Table 2).

Table 2. Traditional ALB assignment.

	a_{11}	a_{12}	a_{22}	a_{13}	a_{14}	a_{15}	
p_{ij}	1	1	1	1	2	1	$\sum p_{ij} = 7$
Υ_{ij}	40	32	48	30	48	60	$\Upsilon_{line} = 30$

The assignment of workers in a step-by-step manner uses considerable time to obtain the following final result (Table 3).

Table 3. Traditional ALB assignment.

	a_{11}	a_{12}	a_{22}	a_{13}	a_{14}	a_{15}	
p_{ij}	74	92	62	99	123	50	$\sum p_{ij} = 500$
Υ_{ij}	2960	2944	2976	2970	2952	3000	$\Upsilon_{line} = 2944$

Considerable time was spent and many steps performed to determine the optimal arrangement of operators at the assembly line's workstations. Although some studies have discussed and tried to solve the ALBP, the question remains: Is there another method for arranging the workers, instead of using the step-by-step method? This paper presents a novel approach for solving this problem.

Almost all workers require training before operating a workstation. Ideally, the operators on an assembly line would have all of the skills required for each workstation, but that would be very costly. Thus, the operators' skills must be determined so that they can be placed at a workstation that requires their skills. For example, consider an assembly line that has welding and drilling workstations. If operator κ possesses welding skills, assigning operator κ to the welding workstation rather than to the drilling workstation will reduce the training time for the welding workstation.

Suppose that there is a balanced assembly line with seven workstations (Figure 3), and workstations 1–7 require one, three, four, two, two, two, and one operator, respectively.

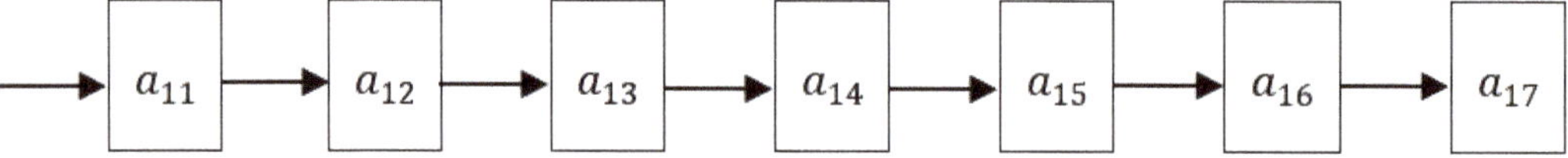

Figure 3. Assembly line with seven workstations.

Each workstation requires a skill, which is denoted as $s_1, s_2, ..., s_7$. A matrix $SO = \left\{so_{ij}\right\}_{uv}$ describes the relationship between skills and operators. For example, $so_{ij} = 1$ indicates that operator j possesses skill i. For the assembly line in Figure 3 to function, the minimum matrix SO must be as follows:

$SO = $

	o_1	o_2	o_3	o_4	o_5	o_6	o_7	o_8	o_9	o_{10}	o_{11}	o_{12}	o_{13}	o_{14}	o_{15}
s_1	1														
s_2		1	1	1											
s_3					1	1	1	1							
s_4									1	1					
s_5											1	1			
s_6													1	1	
s_7															1

Suppose it is known that:

$SO = $

	o_1	o_2	o_3	o_4	o_5	o_6	o_7	o_8	o_9	o_{10}	o_{11}	o_{12}	o_{13}	o_{14}	o_{15}
s_1					1		1	1			1			1	
s_2	1	1										1		1	
s_3			1	1			1				1		1		1
s_4	1						1	1		1		1			
s_5				1			1		1				1		1
s_6					1	1		1					1		
s_7		1								1		1		1	

On the basis of the first-come, first-served rule, operator 1 is assigned to s_1; operators 2, 3, and 4 are assigned to s_2; operators 5, 6, 7, and 8 are assigned to s_3; operators 9 and 10 are assigned to s_4; operators 11 and 12 are assigned to s_5; operators 13 and 14 are assigned to s_6; and operator 15 is assigned to s_7. Thus, at least 11 operators must take the training course, namely operators 1, 3, 4, 5, 6, 8, 9, 11, 12, 14, and 15, as illustrated in the following matrix where an "X" denotes that the operator needs the training course:

$SO = $

	o_1	o_2	o_3	o_4	o_5	o_6	o_7	o_8	o_9	o_{10}	o_{11}	o_{12}	o_{13}	o_{14}	o_{15}
s_1	X				1		1	1			1			1	
s_2	1	1	X	X								1		1	
s_3			1	1	X	X	1	X			1		1		1
s_4	1						1	1	X	1		1			
s_5				1			1		1		X	X	1		1
s_6					1	1		1					1	X	
s_7		1								1		1		1	X

A cluster algorithm is introduced in this paper to reduce the training time and increase the speed at which an assembly line functions. Workers' absences are also a problem for managers [48,49], but the proposed cluster algorithm can suggest which worker to rotate to a workstation with an absent operator.

Instead of using the traditional step-by-step method, this study aims to use the proposed algorithm to assign operators to each workstation. Before assigning the operators, some basic data must first be known about each workstation, such as the required skills, the standardization time, the work time per day, and the total number of operators. Then, the approximate number of operators needed at each workstation can be found by using Equations (1)–(7) in Section 3.1. Finally, the number of operators at each workstation can be adjusted by using the traditional step-by-step method. The training time can be reduced by assigning an operator with suitable skills to a suitable workstation, or by sorting the skill–operator matrix by using the cluster algorithm proposed in Section 3.2. The concept is shown in Figure 4.

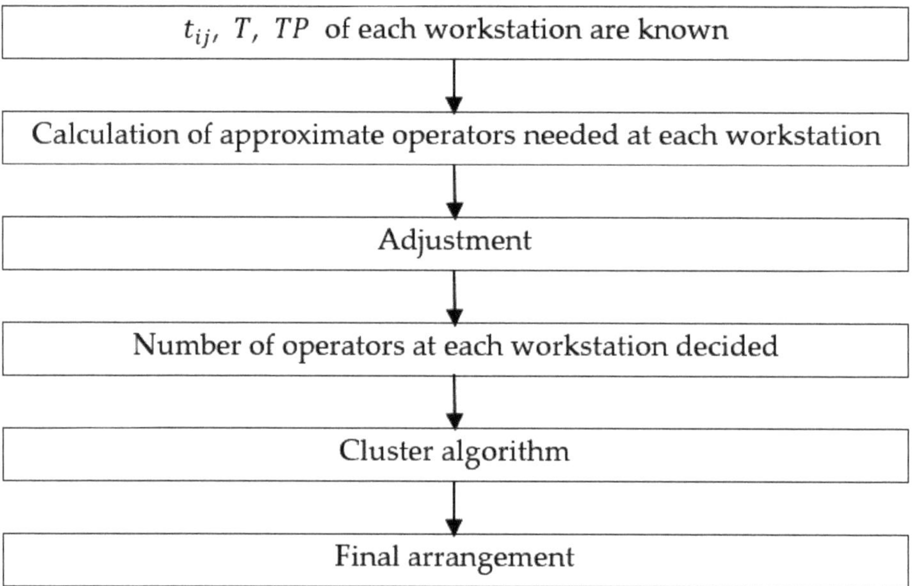

Figure 4. The novel algorithm proposed in this study.

3.2. Number of Operators at Each Workstation Decided

Suppose an assembly line has $m \times n$ workstations (Figure 5).

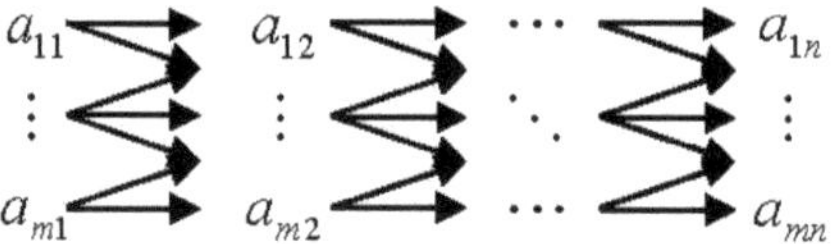

Figure 5. An assembly line with $m \times n$ workstations.

The matrix $A = \left\{ a_{ij} \right\}_{mn}$ can be used instead of the previous assembly line.

$$A = \begin{bmatrix} a_{11} & a_{12} & \cdots & a_{1n} \\ \vdots & \vdots & \ddots & \vdots \\ a_{m1} & a_{m2} & \cdots & a_{mn} \end{bmatrix}$$

The output rate of the assembly line is λ_{line} where

$$\lambda_{line} = min.\{\lambda_{ij}\},\ 1 \leq i \leq m,\ 1 \leq j \leq n$$

The ideal assembly line means that

$$\lambda_{line} = \lambda_{11} = \lambda_{21} = \cdots = \lambda_{mn} \tag{1}$$

because

$$\lambda_{ij} = p_{ij}\gamma_{ij} \tag{2}$$

The term p_{ij} can be easily determined, as follows:

$$p_{ij} = \frac{\lambda_{ij}}{\gamma_{ij}} \tag{3}$$

Moreover, the total number of operators is $P.$, where

$$P = \sum_{i=1}^{m} \sum_{j=1}^{n} p_{ij} \tag{4}$$

Thus, the following equations are obtained:

$$P = \sum_{i=1}^{m} \sum_{j=1}^{n} \frac{\lambda_{ij}}{\gamma_{ij}} = \lambda_{line} \cdot \sum_{i=1}^{m} \sum_{j=1}^{n} \frac{1}{\gamma_{ij}} = \lambda_{line} \cdot \frac{1}{T} \sum_{i=1}^{m} \sum_{j=1}^{n} t_{ij} \tag{5}$$

and,

$$\lambda_{line} = \frac{P}{\sum_{i=1}^{m} \sum_{j=1}^{n} \frac{1}{\gamma_{ij}}} = \frac{P \cdot T}{\sum_{i=1}^{m} \sum_{j=1}^{n} t_{ij}} \tag{6}$$

Replace λ_{ij} from Equation (3) with λ_{line} from Equation (6) to obtain the following equation:

$$p_{ij} = \frac{\lambda_{ij}}{\gamma_{ij}} = \frac{P}{\gamma_{ij} \sum_{i=1}^{m} \sum_{j=1}^{n} \frac{1}{\gamma_{ij}}} = \frac{P \cdot T}{\gamma_{ij} \sum_{i=1}^{m} \sum_{j=1}^{n} t_{ij}} \tag{7}$$

As term p_{ij} is not always an integer, it may need to be chopped off.

Almost all of the operators are now assigned. The traditional step-by-step method is adjusted (as demonstrated in the previous section) and used to assign the remaining workers.

3.3. Cluster Algorithm

The proposed cluster algorithm is based on the concept of GT, which has been applied to cellular manufacturing systems [50,51]. The concept involves grouping machines into machine cells and parts into part families [52–59]. Several algorithms, such as heuristic algorithms [60,61], genetic algorithms [62,63], and neural network algorithms [64], have been derived for solving the GT problem. The proposed algorithm begins with the closeness matrix and a comparison with other algorithms in example 3.

First, the relationship between u skills and v operators, which is represented by the matrix $SO = \{so_{ij}\}_{uv}$ is determined as follows:

$$so_{ij} = \begin{cases} 1, & \textit{if operator } j \textit{ has } i \textit{ skill} \\ 0, & \textit{otherwise} \end{cases}$$

The closeness matrix of skills is defined as $B_u = \{b_{rs}\}_{u \times u}$, where

$$b_{rs} = \begin{cases} \sum_{k=1}^{v} so_{rk} so_{sk}, & \textit{if } r \neq s \\ 0 & , \textit{ otherwise.} \end{cases} \tag{8}$$

For example, if there are three workers that know how to operate the welding and milling machines, the closeness between the welding and milling skills is three.

3.3.1. Skills Sorting

Whether two skills are grouped depends on their closeness. Closeness denotes how many operators have both of the skills. The more operators have the two skills, the closer the skills are. First, the skill with the highest closeness value is found and a new group is formed. Second, an unsorted skill follows a sorted skill, based on the highest value in its row in the closeness matrix. If there is no sorted skill to follow, the skill is placed in a new group. Finally, all skills are grouped.

Thus, the skills can be sorted in three steps:

Step 1. Let $f \in G_1$, where $f = \{ x \mid b_{xk} = max.\{b_{rs}\}, 1 \leq k \leq v \}$. A tie is broken by selecting the smaller one.

Step 2. Ignore all the grouped rows and form a new closeness matrix $B'u = \{b_{ij}\}$. Let

$$\begin{cases} f \in G_k, & if = \{b_{ij} = max\{b_{ij}\} \wedge y \in G_k \\ h \in G_{new\ group}, & otherwise \end{cases}$$

Step 3. Repeat Step 2 until u skills are grouped.

3.3.2. Operator Sorting

Operator sorting aims to place operators in the group that will use the highest number of their skills.

Thus, let $v \in G_y$, $y = \left\{ x \mid \sum_{u \in G_x} so_{uv} = max.\left(\sum_{u \in G_i} so_{uv} \right), 1 \leq i \leq R \right\}$. A tie is broken by choosing the group with the lowest number of machines.

3.4. GT Efficiency

This work uses the GT efficiency measurement equation defined by Chandrasekharan and Rajagopalan [65], while letting q be equal to 0.5, as follows:

$$\eta = q\eta_1 + (1-q)\eta_2 = q\frac{e_b}{\sum Q_i P_i} + (1-q)\left(1 - \frac{e_0}{uv - \sum Q_i P_i}\right) \tag{9}$$

There are two concerns: the "1" in the group, and the "1" not in the group. The higher the GT efficiency, the more 1s are in the groups, and the fewer 1s are not in the groups. Thus, $\frac{e_b}{\sum NS_i NO_i}$ is the ratio of the number of 1s in the groups to the number of 1s and 0s in all groups, while $\frac{e_0}{uv - \sum NS_i NO_i}$ is the ratio of the number of 1s not in the groups to the number of 1s and 0s not in the groups.

4. Numerical Examples

4.1. Example 1: Number of Operators at Each Workstation Decided

On an assembly line, 500 operators must be assigned to workstations (Figure 6).

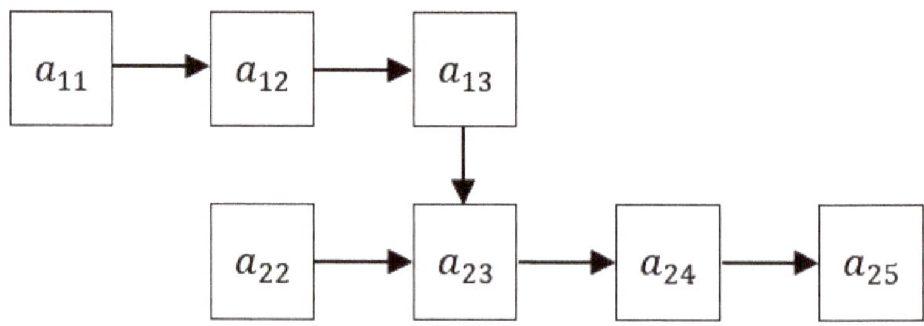

Figure 6. Assembly line with seven workstations.

Let

$$A = \begin{bmatrix} a_{11} & a_{12} & a_{13} & & \\ & a_{22} & a_{23} & a_{24} & a_{25} \end{bmatrix}$$

and

$$\{t_{ij}\} = \begin{bmatrix} 10 & 30 & 24 & 0 & 0 \\ 0 & 5 & 40 & 20 & 15 \end{bmatrix}$$

Let the work time per day be 8 h.

Thus,

$$\{\gamma_{ij}\} = \begin{bmatrix} 48 & 16 & 20 & 0 & 0 \\ 0 & 96 & 12 & 24 & 32 \end{bmatrix}$$

Thus, from Equation (6), $\lambda_{line} = 5000/3$, and from Equation (8),

$$\{p_{ij}\} \overset{chop\ off}{\underset{\sim}{}} \begin{bmatrix} 34 & 104 & 83 & & \\ & 17 & 138 & 69 & 52 \end{bmatrix}$$

The approximate operators needed at each workstation are as follows (Table 4).

Table 4. Approximate operators needed at each workstation.

	a_{11}	a_{12}	a_{13}	a_{22}	a_{23}	a_{24}	a_{25}	
p_{ij}	34	104	83	17	138	69	52	$\sum p_{ij} = 497$
γ_{ij}	1632	1664	1660	1632	1656	1656	1664	$\gamma_{line} = 1632$

The traditional step-by-step method is adjusted to obtain Table 5. Stations a_{11} and a_{22} are the bottleneck stations. Thus, if only one worker is assigned (the 498th one) to station a_{11} or station a_{22}, the assembly line cannot function. The production rate is increased if another worker is assigned to station a_{11} and another is assigned to station a_{22}. The final ALB solution is as follows (Table 5).

Table 5. Final ALB solution proposed by this study.

	a_{11}	a_{12}	a_{13}	a_{22}	a_{23}	a_{24}	a_{25}	
p_{ij}	35	104	83	18	138	69	52	$\sum p_{ij} = 499$
γ_{ij}	1680	1664	1660	1728	1656	1656	1664	$\gamma_{line} = 1656$

4.2. Example 2: Auditing Production Rate

The proposed method can be used to audit the production rate, which could be increased by rearranging the operators at each station. Consider this numerical example: An assembly line has seven workstations and 1100 workers (Figure 7).

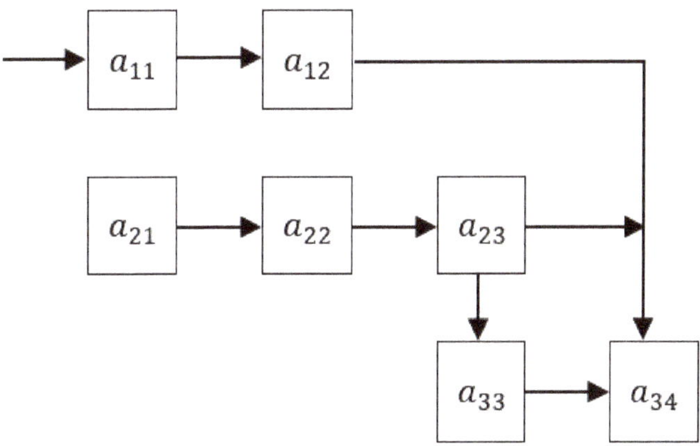

Figure 7. Assembly line with seven workstations.

Let

$$A = \begin{bmatrix} a_{11} & a_{12} & & \\ a_{21} & a_{22} & a_{23} & \\ & & a_{33} & a_{34} \end{bmatrix}$$

$$\{t_{ij}\} = \begin{bmatrix} 10 & 30 & & \\ 20 & 40 & 15 & \\ & & 24 & 48 \end{bmatrix}$$

and

$$\{p_{ij}\} = \begin{bmatrix} 70 & 180 & & \\ 120 & 200 & 80 & \\ & & 150 & 300 \end{bmatrix}$$

Let the work hours per day be 8 h. Thus, the following matrix is obtained:

$$\{\gamma_{ij}\} = \begin{bmatrix} 48 & 16 & & \\ 24 & 12 & 32 & \\ & & 20 & 10 \end{bmatrix}$$

From Equation (2), the following matrix is obtained:

$$\{\lambda_{ij}\} = \begin{bmatrix} 3360 & 2880 & & \\ 2880 & 2400 & 2560 & \\ & & 3000 & 3000 \end{bmatrix}$$

where $\lambda_{line} = 2400$. If the workers are rearranged, the output rate is increased without any increase in cost. Equation (6) is used in the method proposed in this study: $\lambda_{line} = 528{,}000/187$. Thus, the following matrix is obtained:

$$A =$$

	p_1	p_2	p_3	p_4	p_5	p_6	p_7	p_8	p_9	p_{10}	p_{11}	p_{12}	p_{13}	p_{14}	p_{15}	p_{16}	p_{17}	p_{18}	p_{19}	p_{20}	p_{21}	p_{22}	p_{23}	p_{24}	p_{25}	p_{26}	p_{27}	p_{28}	p_{29}	p_{30}	p_{31}	p_{32}	p_{33}	p_{34}	p_{35}
m_1	1				1				1		1	1			1		1			1	1	1						1		1		1	1		
m_2		1							1		1	1				1	1				1		1				1								
m_3	1	1	1							1														1		1									
m_4		1				1					1	1									1		1												
m_5					1								1						1								1					1			
m_6							1						1				1			1			1				1								
m_7	1		1		1		1				1			1		1			1	1		1	1	1		1			1	1	1	1			
m_8	1				1			1					1			1				1			1		1										
m_9							1						1						1							1						1			1
m_{10}							1						1		1				1				1				1								
m_{11}				1		1		1							1							1				1		1		1			1		
m_{12}				1		1		1		1										1											1				
m_{13}		1									1	1										1													
m_{14}		1				1			1		1	1					1					1			1		1								
m_{15}				1		1		1		1										1						1		1							
m_{16}				1		1		1			1									1					1			1		1					
m_{17}	1		1		1											1		1		1					1										1
m_{18}		1					1				1	1								1					1					1					
m_{19}				1				1	1											1						1	1	1							
m_{20}						1							1							1					1			1							

The traditional step-by-step method is adjusted to derive the final arrangement (Table 6):

Table 6. Final ALB solution proposed by this study.

	a_{11}	a_{12}	a_{21}	a_{22}	a_{23}	a_{33}	a_{34}	
p_{ij}	59	176	118	235	88	141	282	$\sum p_{ij} = 1099$
γ_{ij}	2832	2816	2832	2820	2816	2820	2820	$\gamma_{line} = 2816$

4.3. Example 3: Comparison with Boe and Cheng's Algorithm

This example compares the proposed algorithm with Boe and Cheng's close neighbor algorithm [66]. The original part–machine incidence matrix is taken from Boe and Cheng as follows:

4.3.1. Machine Sorting

First, the relationship between the two machines must be obtained. For example, Equation (8) is used to obtain b_{37}.

$$
\begin{array}{c}
\quad p_1\ p_2\ p_3\ p_4\ p_5\ p_6\ p_7\ p_8\ p_9\ p_{10}\ p_{11}\ p_{12}\ p_{13}\ p_{14}\ p_{15}\ p_{16}\ p_{17}\ p_{18}\ p_{19}\ p_{20}\ p_{21}\ p_{22}\ p_{23}\ p_{24}\ p_{25}\ p_{26}\ p_{27}\ p_{28}\ p_{29}\ p_{30}\ p_{31}\ p_{32}\ p_{33}\ p_{34}\ p_{35} \\[4pt]
\begin{array}{c} m_3 \\ m_7 \end{array}
\left[
\begin{array}{ccccccccccccccccccccccccccccccccccc}
1 & 1 & 1 &   &   &   &   &   &   &   &   &   &   &   & 1 &   &   &   &   &   &   &   &   &   &   &   &   &   &   & 1 & 1 &   &   &   &   \\
1 & 1 & 1 & 1 &   &   &   &   &   &   & 1 &   &   & 1 &   & 1 &   & 1 & 1 &   & 1 & 1 & 1 &   & 1 &   &   &   &   & 1 & 1 & 1 & 1 &   &   \\
\end{array}
\right]
\end{array}
$$

$$b_{37} = 1 \times 1 + 0 \times 0 + 1 \times 1 + \ldots + 0 \times 0 = 6$$

Thus, the closeness matrix of skills is as follows.

$$
B_u =
\begin{array}{c}
\quad m_1\ m_2\ m_3\ m_4\ m_5\ m_6\ m_7\ m_8\ m_9\ m_{10}\ m_{11}\ m_{12}\ m_{13}\ m_{14}\ m_{15}\ m_{16}\ m_{17}\ m_{18}\ m_{19}\ m_{20}
\end{array}
$$

	m_1	m_2	m_3	m_4	m_5	m_6	m_7	m_8	m_9	m_{10}	m_{11}	m_{12}	m_{13}	m_{14}	m_{15}	m_{16}	m_{17}	m_{18}	m_{19}	m_{20}
m_1	0	2	2	1	2	3	7	5	4	1	5	2	1	2	3	4	5	2	3	2
m_2	2	0	1	5	0	1	4	0	1	1	1	0	4	8	0	2	0	5	0	1
m_3	2	1	0	0	1	0	6	3	0	0	0	0	0	1	0	0	4	1	0	0
m_4	1	5	0	0	0	0	3	0	0	0	1	0	4	6	0	1	0	5	0	0
m_5	2	0	1	0	0	1	4	3	2	2	0	0	0	0	0	1	2	0	0	3
m_6	3	1	0	0	1	0	3	1	3	4	1	0	0	0	1	1	1	0	1	3
m_7	7	4	6	3	4	3	0	6	3	3	3	0	2	4	1	4	6	5	2	3
m_8	5	0	3	0	3	1	6	0	1	0	1	1	0	0	1	1	6	1	1	1
m_9	4	1	0	0	2	3	3	1	0	3	2	0	0	0	0	1	1	0	1	4
m_{10}	1	1	0	0	2	4	3	0	3	0	1	0	0	0	0	1	0	0	0	4
m_{11}	5	1	0	1	0	1	3	1	2	1	0	3	1	1	5	5	1	1	5	0
m_{12}	2	0	0	0	0	0	0	1	0	0	3	0	0	0	5	3	0	0	4	0
m_{13}	1	4	0	4	0	0	2	0	0	0	1	0	0	4	0	1	0	4	0	0
m_{14}	2	8	1	6	0	0	4	0	0	0	1	0	4	0	0	2	0	6	0	0
m_{15}	3	0	0	0	0	1	1	1	0	0	5	5	0	0	0	4	0	0	6	0
m_{16}	4	2	0	1	1	1	4	1	1	1	5	3	1	2	4	0	0	1	4	1
m_{17}	5	0	4	0	2	1	6	6	1	0	1	0	0	0	0	0	0	1	0	0
m_{18}	2	5	1	5	0	0	5	1	0	0	1	0	4	6	0	1	1	0	0	0
m_{19}	3	0	0	0	0	1	2	1	1	0	5	4	0	0	6	4	0	0	0	0
m_{20}	2	1	0	0	3	3	3	1	4	4	0	0	0	0	0	1	0	0	0	0

In step 1 of machine sorting, the first machine, belonging to the first group, is m_2. Thus, $G_1 = \{m_2\}$. Ignoring m_2, the next step is to form a new closeness matrix of skill $B'u = \{b_{kl}\}$, as follows:

$B'_u =$

	m_1	m_2	m_3	m_4	m_5	m_6	m_7	m_8	m_9	m_{10}	m_{11}	m_{12}	m_{13}	m_{14}	m_{15}	m_{16}	m_{17}	m_{18}	m_{19}	m_{20}
m_1	0	2	2	1	2	3	7	5	4	1	5	2	1	2	3	4	5	2	3	2
m_3	2	1	0	0	1	0	6	3	0	0	0	0	0	1	0	0	4	1	0	0
m_4	1	5	0	0	0	0	3	0	0	0	1	0	4	6	0	1	0	5	0	0
m_5	2	0	1	0	0	1	4	3	2	2	0	0	0	0	0	1	2	0	0	3
m_6	3	1	0	0	1	0	3	1	3	4	1	0	0	0	1	1	1	0	1	3
m_7	7	4	6	3	4	3	0	6	3	3	3	0	2	4	1	4	6	5	2	3
m_8	5	0	3	0	3	1	6	0	1	0	1	1	0	0	1	1	6	1	1	1
m_9	4	1	0	0	2	3	3	1	0	3	2	0	0	0	0	1	1	0	1	4
m_{10}	1	1	0	0	2	4	3	0	3	0	1	0	0	0	0	1	0	0	0	4
m_{11}	5	1	0	1	0	1	3	1	2	1	0	3	1	1	5	5	1	1	5	0
m_{12}	2	0	0	0	0	0	0	1	0	0	3	0	0	0	5	3	0	0	4	0
m_{13}	1	4	0	4	0	0	2	0	0	0	1	0	0	4	0	1	0	4	0	0
m_{14}	2	8	1	6	0	0	4	0	0	0	1	0	4	0	0	2	0	6	0	0
m_{15}	3	0	0	0	0	1	1	1	0	0	5	5	0	0	0	4	0	0	6	0
m_{16}	4	2	0	1	1	1	4	1	1	1	5	3	1	2	4	0	0	1	4	1
m_{17}	5	0	4	0	2	1	6	6	1	0	1	0	0	0	0	0	0	1	0	0
m_{18}	2	5	1	5	0	0	5	1	0	0	1	0	4	6	0	1	1	0	0	0
m_{19}	3	0	0	0	0	1	2	1	1	0	5	4	0	0	6	4	0	0	0	0
m_{20}	2	1	0	0	3	3	3	1	4	4	0	0	0	0	0	1	0	0	0	0

The next machine is m_{14}, which belongs to G_1 because $b_{m_{14}m_2}$ is the largest number in $B'u$. Thus, $G_1 = \{m_2,\ m_{14}\}$. The new closeness matrix of skill $B'u = \{b_{kl}\}$ is as follows:

$B'_u =$

	m_1	m_2	m_3	m_4	m_5	m_6	m_7	m_8	m_9	m_{10}	m_{11}	m_{12}	m_{13}	m_{14}	m_{15}	m_{16}	m_{17}	m_{18}	m_{19}	m_{20}
m_1	0	2	2	1	2	3	7	5	4	1	5	2	1	2	3	4	5	2	3	2
m_3	2	1	0	0	1	0	6	3	0	0	0	0	0	1	0	0	4	1	0	0
m_4	1	5	0	0	0	0	3	0	0	0	1	0	4	6	0	1	0	5	0	0
m_5	2	0	1	0	0	1	4	3	2	2	0	0	0	0	0	1	2	0	0	3
m_6	3	1	0	0	1	0	3	1	3	4	1	0	0	0	1	1	1	0	1	3
m_7	7	4	6	3	4	3	0	6	3	3	3	0	2	4	1	4	6	5	2	3
m_8	5	0	3	0	3	1	6	0	1	0	1	1	0	0	1	1	6	1	1	1
m_9	4	1	0	0	2	3	3	1	0	3	2	0	0	0	0	1	1	0	1	4
m_{10}	1	1	0	0	2	4	3	0	3	0	1	0	0	0	0	1	0	0	0	4
m_{11}	5	1	0	1	0	1	3	1	2	1	0	3	1	1	5	5	1	1	5	0
m_{12}	2	0	0	0	0	0	0	1	0	0	3	0	0	0	5	3	0	0	4	0
m_{13}	1	4	0	4	0	0	2	0	0	0	1	0	0	4	0	1	0	4	0	0
m_{15}	3	0	0	0	0	1	1	1	0	0	5	5	0	0	0	4	0	0	6	0
m_{16}	4	2	0	1	1	1	4	1	1	1	5	3	1	2	4	0	0	1	4	1
m_{17}	5	0	4	0	2	1	6	6	1	0	1	0	0	0	0	0	0	1	0	0
m_{18}	2	5	1	5	0	0	5	1	0	0	1	0	4	6	0	1	1	0	0	0
m_{19}	3	0	0	0	0	1	2	1	1	0	5	4	0	0	6	4	0	0	0	0
m_{20}	2	1	0	0	3	3	3	1	4	4	0	0	0	0	0	1	0	0	0	0

The third machine is m_1, which belongs to a new group, G_2, because $b_{m_1m_7}$ is the largest number in $B'u$, but machine m_7 does not belong to any group. Thus, $G_2 = \{m_1\}$. Applying steps 2–3 in Section 3.3.1 gives four groups: $G_1 = \{m_2,\ m_{14},\ m_4,\ m_{18},\ m_{13}\}$, $G_2 = \{m_1,\ m_7,\ m_3,\ m_8,\ m_{17},\ m_5,\ m_9\}$, $G_3 = \{m_{15},\ m_{19},\ m_{11},\ m_{12},\ m_{16}\}$, and $G_4 = \{m_6,\ m_{10},\ m_{20}\}$.

4.3.2. Part Sorting

By using the operator sorting step in Section 3.3.2, the parts can be sorted. For example, p_{24}, the $\sum_{u \in G_1} so_{u p_{24}} = 5$, and 2, 1, and 0 in G_2, G_3, and G_4, respectively. Thus, p_{24} belongs to G_1.

Finally, the four groups are ordered (Table 7).

Table 7. The four groups in example 3.

Groups	Group Members
G_1	$m_2, m_{14}, m_4, m_{18}, m_{13}, p_2, p_7, p_{10}, p_{12}, p_{13}, p_{18}, p_{24}, p_{27}, p_{31}$
G_2	$m_1, m_7, m_3, m_8, m_{17}, m_5, m_9, p_1, p_3, p_5, p_{15}, p_{17}, p_{20}, p_{23}, p_{25}, p_{29}, p_{34}, p_{35}$
G_3	$m_{15}, m_{19}, m_{11}, m_{12}, m_{16}, p_4, p_6, p_9, p_{11}, p_{21}, p_{28}, p_{30}, p_{32}, p_{33}$
G_4	$m_6, m_{10}, m_{20}, p_8, p_{14}, p_{16}, p_{19}, p_{22}, p_{26},$

The sorted incidence matrix is as follows:

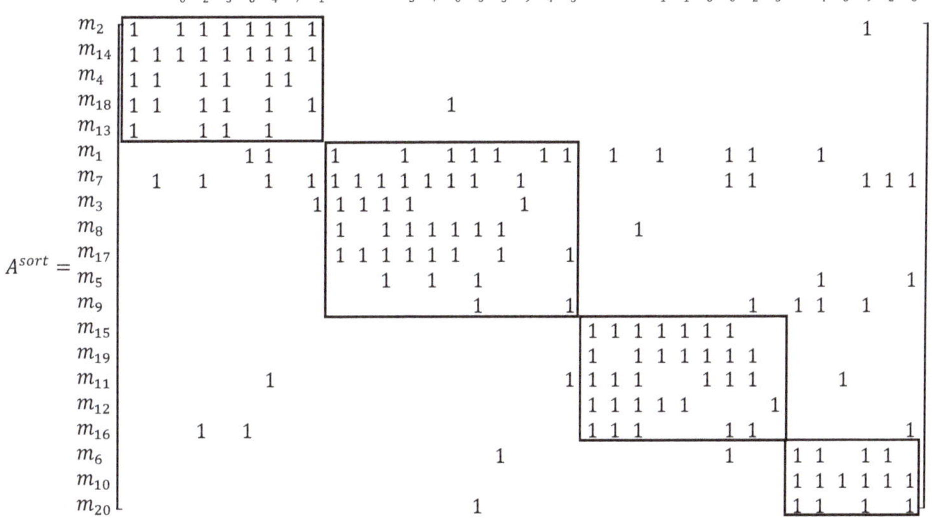

Boe and Cheng's solution of this example is as follows:

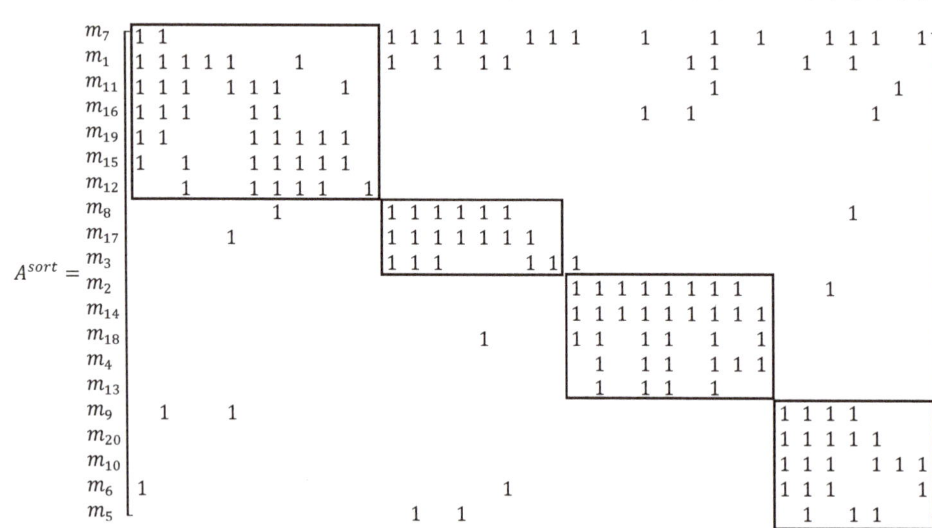

The GT efficiency of the proposed algorithm is obtained using Equation (9), as follows:

$$\eta = 0.5 \times \frac{33 + 40 + 31 + 14}{5 \times 9 + 7 \times 11 + 5 \times 9 + 3 \times 6} + 0.5 \times \left(1 - \frac{35}{20 \times 35 - (5 \times 9 + 7 \times 11 + 5 \times 9 + 3 \times 6)}\right) = 78.5\%$$

In the same way, GT efficiency is calculated using Boe and Cheng's algorithm, as follows:

$$\eta = 0.5 \times \frac{40 + 18 + 33 + 22}{7 \times 11 + 3 \times 8 + 5 \times 9 + 5 \times 7} + 0.5 \times \left(1 - \frac{40}{20 \times 35 - (7 \times 11 + 3 \times 8 + 5 \times 9 + 5 \times 7)}\right) = 77.4\%$$

4.4. Example 4: Reducing Training Time

This example demonstrates how to find an operator with the training and skills that matches a workstation's skill requirements so that the operator does not need to take any training courses. Consider the problem displayed in Figure 3 (Section 2). The proposed algorithm is used to obtain a sorted skill-operator matrix, as follows:

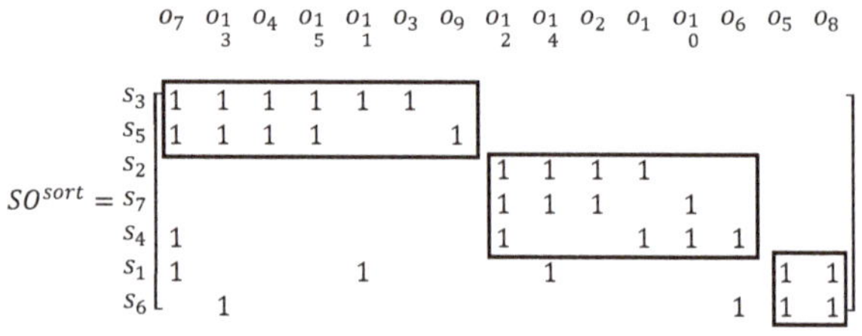

A manager could use the sorted skill–operator matrix as a database for choosing an operator with the right training and skills. Based on the database, the manager could quickly rotate a suitable worker to a workstation. Based on the sorted skill–operator matrix SO^{sort}, the manager could choose operator o_{11} for the workstation that requires skill s_1; o_2,

o_{12}, and o_{14} because of s_2; o_4, o_7, o_{13}, and o_3, because of s_3; o_1 and o_6 because of s_4; o_{15} and o_9 because of s_5; o_5 and o_8 because of s_6; and o_{10} because of s_7, as follows:

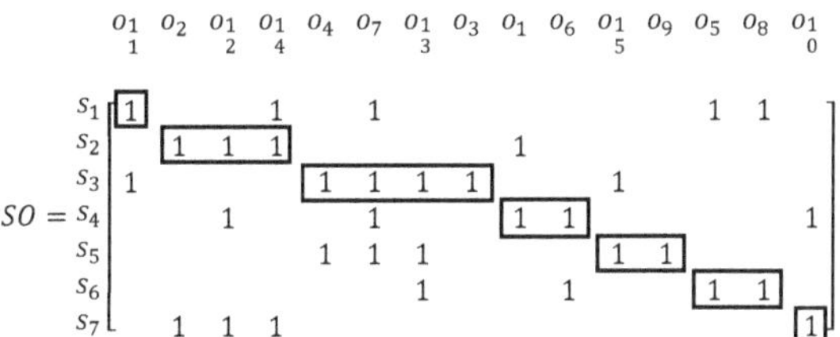

Operators in the same skills group can support other operators in the group.

5. Results and Discussion

In example 1, stations a_{23} and a_{24} are the bottleneck stations. However, only one operator, namely the 500th one, can be assigned. Assigning the 500th operator to either station a_{23} or a_{24} cannot increase the output of the assembly line. This study suggests that the 500th worker should be laid off or shifted to another department. Alternatively, a 501st operator could be hired, which would allow another worker to be assigned to station a_{23} and another to be assigned to station a_{24}. This example demonstrates the proposed method in detail. The maximum output rate and the number of workers at each workstation can be decided. It also shows that multiple workstations can simultaneously become bottleneck stations.

In example 2, stations a_{12} and a_{23} are the bottleneck stations, and only one operator can be assigned. This study suggests that the 1100th worker be laid off or shifted to another department. Alternatively, a 1101st operator could be hired, which would allow another worker to be assigned to station a_{12} and another to be assigned to station a_{23}. With these solutions, a higher production rate than in the original arrangement (2816 > 2400) is achieved without any increase in cost. It shows that a manager can improve productivity by rearranging the number of workers at each workstation, while not increasing costs. It will also cut down on worker complaints because the workers' workloads are balanced.

In example 3, the proposed algorithm is used to categorize four groups, as shown in Section 4.3. Boe and Cheng's algorithm is compared with the GT efficiency measure proposed by Chandrasekharan and Rajagopalan. The GT efficiency of the proposed algorithm is 78.5%, higher than the 77.4% of Boe and Cheng's algorithm. Thus, the proposed GT algorithm works and can perform better than other algorithms in the literature. Workers are assigned to the workstation that best uses their skills, so they feel that their contribution is meaningful and more substantial. Workers at the same workstation also share more skills in common. This makes it easier for the manager to alternate workers.

In example 4, operators within the same skill group can support each other, giving managers more worker rotation solutions. With workers suggested by the sorted skill–operator matrix, managers can quickly match suitable operators with workstations so that productivity remains high. Using the matrix to match workers' training and skills to workstations' skill requirements also enables managers to reduce operator training time and cover workers' absences. For example, if o_4 is absent or busy, then o_{15} can support or share the workload without needing too much training. The risk of an unexpected worker absence or resignation can be mitigated by alternating the workers according to the sorted skill–operator matrix.

6. Conclusions

Company managers aim to achieve sustainable operations, part of which is managing labor costs. To reduce labor costs, an assembly line manager increases the line's efficiency and reduces the training time needed by operators. Inadequate training or distractions, such as an operator focused on a personal problem, can be dangerous, something a plant manager seeks to eliminate, while worker absences hurt efficiency and increase costs. This paper presents an algorithm that helps managers keep assembly lines balanced by arranging operators, so a line's workstations achieve the same output rates. The first example demonstrates how the proposed method can decide the maximum output rate and the number of workers at each workstation given a constant number of operators, especially when setting up a new plant. The second example demonstrates how the method can audit the existing number of operators at each workstation. Using the method to readjust the number of operators at a workstation can increase the output rate, without raising costs. As indicated in the numerical examples, another challenge is that multiple workstations can simultaneously become bottleneck stations. Knowing the number of operators at each workstation, the proposed cluster algorithm can group operators with similar skills and training, which reduces the training time. The third example demonstrates how the proposed clustering algorithm performs better than a previously described algorithm. Within the same group, operators from one workstation can also support operators at other workstations, as demonstrated in example 4. The clustering algorithm could be used by the human resources department to make a database that includes all of the company employees. When an employee is absent, a manger could quickly find a replacement, mitigating the risk of an unexpected absence. The manager could also use such a database to determine which employees should take a training course. By not having all of the employees take the training course, training costs could be reduced. Future applications of the proposed method could include the creation of a course-faculty database, where it could be used to find a replacement instructor for a class when the assigned instructor is absent.

Traditional GT considers only machines and parts, but this study uses the concept of GT to reduce the training time of workstation operators so that managers can achieve sustainable operations. The sorted skill–operator matrix in this study can, similar to a database, provide a manager with suggestions. However, most managers prefer a direct solution to a problem; they want the specific worker whose training and skills match the skill requirements of the problem workstation. Thus, future studies could develop an algorithm to assign specific workers to specific workstations by using a sorted skill–operator matrix, especially for large factories with multiple assembly lines.

Funding: This research received no external funding.

Institutional Review Board Statement: Not applicable.

Informed Consent Statement: Not applicable.

Data Availability Statement: Not applicable.

Conflicts of Interest: The author declares no conflict of interest.

Nomenclature

$A = \left\{ a_{ij} \right\}_{mn}$	Matrix of workstations
t_{ij}	Standard time of workstation a_{ij}
γ_{ij}	Output of workstation a_{ij} per operator per day
p_{ij}	Number of operators at workstation a_{ij}
λ_{ij}	Output of workstation a_{ij} per day
λ_{line}	Output of assembly line per day
TP	Total number of operators
T	Work time per day
$SO = \left\{ so_{ij} \right\}_{uv}$	Matrix of the skill-operator
u	Number of skills
v	Number of operators
$B_u = \left\{ b_{rs} \right\}_{uu}$	Closeness matrix of skills

$B\prime_u = \left\{ b_{ij} \right\}$ New closeness matrix of skills without consideration of sorted rows

G_i — group

R — Total number of groups

η — Group efficiency

e_b — Total number of 1s in the groups

e_0 — Total number of 1s not in the groups

NS_i — Number of skills in the ith group

NO_i — Number of operators in the ith group

— Weighting factor

References

1. Netessine, S.; Taylor, T.A. Product line design and production technology. *Mark. Sci.* **2007**, *26*, 101–117. [CrossRef]
2. Sabuncuoglu, I.; Erel, E.; Tanyer, M. Assembly line balancing using genetic algorithms. *J. Intell. Manuf.* **2000**, *11*, 295–310. [CrossRef]
3. Koo, P.-H. A New Self-Balancing Assembly Line Based on Collaborative Ant Behavior. *Appl. Sci.* **2020**, *10*, 6845. [CrossRef]
4. Adeppa, A. A Study on Basics of Assembly Line Balancing. *Int. J. Emerg. Technol.* **2015**, *6*, 294–297.
5. Nagy, L.; Ruppert, T.; Abonyi, J. Analytic Hierarchy Process and Multilayer Network-Based Method for Assembly Line Balancing. *Appl. Sci.* **2020**, *10*, 3932. [CrossRef]
6. Suer, G.A. Designing parallel assembly lines. *Comput. Ind. Eng.* **1998**, *35*, 467–470. [CrossRef]
7. Kim, S.; Lee, Y.H.; Yang, T.; Park, N. Robust production control policies considering WIP balance and setup time in a semiconductor fabrication line. *Int. J. Adv. Manuf. Technol.* **2008**, *39*, 333–343. [CrossRef]
8. Yuan, M.; Yu, H.; Huang, J.; Ji, A. Reconfigurable assembly line balancing for cloud manufacturing. *J. Intell. Manuf.* **2019**, *30*, 2391–2405. [CrossRef]
9. Rafael, P.; Alberto, G.V.; Manuel, L.; Rafael, M. Metaheuristic procedures for the lexicographic bottleneck assembly line balancing problem. *J. Oper. Res. Soc.* **2015**, *66*, 1815–1825. [CrossRef]
10. Kuroda, M.; Tomita, T. Robust design of a cellular-line production system with unreliable facilities. *Comput. Ind. Eng.* **2005**, *48*, 537–551. [CrossRef]
11. Barthold, J.J.; Eisenstein, D.D. A production line that balances itself. *Oper. Res.* **1996**, *44*, 21–31. [CrossRef]
12. Huang, L.H. Labour assignment and workload balance evaluation for a production line. *Est. J. Eng.* **2009**, *15*, 34–47. [CrossRef]
13. Sawik, T. Mixed integer programming for scheduling flexible flow lines with limited intermediate buffers. *Math. Comput. Mdlng.* **2000**, *31*, 39–52. [CrossRef]
14. Muth, E.J. The reversibility property of production lines. *Manag. Sci.* **1979**, *25*, 152–158. [CrossRef]
15. Zavadlav, E.; McClain, J.O.; Thomas, L.J. Self-buffering, self-balancing, self-flushing production lines. *Manag. Sci.* **1996**, *42*, 1151–1164. [CrossRef]
16. Hazır, O.; Agi, M.A.N.; Guerin, J. A fast and effective heuristic for smoothing workloads on assembly lines: Algorithm design and experimental analysis. *Comput. Oper. Res.* **2020**, *115*, 104857. [CrossRef]
17. Dimitriadis, S.G. Assembly line balancing and group working: A heuristic procedure for workers' groups operating on the same product and workstation. *Comput. Oper. Res.* **2006**, *33*, 2757–2774. [CrossRef]
18. Lian, J.; Liu, C.G.; Li, W.J.; Yin, Y. A multi-skilled worker assignment problem in seru production systems considering the worker heterogeneity. *Comput. Ind. Eng.* **2018**, *118*, 366–382. [CrossRef]
19. Samouei, O.; Fattahi, P.; Ashayeri, J.; Ghazinoory, S. Bottleneck easing-based assignment of work and product mixture determination: Fuzzy assembly line balancing approach. *Appl. Math. Model.* **2016**, *40*, 4323–4340. [CrossRef]
20. Miralles, C.; Garcia-Sabater, J.P.; Andres, C.; Cardos, M. Branch and bound procedures for solving the Assembly Line Worker Assignment and Balancing Problem: Application to Sheltered Work centres for Disabled. *Discrete Appl. Math.* **2008**, *156*, 352–367. [CrossRef]
21. Borba, L.; Ritt, M. A heuristic and a branch-and-bound algorithm for the Assembly Line Worker Assignment and Balancing Problem. *Comput. Oper. Res.* **2014**, *45*, 87–96. [CrossRef]
22. Pearce, B.W.; Antani, K.; Mears, L.; Funk, K.; Mayorga, M.E.; Kurz, M.E. An effective integer program for a general assembly line balancing problem with parallel workers and additional assignment restrictions. *J. Manuf. Syst.* **2019**, *50*, 180–192. [CrossRef]
23. Nourmohammadi, A.; Fathi, M.; Zandieh, M.; Ghobakhloo, M. A Water-Flow Like Algorithm for Solving U-Shaped Assembly Line Balancing Problems. *IEEE Access* **2019**, *7*, 129824–129833. [CrossRef]
24. Pastor, R. LB-ALBP: The lexicographic bottleneck assembly line balancing problem. *Int. J. Prod. Res.* **2011**, *49*, 2425–2442. [CrossRef]
25. Karas, A.; Ozcelik, F. Assembly line worker assignment and rebalancing problem: A mathematical model and an artificial bee colony algorithm. *Comput. Ind. Eng.* **2021**, *156*, 107195. [CrossRef]
26. Oksuz, M.K.; Buyukozkan, K.; Satoglu, S.I. U-shaped assembly line worker assignment and balancing problem: A mathematical model and two meta-heuristics. *Comput. Ind. Eng.* **2017**, *112*, 246–263. [CrossRef]
27. Saif, U.; Guan, Z.; Zhang, L.; Zhang, F.; Wang, B.; Mirza, J. Multi-objective artificial bee colony algorithm for order oriented simultaneous sequencing and balancing of multi-mixed model assembly line. *J. Intell. Manuf.* **2019**, *30*, 1195–1220. [CrossRef]

28. Mutlu, O.; Polat, O.; Supciller, A.A. An iterative genetic algorithm for the assembly line worker assignment and balancing problem of type-II. *Comput. Oper. Res.* **2013**, *40*, 418–426. [CrossRef]
29. Zacharia, P.T.; Nearchou, A.C. A population-based algorithm for the bi-objective assembly line worker assignment and balancing problem. *Eng. Appl. Artif. Intell.* **2016**, *49*, 1–9. [CrossRef]
30. Zacharia, P.T.; Nearchou, A.C. Multi-objective fuzzy assembly line balancing using genetic algorithms. *J. Intell. Manuf.* **2012**, *23*, 615–627. [CrossRef]
31. Vila, M.; Pereira, J. A branch-and-bound algorithm for assembly line worker assignment and balancing problems. *Comput. Oper. Res.* **2014**, *44*, 105–114. [CrossRef]
32. Blum, C.; Miralles, C. On solving the assembly line worker assignment and balancing problem via beam search. *Comput. Oper. Res.* **2011**, *38*, 328–339. [CrossRef]
33. Erel, E.; Sabuncuoglu, I.; Sekerci, H. Stochastic assembly line balancing using beam search. *Int. J. Prod. Res.* **2005**, *43*, 1411–1426. [CrossRef]
34. Sun, B.Q.; Wang, L. A decomposition-based matheuristic for supply chain network design with assembly line balancing. *Comput. Ind. Eng.* **2019**, *131*, 408–417. [CrossRef]
35. Ramezanian, R.; Ezzatpanah, A. Modeling and solving multi-objective mixed-model assembly line balancing and worker assignment problem. *Comput. Ind. Eng.* **2015**, *87*, 74–80. [CrossRef]
36. Zhang, Z.; Tang, Q.H.; Ruiz, R.; Zhang, L. Ergonomic risk and cycle time minimization for the U-shaped worker assignment assembly line balancing problem: A multi-objective approach. *Comput. Oper. Res.* **2020**, *118*, 104905. [CrossRef]
37. Zhang, Z.; Tang, Q.; Chica, M. Multi-manned assembly line balancing with time and space constraints: A MILP model and memetic ant colony system. *Comput. Ind. Eng.* **2020**, *150*, 106862. [CrossRef]
38. Meng, K.; Tang, Q.; Zhang, Z.; Qian, X. An Improved Lexicographical Whale Optimization Algorithm for the Type-II Assembly Line Balancing Problem Considering Preventive Maintenance Scenarios. *IEEE Access* **2020**, *8*, 30421–30435. [CrossRef]
39. Song, B.L.; Wong, W.K.; Chan, S.F. A recursive operator allocation approach for assembly line-balancing optimization problem with the consideration of operator efficiency. *Comput. Ind. Eng.* **2006**, *51*, 585–608. [CrossRef]
40. Mutlu, O.; Ozgormus, E. A fuzzy assembly line balancing problem with physical workload constraints. *Int. J. Prod. Res.* **2012**, *50*, 5281–5291. [CrossRef]
41. Zhong, Y.; Deng, Z.; Xu, K. An effective artificial fish swarm optimization algorithm for two-sided assembly line balancing problems. *Comput. Ind. Eng.* **2019**, *138*, 106121. [CrossRef]
42. Sahin, M.; Kellegoz, T. A new mixed-integer linear programming formulation and particle swarm optimization based hybrid heuristic for the problem of resource investment and balancing of the assembly line with multi-manned workstations. *Comput. Ind. Eng.* **2019**, *133*, 107–120. [CrossRef]
43. Pilati, F.; Ferrari, E.; Gamberi, M.; Margelli, S. Multi-Manned Assembly Line Balancing: Workforce Synchronization for Big Data Sets through Simulated Annealing. *Appl. Sci.* **2021**, *11*, 2523. [CrossRef]
44. Jolai, F.; Jahangoshai Rezaee, M.; Vazifeh, A. Multi-criteria decision making for assembly line balancing. *J. Intell. Manuf.* **2009**, *20*, 113. [CrossRef]
45. Meng, W.; Zhang, X. Optimization of Remanufacturing Disassembly Line Balance Considering Multiple Failures and Material Hazards. *Sustainability* **2020**, *12*, 7318. [CrossRef]
46. Xia, X.; Liu, W.; Zhang, Z.; Wang, L.; Cao, J.; Liu, X. A Balancing Method of Mixed-model Disassembly Line in Random Working Environment. *Sustainability* **2019**, *11*, 2304. [CrossRef]
47. Cao, J.; Xia, X.; Wang, L.; Zhang, Z.; Liu, X. A Novel Multi-Efficiency Optimization Method for Disassembly Line Balancing Problem. *Sustainability* **2019**, *11*, 6969. [CrossRef]
48. Szwarc, E.; Wikarek, J. Proactive Planning of Project Team Members' Competences. *Found. Manag.* **2020**, *12*, 71–84. [CrossRef]
49. Szwarc, E.; Bocewicz, G.; Bach-Dąbrowska, I.; Banaszak, Z. Declarative Model of Competences Assessment Robust to Personnel Absence. In Proceedings of the International Conference on Information and Software Technologies (ICIST 2019), Vilnius, Lithuania, 10–12 October 2019; pp. 12–23.
50. Shahdi-Pashaki, S.; Teymourian, E.; Kayvanfar, V.; Komaki, G.M.; Sajadi, A. Group technology-based model and cuckoo optimization algorithm for resource allocation in cloud computing. *IFAC* **2015**, *48*, 1140–1145. [CrossRef]
51. Zohrevand, A.M.; Rafiei, H.; Zohrevand, A.H. Multi-objective dynamic cell formation problem: A stochastic programming approach. *Comput. Ind. Eng.* **2016**, *98*, 323–332. [CrossRef]
52. McAuley, J. Machine grouping for efficient production. *Prod. Engr.* **1972**, *51*, 53–57. [CrossRef]
53. Li, M.-L.; Parkin, R.E. Group technology revisited: A simple and robust algorithm with enhanced capability. *Int. J. Prod. Res.* **1997**, *35*, 1969–1992. [CrossRef]
54. Boulif, M.; Atif, K. A new branch-&-bound-enhanced genetic algorithm for the manufacturing cell formation problem. *Comput. Oper. Res.* **2006**, *33*, 2219–2245. [CrossRef]
55. Li, M.-L. Arrangement of Machines Under Spatial Constraint by Using a Novel Algorithm. *IEEE Access* **2020**, *8*, 144565–144574. [CrossRef]
56. Agustin-Blas, L.E.; Salcedo-Sanz, S.; Ortiz-Garcia, E.G.; Portilla-Figueras, A.; Perez-Bellido, A.M.; Jimenez-Fernandez, S. Team formation based on group technology: A hybrid grouping genetic algorithm approach. *Comput. Oper. Res.* **2011**, *38*, 484–495. [CrossRef]

57. Hazarika, M.; Laha, D. Machine-part cell formation for maximum grouping efficacy based on genetic algorithm. In Proceedings of the 2015 IEEE Workshop on Computational Intelligence: Theories, Applications and Future Directions (WCI), Kanpur, India, 14–17 December 2015; pp. 1–6.
58. Laha, D.; Hazarika, M. A heuristic approach based on Euclidean distance matrix for the machine-part cell formation problem. *Matl. Tdy. Proc.* **2017**, *4*, 1442–1451. [CrossRef]
59. Li, M.-L. A novel algorithm of cell formation with alternative machines and multiple-operation-type machines. *Comput. Ind. Eng.* **2021**, *154*, 107172. [CrossRef]
60. Suzic, N.; Stevanov, B.; Cosic, I.; Anisic, Z.; Sremcev, N. Customizing Products through Application of Group Technology: A Case Study of Furniture Manufacturing. *J. Mech. Eng.* **2012**, *58*, 724–731. [CrossRef]
61. Kumar, R.; Singh, S.P. A similarity score-based two-phase heuristic approach to solve the dynamic cellular facility layout for manufacturing systems. *Eng. Optim.* **2017**, *49*, 1848–1867. [CrossRef]
62. Feng, Y.L.; Li, G.; Sethi, S.P. A three-layer chromosome genetic algorithm for multi-cell scheduling with flexible routes and machine sharing. *Int. J. Prod. Econ.* **2018**, *196*, 269–283. [CrossRef]
63. Imran, M.; Kang, C.; Lee, Y.H.; Jahanzaib, M.; Aziz, H. Cell formation in a cellular manufacturing system using simulation integrated hybrid genetic algorithm. *Comput. Ind. Eng.* **2017**, *105*, 123–135. [CrossRef]
64. Zeidi, J.R.; Javadian, N.; Tavakkoli-Moghaddam, R.; Jolai, F. A hybrid multi-objective approach based on the genetic algorithm and neural network to design an incremental cellular manufacturing system. *Comput. Ind. Eng.* **2013**, *66*, 1004–1014. [CrossRef]
65. Chandrasekharan, M.P.; Rajagopalan, R. MODROC: An extension of rank order clustering for group technology. *Int. J. Prod. Res.* **1986**, *24*, 1221–1233. [CrossRef]
66. Boe, W.J.; Cheng, C.H. A close neighbour algorithm for designing cellular manufacturing systems. *Int. J. Prod. Res.* **1991**, *29*, 2097–2116. [CrossRef]

Article

A Novel Methodology for Simultaneous Minimization of Manufacturing Objectives in Tolerance Allocation of Complex Assembly

Lenin Nagarajan [1,*], Siva Kumar Mahalingam [1], Sachin Salunkhe [1], Emad Abouel Nasr [2], Jõao Paulo Davim [3] and Hussein M. A. Hussein [4,5]

[1] Department of Mechanical Engineering, Vel Tech Rangarajan Dr. Sagunthala R&D Institute of Science and Technology, Chennai 600 062, India; lawan.sisa@gmail.com (S.K.M.); drsalunkhesachin@veltech.edu.in (S.S.)
[2] Industrial Engineering Department, College of Engineering, King Saud University, Riyadh 11421, Saudi Arabia; eabdelghany@ksu.edu.sa
[3] Department of Mechanical Engineering, Campus Universitário de Santiago, University of Aveiro, 3810-193 Aveiro, Portugal; pdavim@ua.pt
[4] Mechanical Engineering Department, Faculty of Engineering, Helwan University, Cairo 11732, Egypt; hussein@h-eng.helwan.edu.eg
[5] Mechanical Engineering Department, Faculty of Engineering, Ahram Canadian University, 6th of October City 19228, Egypt
* Correspondence: n.lenin@gmail.com

Abstract: Tolerance cost and machining time play crucial roles while performing tolerance allocation in complex assemblies. The aim of the proposed work is to minimize the above-said manufacturing objectives for allocating optimum tolerance to the components of complex assemblies, by considering the proper process and machine selections from the given alternatives. A novel methodology that provides a two-step solution is developed for this work. First, a heuristic approach is applied to determine the best machine for each process, and then a combined whale optimization algorithm with a univariate search method is used to allocate optimum tolerances with the best process selection for each sub-stage/operation. The efficiency of the proposed novel methodology is validated by solving two typical tolerance allocation problems of complex assemblies: a wheel mounting assembly and a knuckle joint assembly. Compared with previous approaches, the proposed methodology showed a considerable reduction in tolerance cost and machining time in relatively less computation time.

Keywords: tolerance allocation; machine and process selection; heuristic approach; univariate search method; whale optimization algorithm

Citation: Nagarajan, L.; Mahalingam, S.K.; Salunkhe, S.; Nasr, E.A.; Davim, J.P.; Hussein, H.M.A. A Novel Methodology for Simultaneous Minimization of Manufacturing Objectives in Tolerance Allocation of Complex Assembly. *Appl. Sci.* **2021**, *11*, 9164. https://doi.org/10.3390/app11199164

Academic Editor: Vladimir Modrak

Received: 6 September 2021
Accepted: 27 September 2021
Published: 2 October 2021

Publisher's Note: MDPI stays neutral with regard to jurisdictional claims in published maps and institutional affiliations.

1. Introduction

All aspects of manufacturing such as machine investment cost, manufacturing cost, the functionality of the product, quality of manufacturing, and the reliability of the product directly connect with tolerance allocation. Hence, most of the researchers are still focusing on this research topic to improve manufacturing efficiency. It involves the allocation of critical dimensions of an assembly, known from the product's functional requirements. As per the literature, more equations are available based on the combinations of the component's tolerance values; however, few equations provide better results. Tolerance allocation requires discovering the best possible combination of the component's tolerances by considering the manufacturing objective(s) and the associated constraints. The researchers proposed different tolerance allocation strategies in different periods. The summary of those strategies is explained here. Further, the detailed comparison of the strategies is given in the Supplementary File as Table S1.

Meta-Heuristic Method (MHM): In this method, the tolerance accumulation model, namely the root sum square (RSS) method [1–6], was used to calculate the assembly tolerance value by summing the randomly picked discrete tolerance values. Initially, the discrete tolerance values were identified by splitting the process tolerance limits. Most of the researchers had applied the MHM such as simulated annealing algorithm [7] and genetic algorithm [8–17] for this purpose.

Heuristic Method (HM): The application of this method is seldom found in literature, since the usage of thumb rules, previous experiences, and standards [18,19] in solving the tolerance allocation problems. Nevertheless, the alternative methods, namely the branch and bound algorithm [20] and design of experiments [21], were used to identify the effective tolerance allocation models. Further, a new method was developed by integrating HM with Tabu search for optimal tolerance allocation and subsequent manufacturing cost reduction [22]. Armillotta (2020) [23] minimized the manufacturing cost of the mechanical assemblies by properly allocating the tolerances and choosing the right dimensional properties. Korbi et al. (2021) [24] proposed a computer-aided design model for analyzing the tolerance in manufacturing the mechanical assemblies.

Discrete Cost Function (DCF) and Continuous Cost Function (CCF) Models: The researchers have used different cost function models [25–30] in the various periods to evaluate the manufacturing cost. These models were classified as DCF and CCF based on their nature. However, the researchers mainly used the CCF model, since it yields closed-form solutions to the tolerance allocation problems. On the other hand, the DCF models [4,26,31] were not preferred due to the model fitting errors during the manual formulation. Further, several studies have been carried out by considering the objective function as the sum of quality loss (as per Taguchi quality loss concept) and manufacturing cost [1,8,32–41].

Simple, Complex, and Non-Linear Assembly: The researchers used different tolerance allocation methods to obtain a solution concerning product type. For instance, the LMM was only used for simple products [1,8,42,43], which have only two mating components. Several works have been reported on finding the optimum allocated tolerance values for the components of complex assembly [44–51]. A limited number of authors have concentrated on non-linear assembly products that consist of more than two components [20,26,38].

Alternative Process Selection (APS): In practice, it is possible to produce components using more than one alternative process. It is necessary to select the proper process for the correct component to reduce the manufacturing cost. Some authors have considered alternative process selection (i.e., every combination of the process has a feasible tolerance range, and for a given process combination, the cost of machining is the function of the tolerance value) for optimum tolerance allocation of both simple and complex assemblies [4,10,13,15]. Kumar et al. (2009) [52] dealt with the distribution of tolerance on the component dimension of a complex assembly with alternative process selection. Authors have attempted to reduce the manufacturing cost of a product using the Lagrange multiplier method for complex assemblies with the bottom curve follower method.

Alternative Machine Selection (AMS): It is possible to reduce the manufacturing cost of a product by choosing the suitable machine for the correct process. Several researchers have considered machine selection as one of the criteria for minimizing manufacturing costs in recent years. However, very few authors [9,17,34] have discussed alternative machine selection for optimum tolerance allocation, and even fewer studies [34] on minimizing both cost and machining time with process and machine selection for optimum allocation of tolerance have been reported in the literature.

From this literature review, it appears that no significant effort has been made to consider machine time as an objective in optimum tolerance allocation, even though machine time is a crucial manufacturing parameter. Therefore, in the present study, both tolerance cost and machining time are optimized. Realizing the complexity of the problem, a combined heuristic and univariate search method is introduced to select the best machine for each process and the optimum tolerance for each component using alternative process and machine selection, with available machine time as a constraint.

2. Problem Definition

The biggest challenge facing today's manufacturing companies is to reduce production costs while maintaining better quality and higher productivity. Tolerance allocation plays a vital role in achieving those goals. The production of a component involves selecting processes as well as selecting machines. These decisions directly influence the allocated tolerance values of components. The optimum allocated tolerance values govern the manufacturing costs and machining time of the product. Therefore, an operation may be possible with multiple alternatives, and as such, is treated as a non-polynomial hard problem.

3. Mathematical Formulation

The proposed work aims to simultaneously minimize tolerance cost and machining time, represented in Equation (1). Tolerance cost (TC_i) and machining time (MT_{ik}) are calculated using Equations (2) and (3). Equations (4) and (5), respectively, determine the critical dimension of the sub-assembly (Y) and its tolerance (t_Y). Machine engagement time (met) is estimated using Equation (6). The constraints considered in this work are expressed by Equations (7)–(9). The allocated tolerance (t_i) should be within the process limits, represented by Equation (7). The calculated sub-assembly tolerance within the given sub-assembly tolerance; this constraint is given by Equation (8). The total individual machine engagement time to manufacture the product should be less than the given available machine time (amt) represented in Equation (9). In addition, it is assumed that the following data are known well in advance before manufacturing the product:

- Number of sub-assemblies and their dimensional chain details;
- Allowable tolerance for each sub-assembly (at_y);
- Number of parts in the assembly (nc);
- Number of sub-stages required to make the parts (no);
- Number of possible processes for each sub-stage of the components (np);
- Number of possible machines for each process (nm);
- Number of hours available for each machine (amt);
- Possible process—machine combination details ($opno$);
- Mathematical model to compute tolerance cost and machining time;
- Constants to compute tolerance cost (A and B) and machining time ($X1$ and $Y1$);
- Process tolerance limits ($tmin$ and $tmax$).

$$Z = \min(TC, MT) \tag{1}$$

$$TC_{ijk} = \eta_{jk}\left(A_j + \frac{B_j}{t_{ijk}}\right) \tag{2}$$

$$MT_{ijk} = \eta_{jk}\left(X1_j + \frac{Y1_j}{t_{ijk}}\right) \tag{3}$$

$$Y = \sum_{l=1}^{nc} \pm d_l \tag{4}$$

$$t_Y = \sum_{i=1}^{no} t_{ijk} \tag{5}$$

$$met_k = \sum_{i=1}^{no} MT_{ijk} \tag{6}$$

$$t_{\min} \leq t_i \leq t_{\max} \tag{7}$$

$$t_{aY} \geq t_Y \tag{8}$$

$$amt_k \geq met_k \tag{9}$$

Z Objective function
TC_{ijk} Tolerance cost of ith sub-stage using jth process on kth machine
MT_{ijk} Machining time of ith sub-stage using jth process on kth machine
t_{ijk} Allocated tolerance of ith sub-stage in jth process and kth machine
A_j, B_j Tolerance cost function constants for jth process
$X1_j, Y1_j$ Tolerance machining time function constants for jth process
η_{jk} Efficiency factor using jth process on kth machine
d_l Dimension of lth component
Y Critical dimension of sub-assembly
t_y Calculated tolerance of critical dimension
t_{ay} Specified tolerance of critical dimension
i Sub-stage number index
j Process number index
k Machine number index
l Component number index
met_k kth machine engagement time
amt_k Available time of kth machine

4. Methodology

The proposed method consists of two stages: (i) selection of the best machine for each process by applying a heuristic approach; (ii) selection of the best process and optimum allocated tolerance for each component using combined whale optimization algorithm and univariate search method. In the first stage, the process tolerance is divided into nd number of discrete values using Equation (10) and the allocated tolerance (t_{ejk}) is calculated using Equation (11). The tolerance cost (TC_{ejk}) and machining time (MT_{ejk}) for t_{ejk} are calculated using Equations (2) and (3), respectively. Nagarajan et al. (2018) explained that the distance method is used to combine the two different objective functions into a single one. For each discrete value, points are plotted on a graph where the x-axis and y-axis represent tolerance cost (TC_{ejk}) and machining time (MT_{ejk}), respectively. Assuming point ($x1, y1$) as the origin and point ($x2, y2$) as discrete tolerance cost and machining time, and substituting ($x1, y1$) as (0,0) and ($x2, y2$) as (TC_{ejk}, MT_{ejk}) in Equation (12), then the distance equation becomes Equation (13). The detailed steps of the heuristic and univariate search methods are shown in Figures 1 and 2. The pseudocode for the combined whale optimization algorithm and a univariate search method is presented in Section 5.

$$td_j = \frac{tmax_j - tmin_j}{nd} \tag{10}$$

$$t_{ejk} = tmin_j + (e - 1)td_j \tag{11}$$

where e is the index for discrete point of tolerance and takes from 1,2,3 ... nd.

$$dis = \sqrt{(x2 - x1)^2 + (y2 - y1)^2} \tag{12}$$

$$dis_{ejk} = \sqrt{(TC_{ejk} - 0)^2 + (MT_{ejk} - 0)^2} = \sqrt{TC_{ejk}^2 + MT_{ejk}^2} \tag{13}$$

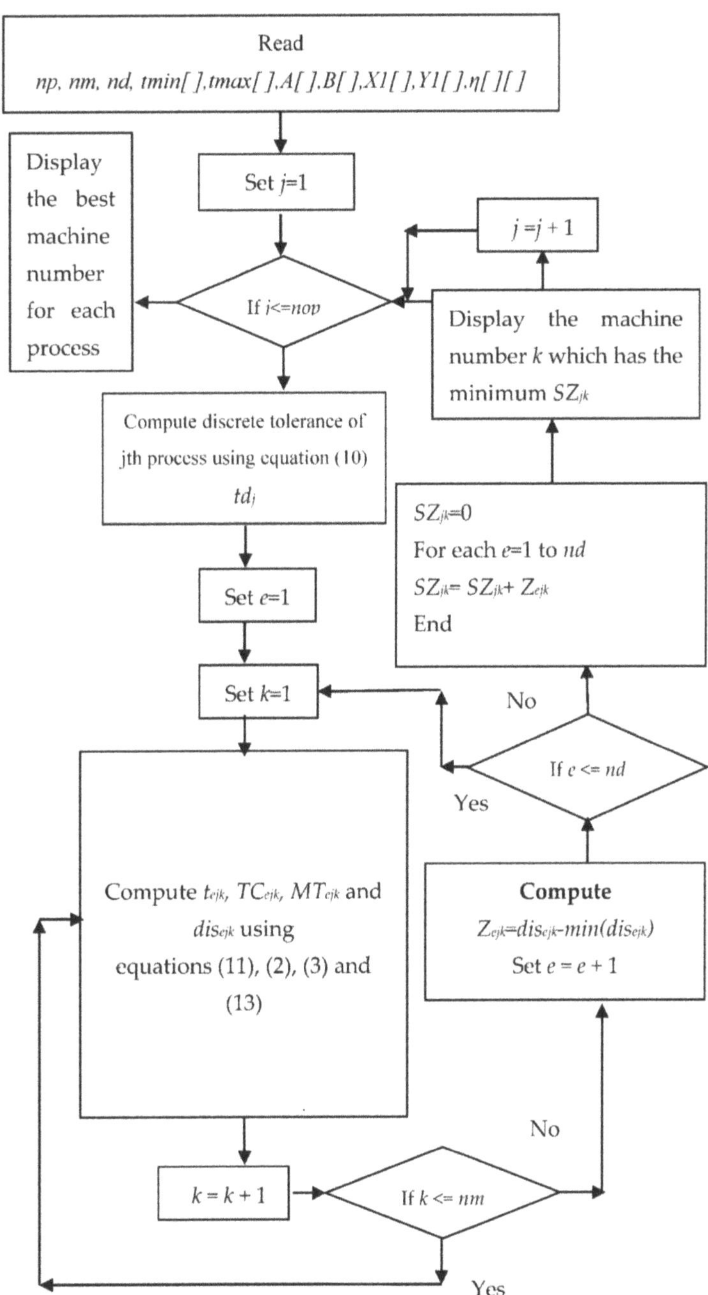

Figure 1. Heuristic approach to determine the best machine for each process.

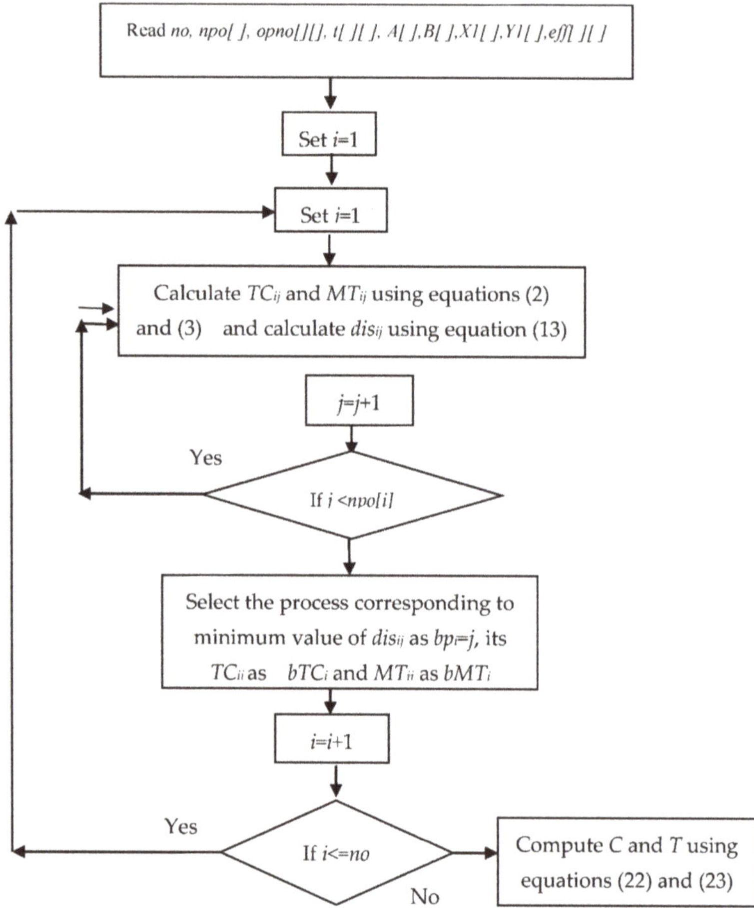

Figure 2. Flow chart of univariate search method.

5. Numerical Illustration

The proposed method is initially implemented in the existing problem (wheel mounting assembly) discussed by Geetha et al. (2013) [34] to show the method's effectiveness in case study 1. Later, it is implemented in knuckle joint assembly in case study 2.
Case study 1: Wheel Mounting Assembly (WMA)

The components of the wheel mounting assembly are given in the Supplemental File as Figure S1. The operations required to manufacture the components of the assembly are illustrated in Figure S1, starting from O1 to O8. The feasibility of performing the operations through the processes from P1 to P5 is given in Table S2. The infeasibility of performing the operation using the specific process is marked as '0'. The same way, the feasibility of using machines for the specific processes is also given in Table S2. Further, the cost and time function constants are represented in Table S3. The first stage of the proposed work, the procedure to determine the best machine for each process, is discussed below.

Figure 3 represents the graphical representation of possible alternative processes and machines to carry out all the sub-stage operations to complete the manufacturing of the product. In the first stage, the best machine is selected for each process by implementing the heuristic approach.

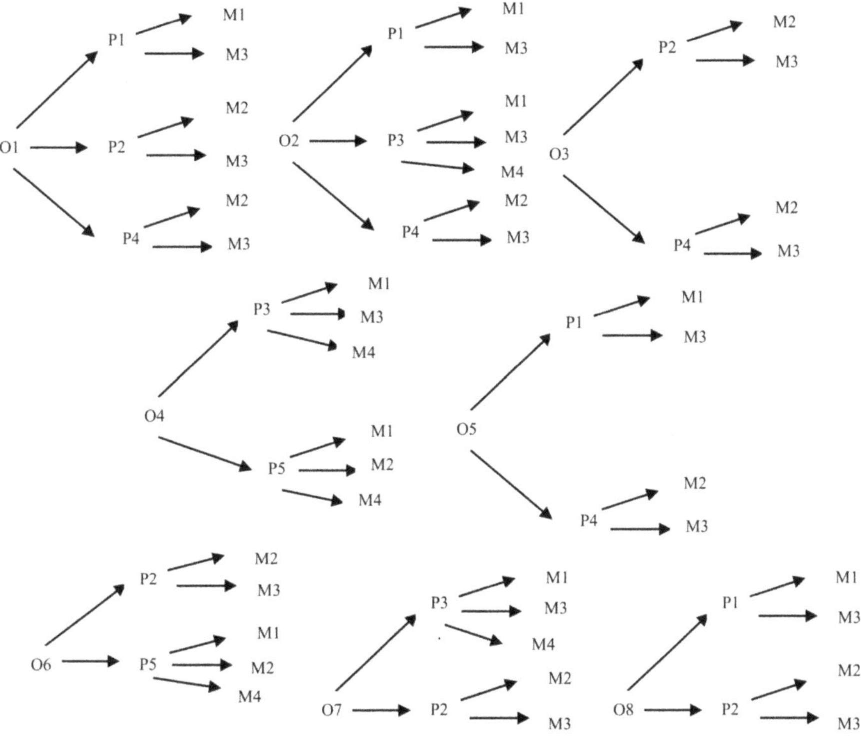

Figure 3. Possible processes and machines for each operation/sub-stage to manufacture WMA.

Using Equations (2), (3), (10), (11) and (13), discrete tolerance (td_j), tolerance (t_{ejk}), tolerance cost (TC_{ejk}), machining time (MT_{ejk}), and distance values (dis_{ejk}) are calculated for each machine and are shown in Table 1. Figure 4 is constructed for a discrete tolerance value of 0.01 mm ($t_{ejk} = 0.01$ mm), assuming TC_{ejk} on the x-axis and MT_{ejk} on the y axis. The distance between the origin and the discrete point of the machine represents the critical factor that decides on selecting the best machine. The value of less distance is more likely to be the best machine for the process.

Table 1. Best machine for process number 1.

T_{ejk}	TC_{e11}	MT_{e11}	dis_{e11}	TC_{e13}	MT_{e13}	dis_{e13}	Min (dis_{ejk})	Z_{e11}	Z_{e13}
0.0100	20.32	33.60	39.27	29.21	48.30	56.45	39.27	0.00	17.18
0.0147	14.21	23.42	27.39	20.43	33.66	39.38	27.39	0.00	11.98
0.0193	11.05	18.15	21.25	15.89	26.09	30.55	21.25	0.00	9.30
0.0240	9.12	14.93	17.50	13.11	21.47	25.15	17.50	0.00	7.66
0.0287	7.82	12.76	14.97	11.24	18.35	21.51	14.97	0.00	6.55
0.0333	6.88	11.20	13.14	9.89	16.10	18.90	13.14	0.00	5.75
0.0380	6.17	10.02	11.77	8.87	14.41	16.92	11.77	0.00	5.15
0.0427	5.62	9.10	10.70	8.08	13.08	15.37	10.70	0.00	4.68
0.0473	5.18	8.36	9.83	7.44	12.02	14.14	9.83	0.00	4.30
0.0520	4.81	7.75	9.13	6.92	11.15	13.12	9.13	0.00	3.99
0.0567	4.51	7.25	8.53	6.48	10.42	12.27	8.53	0.00	3.73
0.0613	4.25	6.82	8.03	6.11	9.80	11.55	8.03	0.00	3.51
0.0660	4.03	6.45	7.60	5.79	9.27	10.93	7.60	0.00	3.33
0.0707	3.84	6.13	7.23	5.52	8.81	10.39	7.23	0.00	3.16
0.0753	3.67	5.85	6.90	5.27	8.41	9.92	6.90	0.00	3.02
0.0800	3.52	5.60	6.61	5.06	8.05	9.51	6.61	0.00	2.89

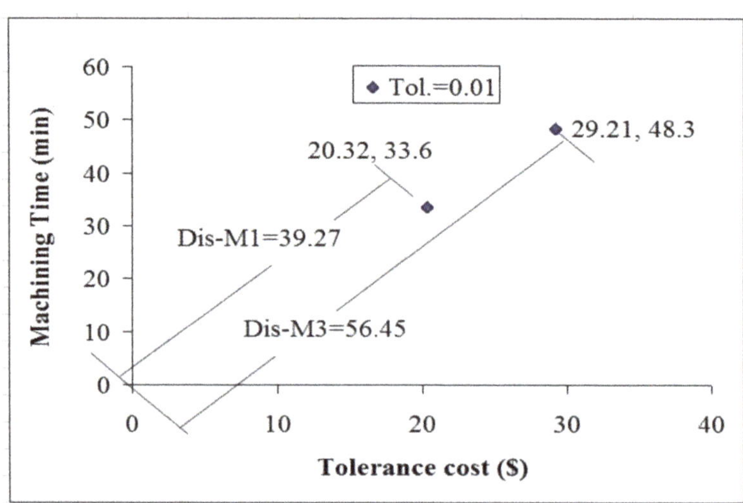

Figure 4. Selection of the best machine based on discrete tolerance.

TC_{e11}—tolerance cost of *e*th discrete tolerance using process number 1 on machine number 1; MT_{e13}—machining time of *e*th discrete tolerance using process number 1 on machine number 3.

As shown in Figure 4, it is clear that *dis-M1* is less than *dis-M3*; therefore, machine 1 is considered the best machine for process 1 for achieving a discrete tolerance value of 0.01 mm. The sum of the difference between minimum discrete distance ($min(dis_{ijk})$) and discrete distance (dis_{ijk}) to that particular machine is calculated to select the best machine suitable for all discrete tolerances of the process tolerance. As shown in Table 1, it can be concluded that machine number 1 is the best machine for process number 1, shown graphically in Figure S2 (Supplementary File). Similarly, the best machine for the other processes is calculated and shown in Figure S2. It is clearly understood that machine numbers 2, 1, 3, and 4 are the best machines for process numbers 2, 3, 4, and 5, respectively.

After implementing the heuristic approach, alternative machines are removed for each process; instead, the best machine is selected, which is shown in Figure 5.

In the second stage, since the same allocated tolerances reported in Geetha et al. (2013) have been used for demonstration purposes, only univariate search method is implemented to get the best process, minimum tolerance cost, and minimum machining time for each sub-stage/operation of WMA. The univariate search method proposed in this work provides the best results in relatively less computation time than the existing method. The search space of the existing method proposed in Geetha et al. (2013) contains 403,200 combinations, whereas, in the proposed method, there are only 11 possible combinations in the search space. Equations (14) and (15) compute the possible combinations in the existing and proposed methods, respectively. Table 2 represents the 11 possible combinations obtained using the univariate search approach, and the first combination is shown as yellow shaded text in Figure 5.

$$ncs_e = \prod_{i=1}^{no} \left(\sum_{j=1}^{np} nm_{ij} \right)$$
$$= (2+2+2)(2+3+2)(2+2)(3+3)(2+2)(2+3)(3+2)(2+2)$$
$$= 6 \times 7 \times 4 \times 6 \times 4 \times 5 \times 5 \times 4 = 403,200 \tag{14}$$

$$ncs_u = 1 + \sum_{i=1}^{no} np_i - no = 1 + 3 + 3 + 2 + 2 + 2 + 2 + 2 + 2 - 8 = 11 \tag{15}$$

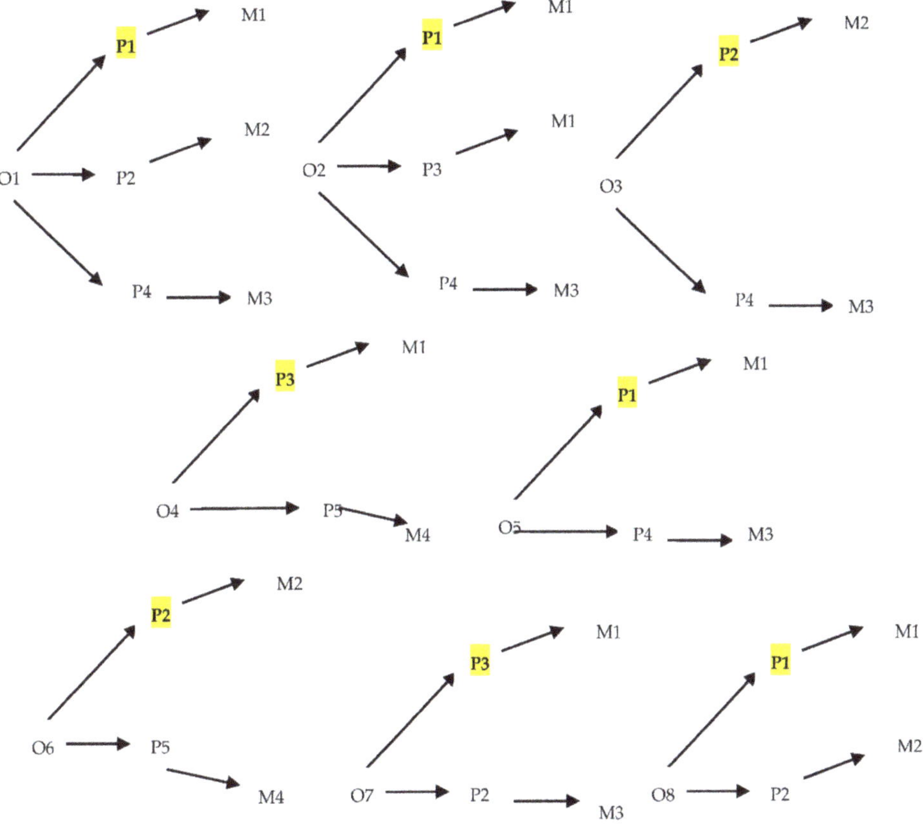

Figure 5. Best machine for process of each operation/sub-stage after implementation of the first stage.

Using Equations (2) and (3), the tolerance cost and machining time are calculated for each operation with alternative processes and its corresponding best machine is shown in Figure 5. In all 11 combinations of alternative processes, the same tolerance mentioned in case 1 and case 2 by Geetha et al. (2013) is assumed for each operation. The alternative process and its corresponding machine-allocated tolerance of each sub-stage/operation and its tolerance cost and machining time are presented in Tables 2 and 3. The best result is shown as shaded text in both Tables 2 and 3.

Figures 6 and 7 show a comparison of the tolerance cost and machining time for the existing problem presented in Geetha et al. (2013) and the proposed method; it can be seen that the proposed method works well, considering tolerance cost, machining time, and both as objective functions.

Table 2. Result of univariate search method for the existing problem.

	Process No. (Machine No.) Table 10 in Geetha et al. (2013) Case 1										
	O1	O2	O3	O4	O5	O6	O7	O8	TC	MT	Dist
	0.060784	0.061882	0.039412	0.041831	0.055569	0.070471	0.062745	0.056824			
1	1(1)	1(1)	2(2)	3(1)	1(1)	2(2)	3(1)	1(1)	35.15	75.64	83.41
2	1(1)	1(1)	2(2)	3(1)	1(1)	2(2)	3(1)	2(2)	35.22	75.66	83.45
3	1(1)	1(1)	2(2)	3(1)	1(1)	2(2)	4(3)	1(1)	37.80	74.10	83.19
4	1(1)	1(1)	2(2)	3(1)	1(1)	5(4)	4(3)	1(1)	37.10	72.11	81.09
5	1(1)	1(1)	2(2)	3(1)	4(3)	5(4)	4(3)	1(1)	38.83	77.58	86.75
6	1(1)	1(1)	2(2)	5(4)	1(1)	5(4)	4(3)	1(1)	37.06	59.53	70.13
7	1(1)	1(1)	4(3)	5(4)	1(1)	5(4)	4(3)	1(1)	38.96	67.29	77.76
8	1(1)	3(1)	2(2)	5(4)	1(1)	5(4)	4(3)	1(1)	36.08	66.30	75.48
9	1(1)	4(3)	2(2)	5(4)	1(1)	5(4)	4(3)	1(1)	36.30	57.74	68.21
10	2(2)	4(3)	2(2)	5(4)	1(1)	5(4)	4(3)	1(1)	36.37	57.93	68.40
11	4(3)	4(3)	2(2)	5(4)	1(1)	5(4)	4(3)	1(1)	37.99	62.97	73.54

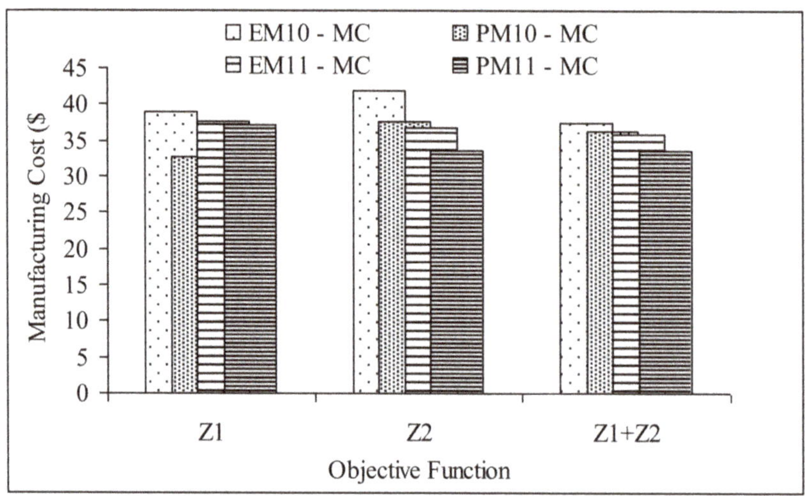

Figure 6. Comparison of tolerance cost of the proposed method with the existing problem presented in Geetha et al. (2013) MC: tolerance cost; EM10 & EM11: existing method, as per data presented in Tables 10 and 11 of Geetha et al.; PM10 & PM11: proposed method, as per data presented in Tables 10 and 11 of Geetha et al. (2013).

Table 3. Details of individual sub-stages of the existing problem presented in Table 11 Geetha et al. (2013) case 2.

O.No.	O1				O2				O3				O4				O5				O6				O7				O8						
Tol	0.0688235				0.0440392				0.0824706				0.0511098				0.0583137				0.0685882				0.0617647				0.0589412				Cmfg	Tmc	Dist
T.No.	P.No.	M.No.	TC	MT	P.No.	M.No.	TC	MT	P.No.	M.No.	TC	MT	P.No.	M.No.	TC	MT	P.No.	M.No.	TC	MT	P.No.	M.No.	TC	MT	P.No.	M.No.	TC	MT	P.No.	M.No.	TC	MT			
1	1	1	3.9	6.2	1	1	5.5	8.9	2	2	3.5	6.3	3	1	3.8	15.9	1	1	4.4	7.1	2	2	4.0	6.7	3	1	3.2	13.6	1	1	4.4	7.0	32.7	71.7	78.8
2	1	1	3.9	6.2	1	1	5.5	8.9	2	2	3.5	6.3	3	1	3.8	15.9	1	1	4.4	7.1	2	2	4.0	6.7	3	1	3.2	13.6	2	2	4.4	7.1	32.8	71.8	78.9
3	1	1	3.9	6.2	1	1	5.5	8.9	2	2	3.5	6.3	3	1	3.8	15.9	1	1	4.4	7.1	2	2	4.0	6.7	4	3	5.9	12.0	1	1	4.4	7.0	35.4	70.1	78.5
4	1	1	3.9	6.2	1	1	5.5	8.9	2	2	3.5	6.3	3	1	3.8	15.9	1	1	4.4	7.1	5	4	3.3	4.7	4	3	5.9	12.0	1	1	4.4	7.0	34.7	68.1	76.4
5	1	1	3.9	6.2	1	1	5.5	8.9	2	2	3.5	6.3	3	1	3.8	15.9	4	3	6.1	12.4	5	4	3.3	4.7	4	3	5.9	12.0	1	1	4.4	7.0	36.4	73.4	81.9
6	1	1	3.9	6.2	1	1	5.5	8.9	2	2	3.5	6.3	5	4	3.9	5.5	1	1	4.4	7.1	5	4	3.3	4.7	4	3	5.9	12.0	1	1	4.4	7.0	34.8	57.8	67.4
7	1	1	3.9	6.2	1	1	5.5	8.9	4	3	5.0	10.0	5	4	3.9	5.5	1	1	4.4	7.1	5	4	3.3	4.7	4	3	5.9	12.0	1	1	4.4	7.0	36.3	61.5	71.4
8	1	1	3.9	6.2	3	1	4.2	18.0	2	2	3.5	6.3	5	4	3.9	5.5	1	1	4.4	7.1	5	4	3.3	4.7	4	3	5.9	12.0	1	1	4.4	7.0	33.5	66.9	74.8
9	1	1	3.9	6.2	5	4	4.2	6.0	2	2	3.5	6.3	5	4	3.9	5.5	1	1	4.4	7.1	5	4	3.3	4.7	4	3	5.9	12.0	1	1	4.4	7.0	33.5	54.9	64.4
10	2	2	4.0	6.7	5	4	4.2	6.0	2	2	3.5	6.3	5	4	3.9	5.5	1	1	4.4	7.1	5	4	3.3	4.7	4	3	5.9	12.0	1	1	4.4	7.0	33.6	55.4	64.8
11	4	3	5.5	11.2	5	4	4.2	6.0	2	2	3.5	6.3	5	4	3.9	5.5	1	1	4.4	7.1	5	4	3.3	4.7	4	3	5.9	12.0	1	1	4.4	7.0	35.2	59.9	69.4

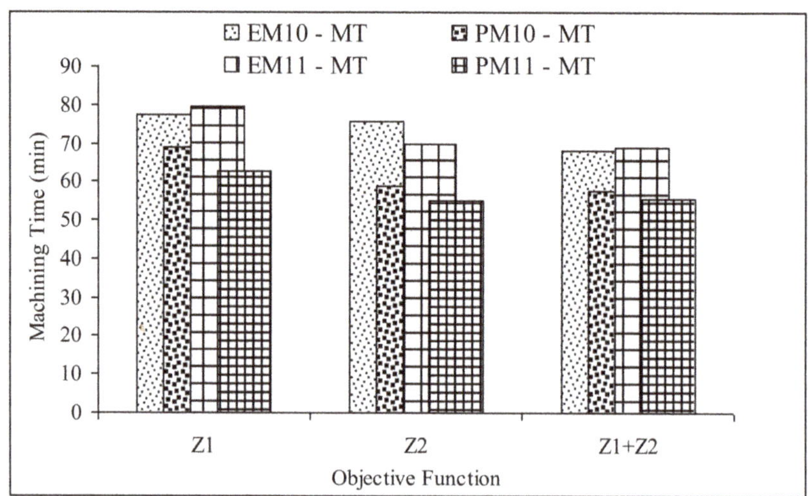

Figure 7. Comparison of machining time of the proposed method with the existing problem presented in Geetha et al. (2013).

Table 4 represents the % of savings in both tolerance cost and machining time as compared to the existing method and the proposed method.

Table 4. % Savings in tolerance cost and machining time.

Details	Case 1		Case 2	
Existing Method	37.32	68.21	35.72	68.90
Proposed Method	36.30	57.74	33.50	54.90
% Savings	2.73	15.35	6.22	20.32

Case Study 2: Knuckle Joint Assembly

In practice, the availability of machine time will restrict the selection of the machine for performing a process, which will influence the tolerance cost and machining time. Therefore, in this work, the available machine time is considered as a constraint in selecting the machine. The methodology is demonstrated using a knuckle joint assembly (Figure 8), consisting of six components performed in ten sub-stages. Two critical dimensions are considered for the proper functioning of the product. It is assumed that nine processes are performed using ten machines. Equations (16) and (17) are used to determine the critical dimension, and the tolerances of the critical dimensions are estimated using Equations (18)–(21). Table 5 shows the dimensions, sub-stages, tolerance symbols, and tolerance stake-up of the knuckle joint assembly.

$$Y1 = d5 - d1 - d2 - d3 - d4 \tag{16}$$

$$Y2 = d2 - d6 \tag{17}$$

$$t_{Y1} = t_{d5} + t_{d1} + t_{d2} + t_{d3} + t_{d4} \tag{18}$$

$$t_{Y1} = t_{O7} + t_{O8} + t_{O1} + t_{O2} + t_{O3} + t_{O4} + t_{O5} + t_{O6} \tag{19}$$

$$t_{Y2} = t_{d2} + t_{d6} \tag{20}$$

$$t_{Y2} = t_{O2} + t_{O3} + t_{O9} + t_{O10} \tag{21}$$

$$C = \sum_{i=1}^{no} TC_i \tag{22}$$

$$T = \sum_{i=1}^{no} MT_i \tag{23}$$

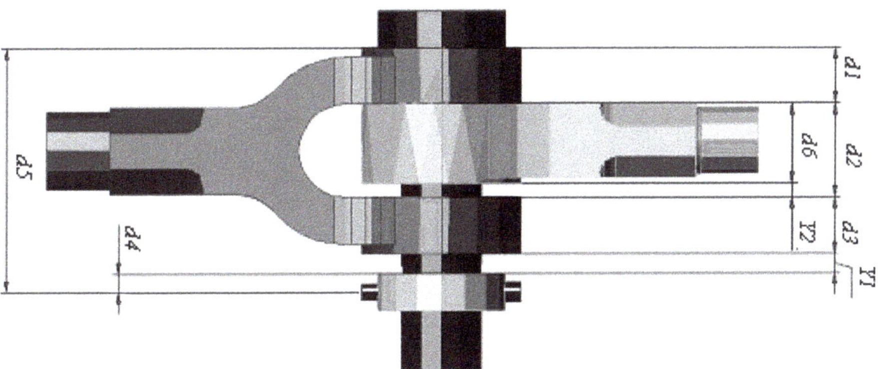

Figure 8. Knuckle joint assembly (KJA).

Table 5. Dimension and tolerance symbol of knuckle joint assembly.

Name of the Component	Dimension No.	Operation/Sub-Stage No.	Tolerance Symbol	Tolerance Stack-Up
Width of fork top	$d1$	$O1$	t_{d1}	t_{O1}
Dimension of fork space	$d2$	$O2$ and $O3$	t_{d2}	$t_{O2} + t_{O3}$
Width of bottom fork	$d3$	$O4$	t_{d3}	t_{O4}
Center distance of hole in spacer	$d4$	$O5$ and $O6$	t_{d4}	$t_{O5} + t_{O6}$
Center distance of hole in pin	$d5$	$O7$ and $O8$	t_{d5}	$t_{O7} + t_{O8}$
Width of collar	$d6$	$O9$ and $O10$	t_{d6}	$t_{O9} + t_{O10}$
Critical dimension 1	$Y1$		t_{Y1}	
Critical dimension 2	$Y2$		t_{Y2}	

Table 6 shows the cost and time function constants of each process of the knuckle joint assembly. The possible operation–process and process–machine feasibility matrix and the available time of the individual machines are presented in Tables 7 and 8, respectively.

Table 6. Cost and time function constants of each process of knuckle joint assembly.

P.No.	A	B	X1	Y1	*tmin*	*tmax*	M1	M2	M3	M4	M5	M6	M7	M8	M9	M10
$P1$	2.5	0.22	4.7	1.3	0.030	0.09	1.3	1.2	0	0.8	0.9	0	1.3	0	0.7	0
$P2$	3.9	0.18	3.8	0.8	0.020	0.07	0.9	0	0.75	0.82	0	1.2	1	0	1.8	1.6
$P3$	2.9	1.23	4.1	1.5	0.030	0.13	0	1.3	1.1	0	1.4	0	1.4	0	0	0.9
$P4$	3.9	2.15	3.9	2	0.009	0.10	0	1.1	0	1.7	0	1.6	1	1.7	0	0
$P5$	1.7	1.09	3.1	0.6	0.020	0.14	0	0.7	1.1	0	0	1.8	0	0.8	0	1.5
$P6$	2.3	1.22	2.1	0.9	0.010	0.13	1.8	0	1.7	0	0.9	0	1.3	0	0.9	0
$P7$	1.7	0.4	4.2	1.2	0.007	0.15	1.7	0	1.2	0	1	1.8	1.7	1.1	0.9	0
$P8$	3.2	1.1	2.9	1.1	0.010	0.15	0.8	0	1	1.7	0	1.4	1.1	0	1	1.6
$P9$	2.1	0.08	5.2	1.7	0.030	0.17	0	1.1	0	0.7	1.8	0	1.1	0	0	0.8

Table 7. Possible operation—process and process—machine feasibility matrix of knuckle joint assembly.

P.No.	Operation/Sub-Stage Number										Machine Number									
	O1	O2	O3	O4	O5	O6	O7	O8	O9	O10	M1	M2	M3	M4	M5	M6	M7	M8	M9	M10
P1	1	0	0	1	1	1	0	0	0	1	1	1	0	1	1	0	1	0	1	0
P2	0	1	1	0	1	0	0	0	1	0	1	0	1	1	0	1	1	0	1	1
P3	1	1	0	0	1	0	1	1	1	1	0	1	1	0	1	0	1	0	0	1
P4	1	1	0	1	1	1	1	0	1	1	0	1	0	1	0	1	1	1	0	0
P5	0	0	0	0	0	1	0	1	1	1	0	1	1	0	0	1	0	1	0	1
P6	1	1	0	1	1	1	0	1	0	1	1	0	1	0	1	0	1	0	1	0
P7	0	0	0	1	1	0	0	1	0	1	1	0	1	0	1	1	1	1	1	0
P8	0	0	1	0	1	1	1	1	0	0	1	0	1	1	0	1	1	0	1	1
P9	0	0	0	0	1	0	1	1	0	0	0	1	0	1	1	0	1	0	0	1

Table 8. Available machine time of each machine in the knuckle joint assembly.

Machine Number	M1	M2	M3	M4	M5	M6	M7	M8	M9	M10
Available machine time (AMT)	75	80	60	40	30	70	30	30	90	50

As per stage 1, the best machine for each process is determined using the heuristic method and is presented in Table 9 and Figure S3 (in Supplementary File).

Table 9. Best machine for each process of the knuckle joint assembly.

Process Number	P1	P2	P3	P4	P5	P6	P7	P8	P9
Best machine number	9	3	10	7	2	5	9	1	4

In the second stage, the whale optimization algorithm is implemented with a univariate search method. The optimum allocated tolerance for each sub-stage/operation, tolerance cost, machining time, and best process is obtained by assuming 100 whales, 100 iterations as stopping criteria with 20 runs. The parameters involved in the algorithm are given in Table 10. Further, the pseudocode of this algorithm is given in the Supplementary File.

Table 10. Terms involved in whale optimization algorithm.

Whale Optimization Algorithm	Optimization Problem
Number of whales ($i = 1,2, \ldots nw$)	Number of solutions = 100
Position of a whale (X_i)	Combination of allocated tolerance of each sub-stage/operation
Number of dimensions involved in defining the position of whale ($j = 1,2, \ldots nd$)	Number of operations = 10
Position of Prey (X_p)	Value of optimum tolerance of each operation (*Tol.*)
Fitness of whale (F_i)	Combined optimum tolerance cost (*TC*) and machining time (*MT*) in terms of distance (*dis*)
Stopping criteria	Maximum number of iteration = 100

The convergence plot for tolerance cost and machining time for stopping criteria considered as 100 iterations are shown in Figure 9. The Pareto optimal solution for a sample run is shown in Figure 10. The tolerance cost and machining time for 31 runs are presented in Table 11. Out of these 31 runs, the best value is calculated using EDAS and CODAS multi-criteria decision-making techniques implemented by Adali & Tuş (2019) [53] The appraisal score (APS) and the assessment score (AS) obtained by implementing EDAS and CODAS method are presented in Table 11. Run number (*R.No.*) 11 has the highest APS value in EDAS, and run number 7 has the highest AS value in CODAS (Highlighted in Table 11). The optimum allocated tolerances of each sub-stage are presented for both methods and are listed in Table 12, along with the tolerance cost and machining time for the required sub-assembly tolerances of 0.55 mm and 0.22 mm. The optimum total tolerance cost and total machining time are presented in Table 13 for the given sub-assembly tolerance values with the constraint of available machine time.

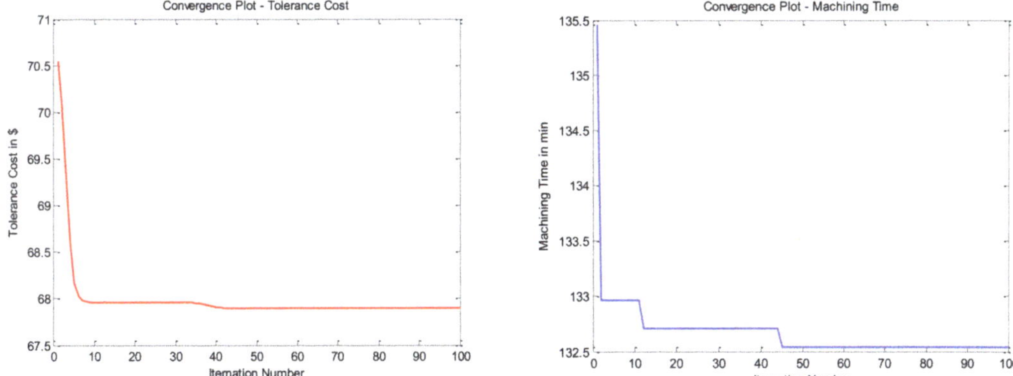

Figure 9. Convergence plot for tolerance cost and machining time.

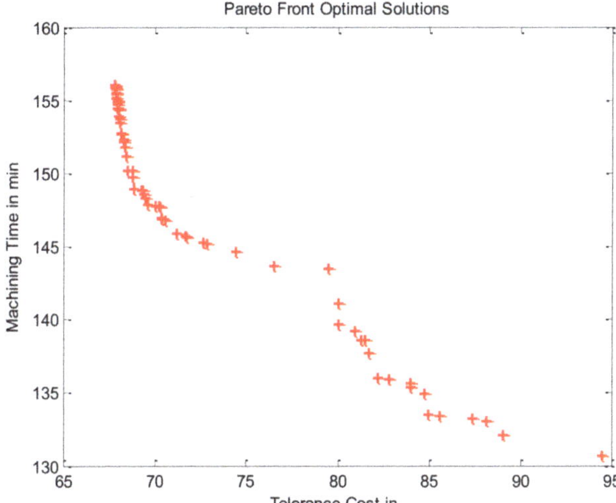

Figure 10. Optimum Pareto front solutions.

Table 14 shows the machine engagement times of the individual machines to manufacture the knuckle joint assembly within the available machine time.

For supporting the proposed method, the statistical analysis for EDAS and CODAS methods are executed through Minitab software. The statistical analysis results and probability plots for both the methods are presented in Figure 11. The probability values are 0.329 and 0.231 for the EDAS and CODAS methods, respectively. Since the value of probability is greater than 0.005 in both cases, it is clearly understood that the results obtained in 31 runs are from normally distributed data.

Table 11. Appraisal (APS) and assessment (AS) scores of EDAS and CODAS.

R.No.	EDAS			CODAS		
	TC	MT	APS	TC	MT	AS
1	68.08	153.97	0.5009	68.01	154.60	−0.0245
2	68.83	149.68	0.5346	67.92	155.39	−0.0106
3	69.75	147.60	0.5308	67.92	158.54	−0.0275
4	69.24	149.31	0.5185	67.93	155.59	−0.0147
5	69.07	149.32	0.5272	67.91	155.64	−0.0109
6	68.83	149.18	0.5433	67.82	155.70	0.0111
7	69.11	149.15	0.5283	67.90	156.88	−0.018
8	82.51	129.70	0.519	67.84	155.22	0.0113
9	68.85	149.45	0.5375	67.81	156.30	0.0085
10	69.45	148.01	0.5367	67.90	159.17	−0.0222
11	68.47	150.20	0.5452	67.83	156.08	0.0062
12	68.57	151.45	0.5181	67.90	158.74	−0.0233
13	68.18	151.55	0.5379	67.87	154.75	0.0075
14	69.23	148.28	0.5417	67.82	158.41	−0.0036
15	68.25	152.04	0.5256	67.92	156.45	−0.0197
16	69.29	149.66	0.5093	68.01	157.39	−0.0453
17	68.49	152.20	0.5093	68.03	155.22	−0.0348
18	68.74	149.58	0.5414	67.89	157.46	−0.018
19	69.20	148.34	0.5418	67.83	157.84	−0.0046
20	69.14	148.82	0.5326	67.82	158.67	−0.0045
21	68.30	151.69	0.5287	67.94	154.87	−0.0095
22	68.81	149.56	0.5379	67.81	157.83	0.0004
23	69.00	149.17	0.5337	67.89	158.09	−0.0191
24	69.28	148.94	0.5224	67.88	157.48	−0.0145
25	68.94	149.36	0.5341	67.89	158.21	−0.0192
26	69.58	148.35	0.5201	67.89	156.61	−0.0134
27	69.07	149.18	0.5302	67.90	157.29	−0.0193
28	69.00	149.61	0.5262	67.98	154.50	−0.0152
29	68.23	151.46	0.5367	67.80	156.40	0.0095
30	68.14	152.32	0.5266	67.94	154.21	−0.0031
31	68.24	152.13	0.5245	67.93	155.66	−0.0167

Table 12. The optimum allocated tolerance of knuckle joint assembly.

O.No.	EDAS					CODAS				
	Tol.	P.No.	M.No.	TC	MT	Tol.	P.No.	M.No.	TC	MT
1	0.074144	1	9	4.58	16.97	0.074438	1	9	4.57	16.93
2	0.040709	2	9	7.22	18.54	0.036949	2	9	7.55	20.04
3	0.039105	2	9	7.35	19.14	0.041246	2	9	7.17	18.35
4	0.081597	1	9	4.39	15.85	0.069137	1	9	4.73	17.86
5	0.065967	2	3	5.95	12.90	0.062934	2	3	6.05	13.33
6	0.089941	5	10	10.18	7.77	0.09	5	10	10.18	7.77
7	0.050332	9	7	3.21	28.84	0.045295	9	7	3.34	31.47
8	0.108207	5	3	8.75	6.98	0.13	5	3	7.57	6.33
9	0.051985	2	9	6.50	15.34	0.051806	2	9	6.51	15.38
10	0.088202	5	7	10.35	7.86	0.089999	5	7	10.18	7.77

O.No.—operation/sub-stage number; *P.No.*—process number; *M.No.*—machine number, *Tol.*—allocated tolerance; *TC*—tolerance cost; *MT*—machining time.

Table 13. Sub-assembly tolerance, total cost, and machining time.

Details	EDAS	CODAS
*ta*1	0.55	0.55
*ta*2	0.22	0.22
TC	68.47	67.84
MT	150.20	155.22

*ta*1—Calculated sub-assembly 1 tolerance. *ta*2—Calculated sub-assembly 2 tolerance.

Table 14. Machine engagement time (*met*) of individual machine.

M.No.	1	2	3	4	5	6	7	8	9	10
EDAS	0	0	19.88	0	0	0	36.70	0	85.85	7.77
CODAS	0	0	19.66	0	0	0	39.24	0	88.55	7.77

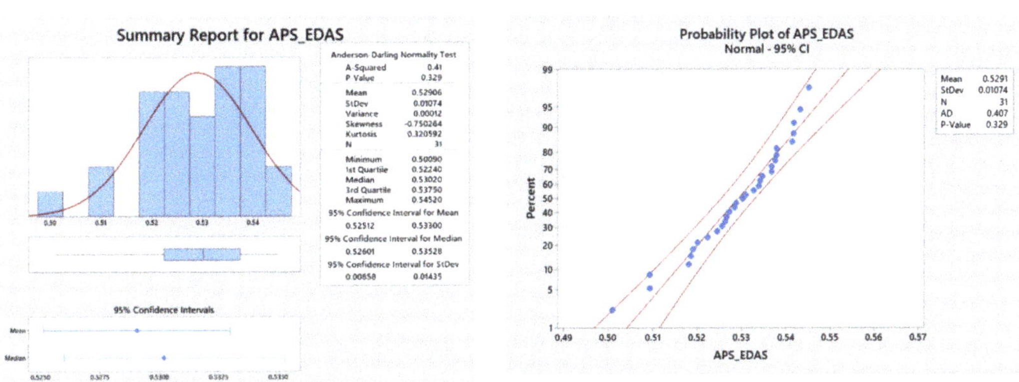

(a) Statistical analysis and probability plot for EDAS method

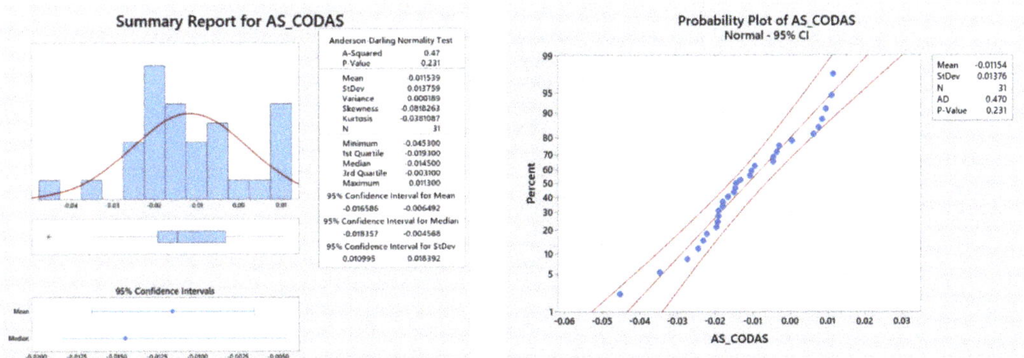

(b) Statistical analysis and probability plot for CODAS method

Figure 11. Statistical analysis and probability plot for EDAS and CODAS methods.

6. Conclusions

Most previous studies on tolerance allocation problems concentrated on minimizing manufacturing costs, quality loss, or combining the two. Machining time, a vital manufacturing objective, has barely been contemplated. In this paper, the machining time was considered along with manufacturing cost in optimum tolerance allocation of complex assemblies, representing a more realistic product development scenario. Alternative machine and process selections with available machine time make this problem cumbersome and complicated. Therefore, a new methodology was developed that applies a heuristic approach and combines whale optimization algorithm with a univariate search method. The total manufacturing cost and machining time of 36.3 USD and 57.74 min reported in this paper for wheel mounting assembly is 2.73% and 15.35% less than the problem dealt with in case 1 by Geetha et al. (2013). Similarly in case 2, there was 6.22% and 20.32% of savings in the tolerance cost and machining time reported by implementing the proposed method. The results presented in this paper demonstrate that the proposed method can reduce tolerance cost and machining time simultaneously with less computation time. The proposed method is also suitable for solving two- and three-dimensional problems. As a further extension of this work, the operation sequence, machine sequence, or both may be considered with additional objectives such as total investment cost of machines, idle time of machines, idle cost of machines, and the number of machines required to manufacture the product.

Supplementary Materials: The following are available online at https://www.mdpi.com/article/10.3390/app11199164/s1, Figure S1: Wheel Mounting Assembly (WMA) [Geetha et al. (2013)], Figure S2: Selection of best machine for process number 1 to 5 to manufacture WMA, Figure S3: Selection of best machine for process number 1 to 9 to manufacture KJA, Table S1: Summary of literature survey, Table S2: Feasibility Matrix for WMA [Geetha et al. (2013)], Table S3: Cost and Time Function Constants for WMA [(Geetha et al. 2013)] [54–69].

Author Contributions: Conceptualization and methodology by L.N. and S.K.M.; writing—original draft preparation by L.N. and S.K.M.; writing—review and editing by S.S., E.A.N., J.P.D. and H.M.A.H. All authors have read and agreed to the published version of the manuscript.

Funding: The authors extend their appreciation to King Saud University for funding this work through Researchers Supporting Project number (RSP-2021/164), King Saud University, Riyadh, Saudi Arabia.

Institutional Review Board Statement: Not applicable.

Informed Consent Statement: Not applicable.

Data Availability Statement: Not applicable.

Conflicts of Interest: The authors declare that they have no conflict of interest.

References

1. Vasseur, H.; Kurfess, T.R.; Cagan, J. Use of a quality loss function to select statistical tolerances. *J. Manuf. Sci. Eng.* **1977**, *119*, 410–416. [CrossRef]
2. Cheng, K.M.; Tsai, J.C. Optimal statistical tolerance allocation of assemblies for minimum manufacturing cost. *Appl. Mech. Mater.* **2011**, *52*, 1818–1823. [CrossRef]
3. Cheng, K.-M.; Jhy-Cherng, T. Optimal statistical tolerance allocation for reciprocal exponential cost–tolerance function. *Proc. Inst. Mech. Eng. Part B J. Eng. Manuf.* **2013**, *227*, 650–656. [CrossRef]
4. Feng, C.X.; Kusiak, A. Robust tolerance design with the integer programming approach. *ASME J. Manuf. Sci. Eng.* **1997**, *119*, 603–610. [CrossRef]
5. Chase, K.W.; Greenwood, W.H.; Loosli, B.G.; Hauglund, L.F. Least cost tolerance allocation for mechanical assemblies with automated process selection. *Manuf. Rev.* **1990**, *3*, 49–59.
6. Jayaprakash, G.; Sivakumar, K.; Thilak, M. FEA compliant parametric tolerance allocation of mechanical assembly using neural network and differential evolution algorithm. *Int. J. Comput. Integr. Manuf.* **2012**, *25*, 636–654. [CrossRef]
7. Zhang, C.; Wang, H.P. The discrete tolerance optimization problem. *Manuf. Rev.* **1993**, *6*, 60–71.
8. Ji, S.; Li, X.; Ma, Y.; Cai, H. Optimal tolerance allocation based on fuzzy comprehensive evaluation and genetic algorithm. *Int. J. Adv. Manuf. Technol.* **2000**, *16*, 461–468. [CrossRef]
9. Uttam, S. Optimal Tolerance Synthesis for Process Planning with Machine selection. Ph.D. Thesis, University of Cincinnati, Cincinnati, OH, USA, 2001.
10. Singh, P.K.; Jain, P.K.; Jain, S.C. Simultaneous optimal selection of design and manufacturing tolerances with different stack-up conditions using genetic algorithms. *Int. J. Prod. Res.* **2003**, *41*, 2411–2429. [CrossRef]
11. Prabhaharan, G.; Asokan, P.; Ramesh, P.; Rajendran, S. Genetic-algorithm-based optimal tolerance allocation using a least-cost model. *Int. J. Adv. Manuf. Technol.* **2004**, *24*, 647–660. [CrossRef]

12. Singh, P.K.; Jain, P.K.; Jain, S.C. A genetic algorithm-based solution to optimal tolerance synthesis of mechanical assemblies with alternative manufacturing processes: Focus on complex tolerancing problems. *Int. J. Prod. Res.* **2004**, *42*, 5185–5215. [CrossRef]
13. Etienne, A.; Dantan, J.Y.; Qureshi, J.; Siadat, A. Variation management by functional tolerance allocation and manufacturing process selection. *Int. J. Interact. Des. Manuf. (IJIDeM)* **2008**, *2*, 207–218. [CrossRef]
14. Wu, F.; Dantan, J.Y.; Etienne, A.; Siadat, A.; Martin, P. Improved algorithm for tolerance allocation based on Monte Carlo simulation and discrete optimization. *Comput. Ind. Eng.* **2009**, *56*, 1402–1413. [CrossRef]
15. Sivakumar, K.; Balamurugan, C.; Ramabalan, S. Optimal Concurrent Dimensional and Geometrical Tolerancing Based on Evolutionary Algorithms. In Proceedings of the 2009 World Congress on Nature & Biologically Inspired Computing (NaBIC), Coimbatore, India, 9–11 December 2009; pp. 300–305.
16. Sivakumar, K.; Balamurugan, C.; Ramabalan, S. Concurrent multi-objective tolerance allocation of mechanical assemblies considering alternative manufacturing process selection. *Int. J. Adv. Manuf. Technol.* **2011**, *53*, 711–732. [CrossRef]
17. Sivasubramanian, R.; Siva Kumar, M. Optimum tolerance synthesis with process and machine selection for minimizing manufacturing cost and machining time by using genetic algorithm. *J. Manuf. Technol. Res.* **2010**, *2*, 93–108.
18. Kim, C.B.; Pulat, P.S.; Foote, B.L.; Lee, D.H. Least cost tolerance allocation and bicriteria extension. *Int. J. Comput. Integr. Manuf.* **1999**, *12*, 418–426. [CrossRef]
19. Gadallah, M.H. Least sensitive tolerance allocation. *Int. J. Qual. Eng. Technol.* **2011**, *2*, 344–356. [CrossRef]
20. Lee, W.J.; Woo, T.C. Optimum selection of discrete tolerances. *ASME J. Mech Transm. Autom. Des.* **1989**, *111*, 243–251. [CrossRef]
21. Feng, C.X.; Kusiak, A. Robust tolerance synthesis with the design of experiments approach. *J. Manuf. Sci. Eng.* **2000**, *122*, 520–528. [CrossRef]
22. Kumar, M.S.; Stalin, B. Optimum tolerance synthesis for complex assembly with alternative process selection using Lagrange multiplier method. *Int. J. Adv. Manuf. Technol.* **2009**, *44*, 405–411. [CrossRef]
23. Armillotta, A. Concurrent optimization of dimensions and tolerances on structures and mechanisms. *Int. J. Adv. Manuf. Technol.* **2020**, *111*, 3141–3157. [CrossRef]
24. Korbi, A.; Tlija, M.; Louhichi, B. Computer-Aided Design/Tolerancing Integration: A Novel Tolerance Analysis Model of Assemblies with Composite Positional Defects and Deformations of Nonrigid Parts. *J. Manuf. Sci. Eng.* **2021**, *143*, 081012. [CrossRef]
25. Singh, P.K.; Jain, S.C.; Jain, P.K. A genetic algorithm based solution to optimum tolerance synthesis of mechanical assemblies with alternate manufacturing processes—Benchmarking with the exhaustive search method using the Lagrange multiplier. *Proc. Inst. Mech. Eng. Part B J. Eng. Manuf.* **2004**, *218*, 765–778. [CrossRef]
26. Chase, K.W. *Minimum Cost Tolerance Allocation*; ADCATS Report No. 99-5; Department of Mechanical Engineering, Bringham Young University: Provo, UT, USA, 1999.
27. Speckhart, F.H. Calculation of tolerance based on a minimum cost approach. *ASME J. Eng. Ind.* **1972**, *94*, 447–453. [CrossRef]
28. Sutherland, G.H.; Roth, B. Mechanism design: Accounting for manufacturing tolerances and costs in function generating problems. *ASME J. Eng. Ind.* **1975**, *97*, 283–286. [CrossRef]
29. Spotts, M.F. How to use wider tolerances, safely, in dimensioning stacked assemblies. *Mach. Des.* **1978**, *50*, 60–63.
30. Chase, K.W.; Greenwood, W.H. Design issues in mechanical tolerance analysis. *Manuf. Rev.* **1988**, *1*, 50–59.
31. Ostwald, P.F.; Huang, J. A method for optimal tolerance selection. *ASME J. Eng. Ind.* **1977**, *99*, 558–565. [CrossRef]
32. Sampath Kumar, R.; Alagumurthi, N.; Ramesh, R. Integrated total cost and Tolerance Optimization with Genetic Algorith. *Int. J. Comput. Intell. Syst.* **2010**, *3*, 325–333. [CrossRef]
33. Rao, Y.S.; Rao CS, P.; Janardhana, G.R.; Vundavilli, P.R. Simultaneous tolerance synthesis for manufacturing and quality using evolutionary algorithms. *Int. J. Appl. Evol. Comput. (IJAEC)* **2011**, *2*, 1–20. [CrossRef]
34. Geetha, K.; Ravindran, D.; Kumar, M.S.; Islam, M.N. Multi-objective optimization for optimum tolerance synthesis with process and machine selection using a genetic algorithm. *Int. J. Adv. Manuf. Technol.* **2013**, *67*, 2439–2457. [CrossRef]
35. Liu, S.G.; Jin, Q. Closed-form optimal tolerance for minimum manufacturing cost and quality loss cost. *Adv. Mater. Res.* **2013**, *655*, 2084–2087. [CrossRef]
36. Hallmann, M.; Schleich, B.; Wartzack, S. From tolerance allocation to tolerance-cost optimization: A comprehensive literature review. *Int. J. Adv. Manuf. Technol.* **2020**, *107*, 4859–4912. [CrossRef]
37. Wu, Z.; Elmaraghy, W.H.; ElMaraghy, H.A. Evaluation of cost-tolerance algorithms for design tolerance analysis and synthesis. *Manuf. Rev.* **1988**, *1*, 168–179.
38. Choi, H.G.R.; Park, M.H.; Salisbury, E. Optimal tolerance allocation with loss functions. *J. Manuf. Sci. Eng.* **2000**, *122*, 529–535. [CrossRef]
39. Chou, C.Y.; Chang, C.L. Minimum-loss assembly tolerance allocation by considering product degradation and time value of money. *Int. J. Adv. Manuf. Technol.* **2001**, *17*, 139–146. [CrossRef]
40. Ye, B.; Salustri, F.A. Simultaneous tolerance synthesis for manufacturing and quality. *Res. Eng. Des.* **2003**, *14*, 98–106. [CrossRef]
41. Gopalakrishnan, B.; Jaraiedi, M.; Iskander, W.H.; Ahmad, A. Tolerance synthesis based on Taguchi philosophy. *Int. J. Ind. Syst. Eng.* **2007**, *2*, 311–326. [CrossRef]
42. Zhang, C.; Wang HP, B. Integrated tolerance optimisation with simulated annealing. *Int. J. Adv. Manuf. Technol.* **1993**, *8*, 167–174. [CrossRef]

43. Ming, X.G.; Mak, K.L. Intelligent approaches to tolerance allocation and manufacturing operations selection in process planning. *J. Mater. Process. Technol.* **2001**, *117*, 75–83. [CrossRef]
44. Li, Z.; Izquierdo, L.E.; Kokkolaras, M.; Hu, S.J.; Papalambros, P.Y. Multiobjective optimization for integrated tolerance allocation and fixture layout design in multistation assembly. *J. Manuf. Sci. Eng.* **2008**, *130*. [CrossRef]
45. Hung, T.C.; Chan, K.Y. Multi-objective design and tolerance allocation for single-and multi-level systems. *J. Intell. Manuf.* **2013**, *24*, 559–573. [CrossRef]
46. Barari, A. Tolerance allocation based on the minimum deformation zone of finite element structural frame analysis. *Comput.-Aided Des. Appl.* **2013**, *10*, 629–641. [CrossRef]
47. Rao, R.V.; More, K.C. Advanced optimal tolerance design of machine elements using teaching-learning-based optimization algorithm. *Prod. Manuf. Res.* **2014**, *2*, 71–94. [CrossRef]
48. Ramesh Kumar, L.; Padmanaban, K.P.; Balamurugan, C. Optimal tolerance allocation in a complex assembly using evolutionary algorithms. *Int. J. Simul. Model. (IJSIMM)* **2016**, *15*. [CrossRef]
49. Nagarajan, L.; Mahalingam, S.K.; Gurusamy, S.; Dharmaraj, V.K. Solution for bi-objective single row facility layout problem using artificial bee colony algorithm. *Eur. J. Ind. Eng.* **2018**, *12*, 252–275. [CrossRef]
50. Vignesh Kumar, D.; Ravindran, D.; Lenin, N.; Siva Kumar, M. Tolerance allocation of complex assembly with nominal dimension selection using Artificial Bee Colony algorithm. *Proc. Inst. Mech. Eng. Part C J. Mech. Eng. Sci.* **2019**, *233*, 18–38. [CrossRef]
51. Rezaei Aderiani, A.; Hallmann, M.; Wärmefjord, K.; Schleich, B.; Söderberg, R.; Wartzack, S. Integrated Tolerance and Fixture Layout Design for Compliant Sheet Metal Assemblies. *Appl. Sci.* **2021**, *11*, 1646. [CrossRef]
52. Kumar, M.S.; Islam, M.N.; Lenin, N.; Kumar, D.V. Optimum tolerance synthesis for complex assembly with alternative process selection using bottom curve follower approach. *Int. J. Eng.* **2009**, *3*, 380–402.
53. Adalı, E.A.; Tuş, A. Hospital site selection with distance-based multi-criteria decision-making methods. *Int. J. Healthc. Manag.* **2021**, *14*, 534–544. [CrossRef]
54. Carfagni, M.; Governi, L.; Fhiesi, F. Development of a Method for Automatic Tolerance Allocation. In Proceedings of the XII ADM International Conference, Rimini, Italy, 5–7 September 2001.
55. Chen, M.S. Optimising tolerance allocation for mechanical components correlated by selective assembly. *Int. J. Adv. Manuf. Technol.* **1996**, *12*, 349–355. [CrossRef]
56. Diplaris, S.C.; Sfantsikopoulos, M.M. Cost–Tolerance function. A new approach for cost optimum machining accuracy. *Int. J. Adv. Manuf. Technol.* **2000**, *16*, 32–38. [CrossRef]
57. Dong, Z.; Hu, W.; Xue, D. New production cost-tolerance models for tolerance synthesis. *ASME J. Eng. Ind.* **1994**, *116*, 199–206. [CrossRef]
58. Janakiraman, V.; Saravanan, R. Concurrent optimization of machining process parameters and tolerance allocation. *Int. J. Adv. Manuf. Technol.* **2010**, *51*, 357–369. [CrossRef]
59. Kumar, M.S.; Kannan, S.M.; Jayabalan, V. A new algorithm for optimum tolerance allocation of complex assemblies with alternative processes selection. *Int. J. Adv. Manuf. Technol.* **2009**, *40*, 819–836. [CrossRef]
60. Loosli, B.G. Manufacturing Tolerance Cost Minimization Using Discrete Optimization for Alternate Process Selection. Master's Thesis, Brigham Young University, Provo, UT, USA, 1987.
61. Nagarwala, M.Y.; Simin Pulat, P.; Raman, S. A slope-based method for least cost tolerance allocation. *Concurr. Eng.* **1995**, *3*, 319–328. [CrossRef]
62. Prabhaharan, G.; Asokan, P.; Rajendran, S. Sensitivity-based conceptual design and tolerance allocation using the continuous ants colony algorithm (CACO). *Int. J. Adv. Manuf. Technol.* **2005**, *25*, 516–526. [CrossRef]
63. Rao, S.S.; Wu, W. Optimum tolerance allocation in mechanical assemblies using an interval method. *Eng. Optim.* **2005**, *37*, 237–257. [CrossRef]
64. Savage, G.J.; Tong, D.; Carr, S.M. Optimal mean and tolerance allocation using conformance-based design. *Qual. Reliab. Eng. Int.* **2006**, *22*, 445–472. [CrossRef]
65. Singh, P.K.; Jain, S.C.; Jain, P.K. Advanced optimal tolerance design of mechanical assemblies with interrelated dimension chains and process precision limits. *Comput. Ind.* **2005**, *56*, 179–194. [CrossRef]
66. Sivakumar, K.; Balamurugan, C.; Ramabalan, S. Evolutionary multi-objective concurrent maximisation of process tolerances. *Int. J. Prod. Res.* **2012**, *50*, 3172–3191. [CrossRef]
67. Zhang, C.; Wang, H.P. Robust design of assembly and machining tolerance allocations. *IIE Trans.* **1998**, *30*, 17–29. [CrossRef]
68. Jeang, A.; Chang, C.L. Combined robust parameter and tolerance design using orthogonal arrays. *Int. J. Adv. Manuf. Technol.* **2002**, *19*, 442–447. [CrossRef]
69. Huang, Y.M.; Shiau, C.S. Optimal tolerance allocation for a sliding vane compressor. *J. Mech. Des.* **2006**, *128*, 98–107. [CrossRef]

*applied
sciences*

MDPI

Article

Harmony Search Algorithm for Minimizing Assembly Variation in Non-linear Assembly

Siva Kumar Mahalingam [1], Lenin Nagarajan [1,*], Sachin Salunkhe [1] Emad Abouel Nasr [2], Jõao Paulo Davim [3] and Hussein M. A. Hussein [4,5]

[1] Department of Mechanical Engineering, Vel Tech Rangarajan Dr. Sagunthala R&D Institute of Science and Technology, Chennai 600062, India; lawan.sisa@gmail.com (S.K.M.); drsalunkhesachin@veltech.edu.in (S.S.)
[2] Industrial Engineering Department, College of Engineering, King Saud University, Riyadh 11421, Saudi Arabia; eabdelghany@ksu.edu.sa
[3] Department of Mechanical Engineering, University of Aveiro, Campus Universitário de Santiago, 3810-193 Aveiro, Portugal; pdavim@ua.pt
[4] Mechanical Engineering Department, Faculty of Engineering, Helwan University, Cairo 11732, Egypt; hussein@h-eng.helwan.edu.eg
[5] Mechanical Engineering Department, Faculty of Engineering, Ahram Canadian University, Giza 12566, Egypt
* Correspondence: n.lenin@gmail.com

Citation: Mahalingam, S.K.; Nagarajan, L.; Salunkhe, S.; Nasr, E.A.; Davim, J.P.; Hussein, H.M.A. Harmony Search Algorithm for Minimizing Assembly Variation in Non-linear Assembly. *Appl. Sci.* **2021**, *11*, 9213. https://doi.org/10.3390/app11199213

Academic Editor: Vladimir Modrak

Received: 6 September 2021
Accepted: 28 September 2021
Published: 3 October 2021

Publisher's Note: MDPI stays neutral with regard to jurisdictional claims in published maps and institutional affiliations.

Abstract: The proposed work aims to acquire the maximum number of non-linear assemblies with closer assembly tolerance specifications by mating the different bins' components. Before that, the components are classified based on the range of tolerance values and grouped into different bins. Further, the manufacturing process of the components is selected from the given and known alternative processes. It is incredibly tedious to obtain the best combinations of bins and the best process together. Hence, a novel approach using the combination of the univariate search method and the harmony search algorithm is proposed in this work. Overrunning clutch assembly is taken as an example. The components of overrunning clutch assembly are manufactured with a wide tolerance value using the best process selected from the given alternatives by the univariate search method. Further, the manufactured components are grouped into three to nine bins. A combination of the best bins is obtained for the various assembly specifications by implementing the harmony search algorithm. The efficacy of the proposed method is demonstrated by showing 24.9% of cost-savings while making overrunning clutch assembly compared with the existing method. The efficacy of the proposed method is demonstrated by showing 24.9% of cost-savings while making overrunning clutch assembly compared with the existing method. The results show that the contribution of the proposed novel methodology is legitimate in solving selective assembly problems.

Keywords: selective assembly; overrunning clutch assembly; univariate search method; harmony search algorithm

1. Introduction

Product quality is the focus of any manufacturing process. In general, two or more components are assembled to create an assembly. The quality of the assembly depends on the quality of the components, which affects the product's functionality. Tolerance plays an essential role in the component's quality, deciding the fit between the mating parts. The components manufactured with closer tolerance make the precise assembly more suitable for functional requirements. Selective assembly is one of the feasible methods for making precise assemblies with lower manufacturing costs. The complete elimination or reduction of secondary operations by forming wide-tolerance components is the reason for lower manufacturing costs. In selective assembly, the components are grouped as bins based on the uniform grouping method, the equal probability method, or the uniform tolerance method, for example. According to the best bin combinations, the precise assemblies are made by mating the components randomly selected from the corresponding bins. Most of the research on linear/radial assembly have mainly focused on minimizing the objectives, namely, clearance variation or surplus parts. Research works focusing on non-linear assembly are seldom found. Moreover, in the existing literature, the tolerance of the components is usually considered for obtaining the best bin combinations rather than the dimension of components. Since different combinations of bins are possible, this problem environment can be treated as an NP-hard problem. Selection of process for making components from the given alternative processes also plays a vital role in further reducing the product's manufacturing cost.

2. Related Research

The literature relevant to the proposed work has been categorized as selective assembly and the harmony search algorithm. The literature related to both topics is detailed below.

2.1. Selective Assembly

Kern (2003) [1] proposed an approach for selecting assembly by considering variations in the dimensional distributions. Further, the author developed closed-form equations for the different techniques of selective assembly. Mease et al. (2004) [2] introduced a method to classify the components into various classes based on the dimensional variations. The assembly was made by pairing the selected components from the different classes. Further, the absolute and squared error loss functions were studied to select the optimum method. Matsuura et al. (2008) [3] explored a method to minimize clearance variation when pairing the components with less variance for making an assembly. Kwon et al. (2009) [4] studied the effectiveness of optimal binning strategies based on the squared error loss function by considering similar normal distributions of the dimensions of two different components. Matsuura et al. (2011) [5] developed the optimal mean shift to reduce surplus components while making selective assembly. The equal width concept was used for grouping the components into different bins. Three shifted means method was used for the fabrication of components.

Yue et al. (2014) [6] proposed a genetic algorithm to minimize the variation in clearance while mating the components available in the selective groups to manufacture hole and shaft assemblies. Babu and Asha (2014) [7] introduced a customized artificial immune system algorithm for mating the components available in selective groups to minimize assembly clearance variation. Further, they applied Taguchi's loss function to identify the deviation from the mean. The count of selective groups was also one of the important concerns in the proposed work. Ju and Li (2014) [8] solved selective assembly problems using a two-stage decomposition procedure. A two-component assembly system with unreliable Bernoulli machines and limited inventories was considered. The effectiveness of the proposed method was analyzed analytically.

Xu et al. (2014) [9] applied a new method with the combination of discarding and binning theorems to minimize the components' variations while making the selective assembly of a hard disk drive. The utilization rate of the components was high with regard to the number of assemblies made using the newly proposed method. Lu and Fei (2015) [10] proposed a grouping method in selective assembly to increase the success rate of manufacturing the assemblies and reducing the surplus parts. A genetic algorithm with a 2D chromosome structure was used for this purpose. The effectiveness of the proposed method was studied based on solving three different cases. Manickam and De (2015) [11] developed a genetic algorithm to identify the better combination of groups containing the components to create the maximum number of assemblies with less total cost. The small number of components with wider tolerance was made available in the individual groups. This was the uniqueness of the presented work. The model was developed using MATLAB software.

Babu and Asha (2015) [12] evaluated the losses in making assemblies by applying the symmetrical interval-based Taguchi loss function. Further, the dual objectives, namely, minimum clearance variation and minimum losses in making assemblies, were considered to identify the better combination of the parts' groups using the sheep flock heredity algorithm. Ju et al. (2016) [13] studied the performance of the selective assembly system using Bernoulli machine reliability models. For this study, the assemblies made through two components with different qualities were considered. The two-stage decomposition procedure was applied to evaluate the effectiveness of the proposed models analytically. Malaichamy et al. (2016A) [14] developed software capable of applying all kinds of selective assembly methods to identify suitable solutions with the least possible time. For instance, equal area and width methods were used to minimize variation and scrap in selective assembly. The random assembly method was inbuilt in the software to select the best method for minimizing the manufacturing cost of the assembly. Since the software was developed using an optional programming language, changes to the input data were extremely easy, and, in no time, the result could be viewed graphically.

Malaichamy et al. (2016B) [15] introduced a software approach to help engineers to visualize the tolerance cost curve for a given process or a good combination of processes in no time. The tolerance cost was calculated using the reciprocal power cost model. The best process could be selected in this work from the given alternative processes for the known tolerance value. A simulated annealing algorithm was implemented to compute the best bin number and its combinations with almost nil surplus parts for the given dimensional distribution of components of a shaft housing assembly. Compared with random assembly, a good number of surplus parts were reduced by implementing an equal width selective assembly method. Liu and Liu (2017) [16] discussed the remanufacturing of the engine through the selective assembly concept. Remanufacturing is a sustainable strategy in which the group numbers and range of components in each group are not kept constant. The assembly accuracy was greatly increased using the proposed strategy. Chu et al. (2018) [17] developed a method for manufacturing gear reducers using a novel strategy called GA-based selective assembly. The backlash of the gear reducer was the major concern in this work to verify the meeting of assembly requirements.

2.2. Harmony Search Algorithm

Geem et al. (2001) [18] described the harmony search optimization algorithm's features, robustness, simplicity, and search efficiency in solving engineering problems. Geem et al. (2002) [19] proposed a harmony search optimization algorithm to solve the problems associated with pipe network design. Lee and Geem (2004) [20] described a novel structural optimization algorithm using the harmony search method. The effectiveness of the proposed algorithm was verified by identifying the solutions to different truss problems. Lee and Geem (2005) [21] solved different engineering optimization problems that included minimizing mathematical functions and optimizing structural problems using a harmonic-search-based method. Flow demand and low head were considered objectives in this work. Wang et al. (2009) [22] analyzed the harmony search algorithm's effectiveness with the colonal selection algorithm. The colonal search method was used to increase the harmony memory members in the harmony search method.

From the literature survey, it is clear that research works have been very limited in the manufacturing of non-linear assemblies with a closer assembly tolerance specification. Further, the usage of the harmony search algorithm for solving selective assembly problems is seldom found. The problem environment is discussed in the next section.

3. Problem Environment

In selective assembly, parts are manufactured with wider tolerance, measured, and partitioned into groups, and assemblies are made by assembling the components within the random combination of groups. Manufacturing tolerance, assembly specification, and the number of groups are the three main factors that play important roles in controlling surplus parts that affect the product's manufacturing cost. After parts are manufactured, it is tedious to obtain the best combination of groups for different assembly specifications. It is laborious work to compute the numbers of closer assembly for each bin number from the manufactured components. Alternative process selection for making components also makes the problem highly complicated. The problems described above are challenging to process/design engineers and reduce implementation in real situations. Moreover, the components involved in making a non-linear assembly require the individual mating of dimensions for each assembly. Compared to making linear assembly, this required additional computation effort is tedious.

4. Solution Methodology

The proposed problem environment can be solved in three stages. In the first stage, the best process for manufacturing each component of an assembly is obtained from the alternative processes using the univariate search method. In the second stage, 1000 simulated dimensions of each component are generated using MATLAB for the different combinations of alternative processes obtained from the first stage. Further, the components are partitioned into different bin numbers according to the equal area method, which is one of the techniques used in the selective assembly method. In the last stage, the harmony search algorithm is implemented to obtain the best bin combinations. Then, the non-linear assemblies are made by mating the components according to the best bin combinations with almost nil surplus parts. To show the effectiveness of the proposed method, the manufacturing cost of assemblies made both from the random assembly [23] and selective assembly are compared. This three-stage approach is explained in a detailed way in the numerical illustration section.

5. Numerical Illustration

Overrunning clutch assembly (OCA), dealt with by Ganesan et al. (2001) [23], shown in Figure 1, is considered an example product to show the efficiency of the proposed method. It consists of a hub, four rollers, and a cage. The nominal dimensions and their allocated tolerance; the minimum, maximum process tolerances; and the tolerance cost function constants of alternative processes of each component are listed in Table 1. The critical dimension Y is determined using Equation (1), and the accepted value of Y is assumed to be $7.0124 \pm 2°$. Equation (2) represents the reciprocal tolerance cost function to calculate the component's tolerance cost. The cost function constants are taken from Ganesan et al. (2001) [23]. The cost of assembly based on allocated and maximum process tolerances is computed using Equations (3) and (4), respectively, and listed in Table 2.

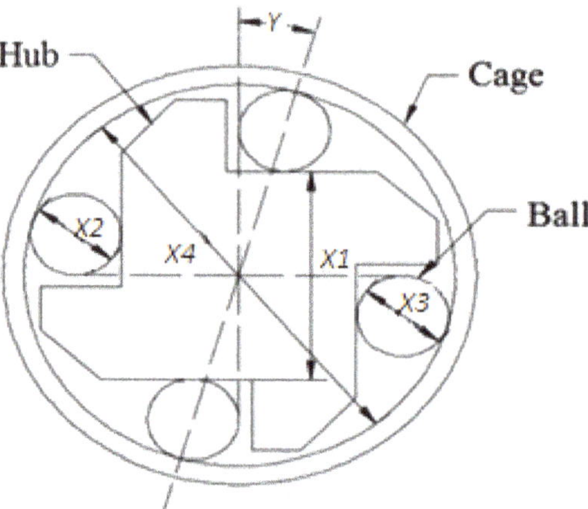

Figure 1. Overrunning clutch assembly.

$$Y = \cos^{-1}\left(\frac{X1 + \left(\frac{X2+X3}{2}\right)}{X4 - \left(\frac{X2+X3}{2}\right)}\right) \tag{1}$$

$$C_i = A_i + \frac{B_i}{t_i} \tag{2}$$

$$C_a = \sum_{i=1}^{nc} A_i + \frac{B_i}{t_{ai}} \tag{3}$$

$$C_m = \sum_{i=1}^{nc} A_i + \frac{B_i}{t_{mi}} \tag{4}$$

where

$X1$	—Dimension of the hub in mm
$X2$ and $X3$	—Dimension of the rollers in mm
$X4$	—Dimension of the cage in mm
Y	—Critical dimension in degrees
A_i	—Fixed cost of the ith component in \$
B_i	—Cost function constants of the ith component
t_i	—ith component's tolerance in mm
C_i	—Cost of the ith component in \$
t_{ai}, t_{mi}	—ith component's allocated and maximum tolerance in mm
C_a, C_m	—Cost of assembly in \$ based on t_{ai} and t_{min}
nc	—Number of components
i	—Component number index

Table 1. Manufacturing details of overrunning clutch assembly.

C.No.	Nominal Dimension (D_i) (mm)	t_{ai} (mm)	P_s	P.No.	A_i	B_i	t_{min} (mm)	t_{max} (mm)
X1	55.29	0.179806	3	P1	10.0	0.015	0.015	0.08
				P2	5.0	0.500	0.060	0.15
				P3	3.5	0.750	0.120	0.25
X2	22.86	0.165358	2	P1	8.0	0.250	0.020	0.15
				P2	3.0	0.650	0.080	0.30
X3	22.86	0.120132	1	P1	2.5	0.300	0.040	0.20
				P2	5.0	0.045	0.120	0.25
X4	101.69	0.200581	3	P1	4.0	0.560	0.080	0.12
				P2	6.0	0.160	0.150	0.25
				P3	0.5	0.880	0.200	0.40

C.No.—component name; P.No.—process number.

Table 2. Manufacturing cost of overrunning clutch assembly for t_{ai}, t_{min}, and t_{max}.

C.No.	C_{ai} (\$)	P.No.	C_{min} (\$)	C_{max} (\$)
X1	7.67	P1	11.00	10.19
		P2	13.33	8.33
		P3	9.75	6.50
X2	6.93	P1	20.50	9.67
		P2	11.13	5.17
X3	5.00	P1	10.00	4.00
		P2	5.38	5.18
X4	4.89	P1	11.00	8.67
		P2	7.07	6.64
		P3	4.90	2.70

5.1. Stage I

The selection of the best process for each component from the given alternative processes using the univariate search method is illustrated in Figure 2. It is understood from Figure 2 that the minimum manufacturing cost of $18.37 can be achieved through making the components *X1*, *X2*, *X3*, and *X4* using process combinations (3213) *P3, P2, P1,* and *P3*, respectively, which is nearly 24.98% less compared with the $24.49 reported in the existing method dealt by Ganesan et al. (2001). It is also understood that the process combinations of 2213 and 1213 for components *X1*, *X2*, *X3*, and *X4*, respectively, can yield 17.52% and 9.92% savings in manufacturing cost. However, in a real situation, the savings may vary slightly because of surplus parts present in the selective assembly method.

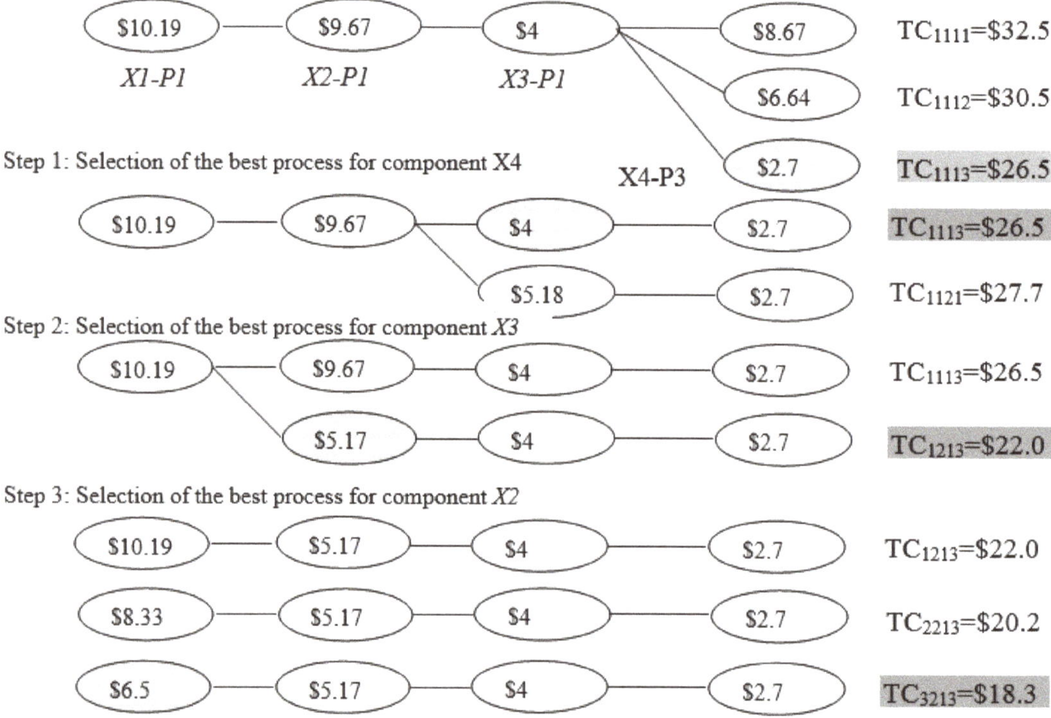

Figure 2. Univariate search method to select the best process for each component. *X1-P1* indicates that *X1* component is produced by the *P1* process; TC1113 = $26.56 indicates that components *X1*, *X2*, *X3*, and *X4* are manufactured using processes *P1*, *P1*, *P1*, and *P3*, respectively, and the total cost to manufacture the same will be $26.56.

5.2. Stage II

As discussed in Section 4, 1000 random values have been generated for each component according to the mean (μ_i) and standard deviation (σ_i) presented in Table 3. The dimensional distribution of 1000 components of *X1*, *X2*, *X3*, and *X4* was generated using the normrnd (*C.No.*) function in MATLAB.

Table 3. Mean and standard deviation of components for different process combinations.

C.No.	μ_i (mm)	PC1–1213			PC2–2213			PC3–3213		
		P.No.	t_{mi} (mm)	$\sigma_i = t_{mi}/3$	P.No.	t_{mi} (mm)	$\sigma_i = t_{mi}/3$	P.No.	t_{mi} (mm)	$\sigma_i = t_{mi}/3$
X1	55.29	P1	0.08	0.02667	P2	0.15	0.05	P3	0.25	0.08333
X2	22.86	P2	0.3	0.1	P2	0.3	0.1	P2	0.3	0.1
X3	22.86	P1	0.25	0.08333	P1	0.25	0.08333	P1	0.25	0.08333
X4	101.69	P3	0.4	0.13333	P3	0.4	0.13333	P3	0.4	0.13333

P.No.—process number; PC1—1st process combinations for components X1, X2, X3, and X4; μ_i—mean dimension of the components; t_{mi}—tolerance of the ith component; σ_i—standard deviation of the ith component.

5.3. Stage III—Implementation of HSA

The harmony search algorithm (HSA) is a meta-heuristic algorithm, and it works based on the identification of good harmony by musicians through a continuous improvisation process. The HSA has the following advantages: (i) quick convergence, (ii) easy to adapt, and (iii) the least computational time. Further, from the literature survey, it is observed that the HSA has outperformed in solving complex optimization problems. Hence, the HSA has been used in this work. Table 4 presents the different terms used in the HSA, its equivalent term in both optimization problems and the present work formulation, and its range of values and examples. The schematic diagram shown in Figure 3 illustrates the implementation of the HSA to obtain the best bin combinations. The technical terms and their meanings used in Figure 3 are presented in Table 5. For demonstration purposes, the components are partitioned into five bins. The step-by-step procedure is given below.

Table 4. Representation of variables in HSA.

HSA Parameters	Equivalent Term in Optimization Problem	Equivalent Term in Present Work	Example
Musical Instrument Pitch Range	Decision Variable Value Range	No. of components Number of bins	X1, X2, X3, and X4 5
Harmony	A Solution Vector	A combination of bins for each component	(X1) 13245 (X2) 43251 (X3) 42135 (X4) 52143
Aesthetics	Objective Function	Number of accepted assemblies	985
Practice	Iteration	No. of iterations	150
Experience	Memory Matrix	Storing of the best solution	
Harmony Memory Size	Size of Solution Vector	No. of initial solutions	20
HMCR (0.7–0.99)	Harmony Memory Considering Rate	0.7	0.7
PAR	Pitch Adjusting Rate	0.3	0.3

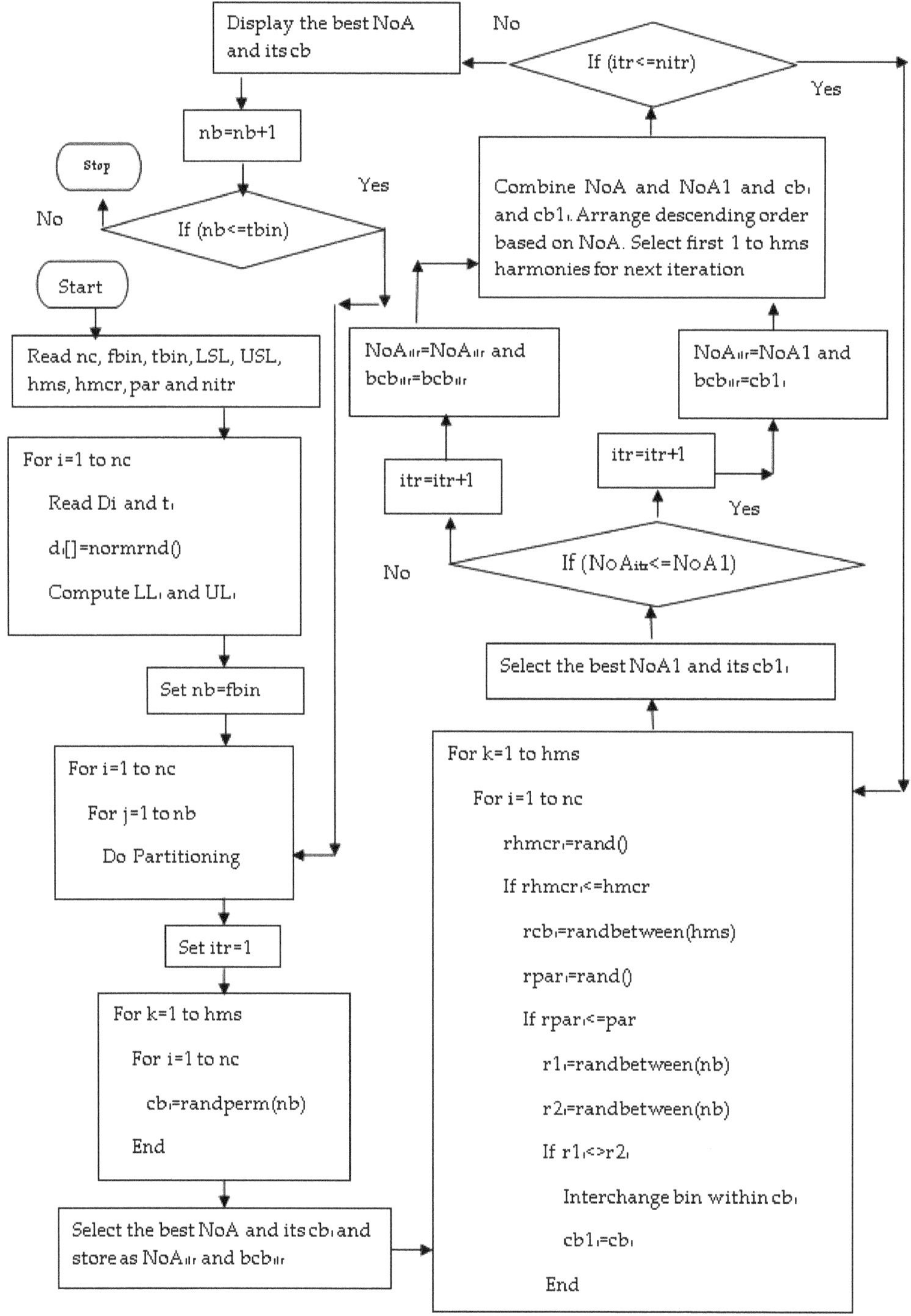

Figure 3. Implementation of HSA.

Table 5. Technical terms and their meanings.

Terms	Meaning
Trimming	Remove the components which do not satisfy $LL_i \leq d_i \leq UL_i$
Partitioning	Arranging the components in ascending order based on its di and placing an equal number of components in each bin starting from bin number 1 to *nb*
Mating	Selecting one component (*X1*, *X2*, *X3*, and *X4*) randomly from the bin based on the bin combination of cb$_i$ and making an assembly. Check the assembly's specification (within *LSL* and *USL*); if it meets the requirement, then treat it as good assembly (*NoA*); otherwise, treat it as a surplus part
normrnd	Matlab function will generate the required number of normally distributed random numbers based on the given mean and standard deviation value
randperm (nb)	Matlab function will generate a permutation combination for the given bin number *nb*

Step 1:   A random combination of bins for each component for the size of harmony memory 10 is generated (listed in Table 6).

Step 2:   The corresponding bin's components are randomly matched with other components for each harmony number, and the assembly is produced. If the assembly meets the given specification limit, it is accepted as an assembly; otherwise, it is treated as a surplus part. This will be carried out until the component exists in each bin. The accepted assemblies are counted and listed in Table 6 as NoA.

Step 3:   Table 7 represents the arrangements of harmony, from maximum NoA to minimum NoA. Table 8 illustrates the best bin combination that will produce maximum NoA in the first iteration.

Table 6. Initial harmony memory.

H.No.(k)	cb_{X1}	cb_{X2}	cb_{X3}	cb_{X4}	NoA_k
1	2 5 1 4 3	2 5 3 1 4	5 3 4 1 2	3 1 2 4 5	308
2	4 2 1 5 3	4 2 5 1 3	4 2 5 3 1	5 3 4 2 1	649
3	4 2 1 5 3	1 5 2 3 4	3 4 2 5 1	3 1 5 4 2	407
4	4 3 1 5 2	3 2 4 5 1	5 2 3 4 1	3 5 2 4 1	508
5	1 5 4 3 2	1 3 4 5 2	3 2 4 5 1	2 4 1 3 5	326
6	2 1 3 4 5	3 2 4 5 1	3 1 5 2 4	3 5 2 4 1	557
7	5 2 3 1 4	2 5 3 1 4	1 2 3 4 5	1 5 4 2 3	354
8	1 2 4 3 5	1 3 5 2 4	5 2 4 3 1	5 3 4 2 1	397
9	2 3 1 4 5	3 1 2 5 4	2 5 1 4 3	5 4 2 1 3	212
10	4 3 1 2 5	4 5 3 2 1	3 2 1 4 5	3 2 4 5 1	458

H.No.—harmony numbers; cb_{X1}, cb_{X2}, cb_{X3}, and cb_{X4} are the combinations of bins for X1, X2, X3, and X4 components.

Table 7. Harmonies after sorting based on NoA.

H.No.(k)	cb_{X1}	cb_{X2}	cb_{X3}	cb_{X4}	NoA_k
1	4 2 1 5 3	4 2 5 1 3	4 2 5 3 1	5 3 4 2 1	649
2	2 1 3 4 5	3 2 4 5 1	3 1 5 2 4	3 5 2 4 1	557
3	4 3 1 5 2	3 2 4 5 1	5 2 3 4 1	3 5 2 4 1	508
4	4 3 1 2 5	4 5 3 2 1	3 2 1 4 5	3 2 4 5 1	458
5	4 2 1 5 3	1 5 2 3 4	3 4 2 5 1	3 1 5 4 2	407
6	1 2 4 3 5	1 3 5 2 4	5 2 4 3 1	5 3 4 2 1	397
7	5 2 3 1 4	2 5 3 1 4	1 2 3 4 5	1 5 4 2 3	354
8	1 5 4 3 2	1 3 4 5 2	3 2 4 5 1	2 4 1 3 5	326
9	2 5 1 4 3	2 5 3 1 4	5 3 4 1 2	3 1 2 4 5	308
10	2 3 1 4 5	3 1 2 5 4	2 5 1 4 3	5 4 2 1 3	212

Table 8. Best combination of bins for the first iteration.

	bcb_{itr}			NoA_{itr}
cb_{X1}	cb_{X2}	cb_{X3}	cb_{X4}	
4 2 1 5 3	4 2 5 1 3	4 2 5 3 1	5 3 4 2 1	649

Step 4:   A random number less than 1 is generated for each component of each harmony number (presented in Table 9 as *RHMCR*).

Step 5:   If this number is less than or equal to *HMCR*, then a random number between 1 and *HMS* (*rcb*) is generated, and others are assumed to be zero. This is presented in Table 9.

Table 9. *RHMCR* and *rcb* values.

H.No.	RHMCR				rcb			
	X1	*X2*	*X3*	*X4*	rcb_{X1}	rcb_{X2}	rcb_{X3}	rcb_{X4}
1	0.0697	0.6581	0.1267	0.0176	1	6	7	7
2	0.1398	0.9349	0.9424	0.3936	8	0	0	1
3	0.4975	0.3145	0.218	0.8001	1	9	3	0
4	0.0639	0.7984	0.1875	0.442	7	0	3	4
5	0.9388	0.0765	0.605	0.1321	0	8	1	1
6	0.2711	0.4453	0.7794	0.9148	9	3	0	0
7	0.6946	0.413	0.216	0.0049	1	6	6	1
8	0.0204	0.9048	0.0267	0.7212	1	0	8	0
9	0.0013	0.4189	0.1216	0.8414	1	6	1	0
10	0.6254	0.8232	0.0301	0.6189	2	0	8	7

RHMCR—a random number decides either to accept or not accept the selection of the existing tune/bin combination of components for improvisation; *rcb*—a random number generated between 1 and hms for each variable.

Step 6:   The cb_{X1}, cb_{X2}, cb_{X3}, and cb_{X4} values corresponding to rcb_{X1}, rcb_{X2}, rcb_X, and rcb_{X4} are taken from Table 6 and listed in Table 10. If the value of *rcb* is zero, then the corresponding harmony's *cbx* value is considered.

Table 10. Harmony after HMCR.

H.No.(k)	cb_{X1}	cb_{X2}	cb_{X3}	cb_{X4}
1′	4 2 1 5 3	1 3 5 2 4	1 2 3 4 5	1 5 4 2 3
2′	1 5 4 3 2	3 2 4 5 1	3 1 5 2 4	5 3 4 2 1
3′	4 2 1 5 3	2 5 3 1 4	5 2 3 4 1	3 5 2 4 1
4′	5 2 3 1 4	4 5 3 2 1	5 2 3 4 1	3 2 4 5 1
5′	4 2 1 5 3	1 3 4 5 2	4 2 5 3 1	5 3 4 2 1
6′	2 5 1 4 3	3 2 4 5 1	5 2 4 3 1	5 3 4 2 1
7′	4 2 1 5 3	1 3 5 2 4	5 2 4 3 1	5 3 4 2 1
8′	4 2 1 5 3	1 3 4 5 2	3 2 4 5 1	2 4 1 3 5
9′	4 2 1 5 3	1 3 5 2 4	4 2 5 3 1	3 1 2 4 5
10′	2 1 3 4 5	3 2 4 5 1	3 2 4 5 1	1 5 4 2 3

Step 7:   A random number (*RPAR*) less than 1 is generated for each value of *RHMCR*, which is not equal to zero and is less than the *HMCR* value (listed in Table 11).

Step 8:   Two random numbers, r1 and r2, within bin numbers, are generated for each value of RPAR, which is not equal to zero (presented in Table 11).

Step 9:   New harmony, i.e., cb_{X1}, cb_{X2}, cb_{X3}, and cb_{X4}, is obtained wherever the bin is located within their bin combinations, according to r1 and r2 (presented in Table 12).

Step 10: Then, NoAs are obtained by mating the components randomly corresponding to the bin combinations given in cb_{X1}, cb_{X2}, cb_{X3}, and cb_{X4} (listed in Table 12).

Step 11:The selection and selected harmonies for the next iteration are presented in Tables 13 and 14.

Step 12: The steps from 4 to 11 are repeated to the specified number of iterations.

Step 13: The above steps, from 1 to 12, can be repeated for various bin numbers, starting from 3 to 9. Figures 4 and 5 represent the iteration number vs. *NoA* and the number of bin vs. *NoA* for various bin numbers.

Step 14: By changing the product specification, starting from $\pm 0.25°$ to $\pm 2°$, the above steps from 1 to 13 can be repeated.

Table 11. *RPAR, r1,* and *r2* values.

H.No.	RPAR				r1				r2			
	$X1$	$X2$	$X3$	$X4$	$r1_{X1}$	$r1_{X2}$	$r1_{X3}$	$r1_{X4}$	$r2_{X1}$	$r2_{X2}$	$r2_{X3}$	$r2_{X4}$
$1'$	0.0917	0.0738	0.4457	0.8808	3	1	0	0	1	2	0	0
$2'$	0.2529	0	0	0.6409	1	0	0	0	3	0	0	0
$3'$	0.5235	0.009	0.344	0	0	1	0	0	0	4	0	0
$4'$	0.6034	0	0.3208	0.6256	0	0	0	0	0	0	0	0
$5'$	0	0.1307	0.599	0.4411	0	2	0	0	0	2	0	0
$6'$	0.5166	0.4696	0	0	0	0	0	0	0	0	0	0
$7'$	0.5402	0.5435	0.8627	0.7751	0	0	0	0	0	0	0	0
$8'$	0.3204	0	0.7643	0	0	0	0	0	0	0	0	0
$9'$	0.1678	0.277	0.0321	0	1	3	1	0	3	3	3	0
$10'$	0.2417	0	0.8962	0.4639	4	0	0	0	1	0	0	0

RPAR—a random number decides the pitch adjustment (interchanging bin numbers within the component); *r1* and *r2*—two random numbers within the bin number (*nb*) to interchange the bin number to generate a new bin combination for a component.

Table 12. Harmony after *HMCR* and *PAR*.

H.No.	After HMCR and PAR				NoA
	$cb1_{X1}$	$cb1_{X2}$	$cb1_{X3}$	$cb1_{X4}$	
$1'$	1 2 4 5 3	3 1 5 2 4	1 2 3 4 5	1 5 4 2 3	709
$2'$	4 5 1 3 2	2 5 4 3 1	4 1 5 3 2	5 3 4 2 1	608
$3'$	4 2 1 5 3	1 5 3 2 4	5 2 3 4 1	3 4 1 5 2	671
$4'$	5 2 3 1 4	1 5 2 3 4	5 2 3 4 1	3 2 4 5 1	555
$5'$	2 1 5 3 4	1 3 4 5 2	4 2 5 3 1	5 3 4 2 1	503
$6'$	2 5 1 4 3	3 2 4 5 1	2 1 5 4 3	1 3 5 2 4	312
$7'$	4 2 1 5 3	1 3 5 2 4	5 2 4 3 1	5 3 4 2 1	519
$8'$	4 2 1 5 3	2 4 5 3 1	3 2 4 5 1	5 3 4 2 1	404
$9'$	1 2 4 5 3	1 3 5 2 4	5 2 4 3 1	3 2 5 1 4	460
$10'$	4 1 3 2 5	1 4 3 5 2	3 2 4 5 1	1 5 4 2 3	549

Table 13. Selection of harmonies for next iteration.

Before Sorting		After Sorting		SHMNo.
HMNo.	NoA	HMNo.	NoA	
1	649	$1'$	709	$1''$
2	557	$2'$	671	$2''$
3	508	1	649	$3''$
4	458	$3'$	608	$4''$
5	407	2	557	$5''$
6	397	$4'$	555	$6''$
7	354	$5'$	549	$7''$
8	326	$6'$	519	$8''$
9	308	3	508	$9''$
10	212	$7'$	503	$10''$
$1'$	709	$8'$	460	NS
$2'$	671	4	458	NS
$3'$	608	5	407	NS
$4'$	555	$9'$	404	NS
$5'$	549	6	397	NS
$6'$	519	7	354	NS
$7'$	503	8	326	NS
$8'$	460	$10'$	312	NS
$9'$	404	9	308	NS
$10'$	312	10	212	NS

NS—not selected; *SHMNo.*—selected harmonies for next iteration.

Table 14. Selected harmonies for next iteration.

IHMno	HMNo	X1	X2	X3	X4	NoA
1″	1′	1 2 4 5 3	3 1 5 2 4	1 2 3 4 5	1 5 4 2 3	709
2″	2′	4 2 1 5 3	1 5 3 2 4	5 2 3 4 1	3 4 1 5 2	671
3″	1	4 2 1 5 3	4 2 5 1 3	4 2 5 3 1	5 3 4 2 1	649
4″	3′	4 5 1 3 2	2 5 4 3 1	4 1 5 3 2	5 3 4 2 1	608
5″	2	2 1 3 4 5	3 2 4 5 1	3 1 5 2 4	3 5 2 4 1	557
6″	4′	5 2 3 1 4	1 5 2 3 4	5 2 3 4 1	3 2 4 5 1	555
7″	5′	4 1 3 2 5	1 4 3 5 2	3 2 4 5 1	1 5 4 2 3	549
8″	6′	4 2 1 5 3	1 3 5 2 4	5 2 4 3 1	5 3 4 2 1	519
9″	3	4 3 1 5 2	3 2 4 5 1	5 2 3 4 1	3 5 2 4 1	508
10″	7′	2 1 5 3 4	1 3 4 5 2	4 2 5 3 1	5 3 4 2 1	503

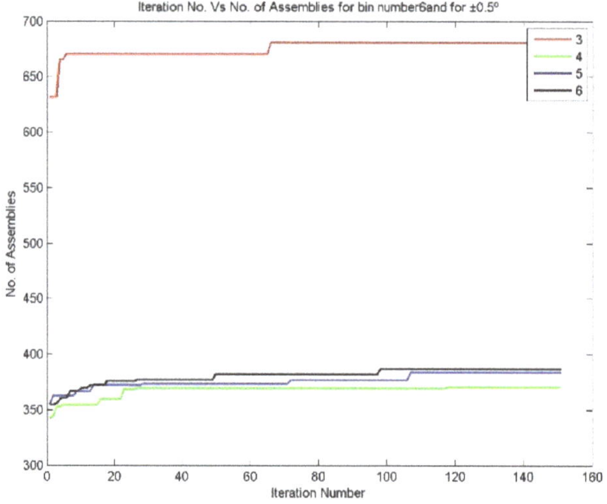

(**a**). *NoA* for ± 0.5°

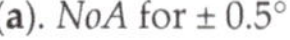

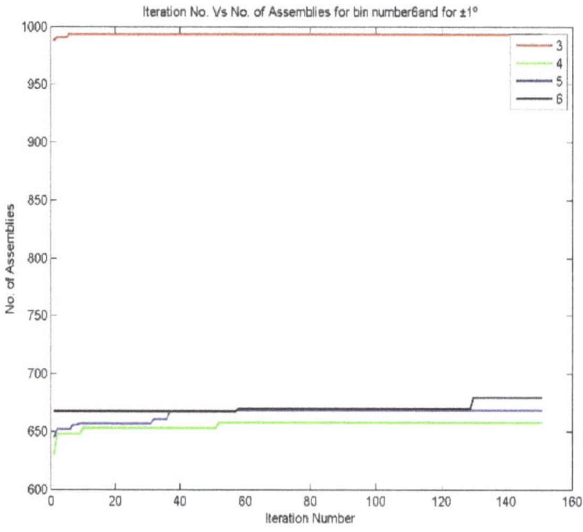

(**b**). *NoA* for ± 1°

Figure 4. *Cont.*

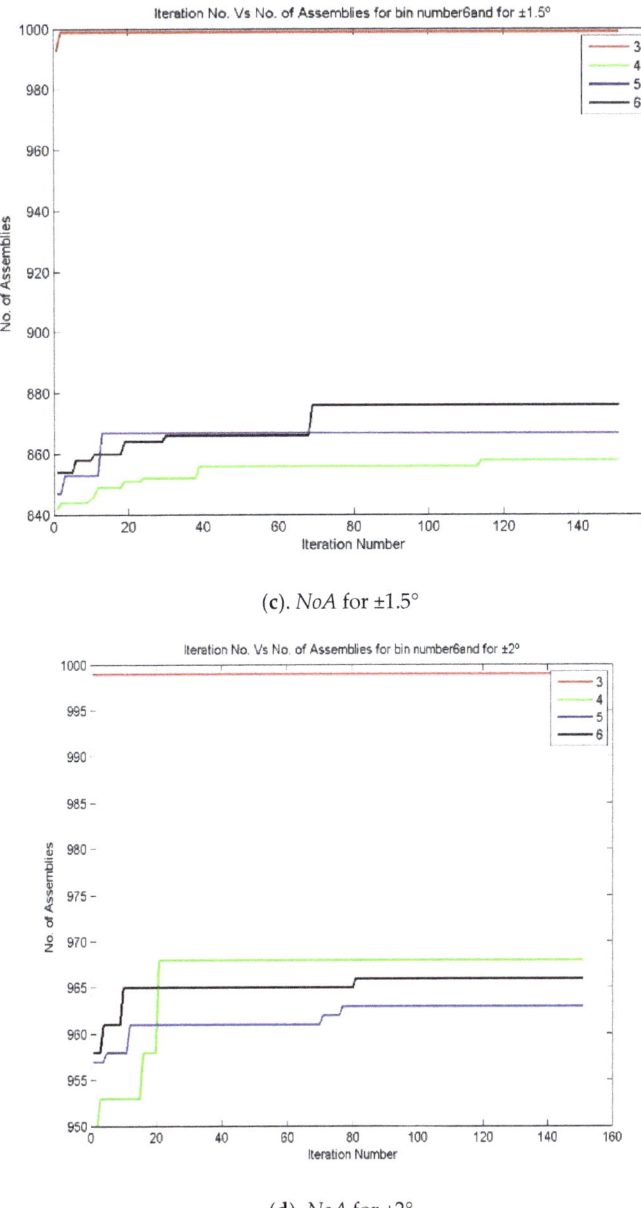

(**c**). *NoA* for ±1.5°

(**d**). *NoA* for ±2°

Figure 4. Iteration number vs. *NoA* for various bin numbers—the 3213 process combination (**a**) *NoA* for ±0.5°, (**b**) *NoA* for ±1°, (**c**). *NoA* for ±1.5° and (**d**) *NoA* for ±2°.

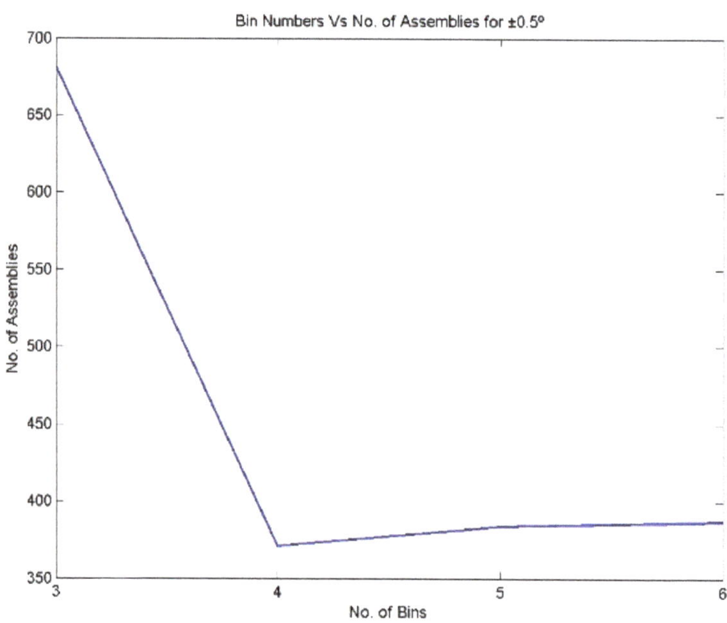

(**a**). *NoA* for ±0.5°

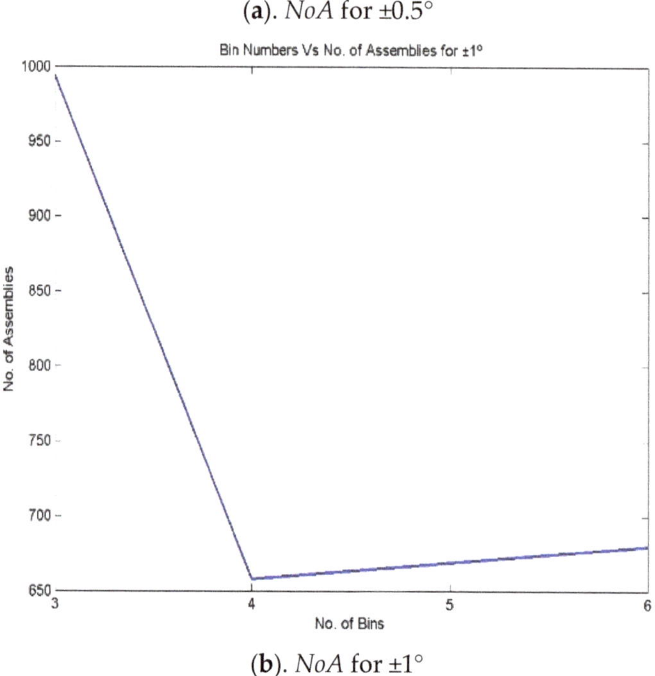

(**b**). *NoA* for ±1°

Figure 5. *Cont.*

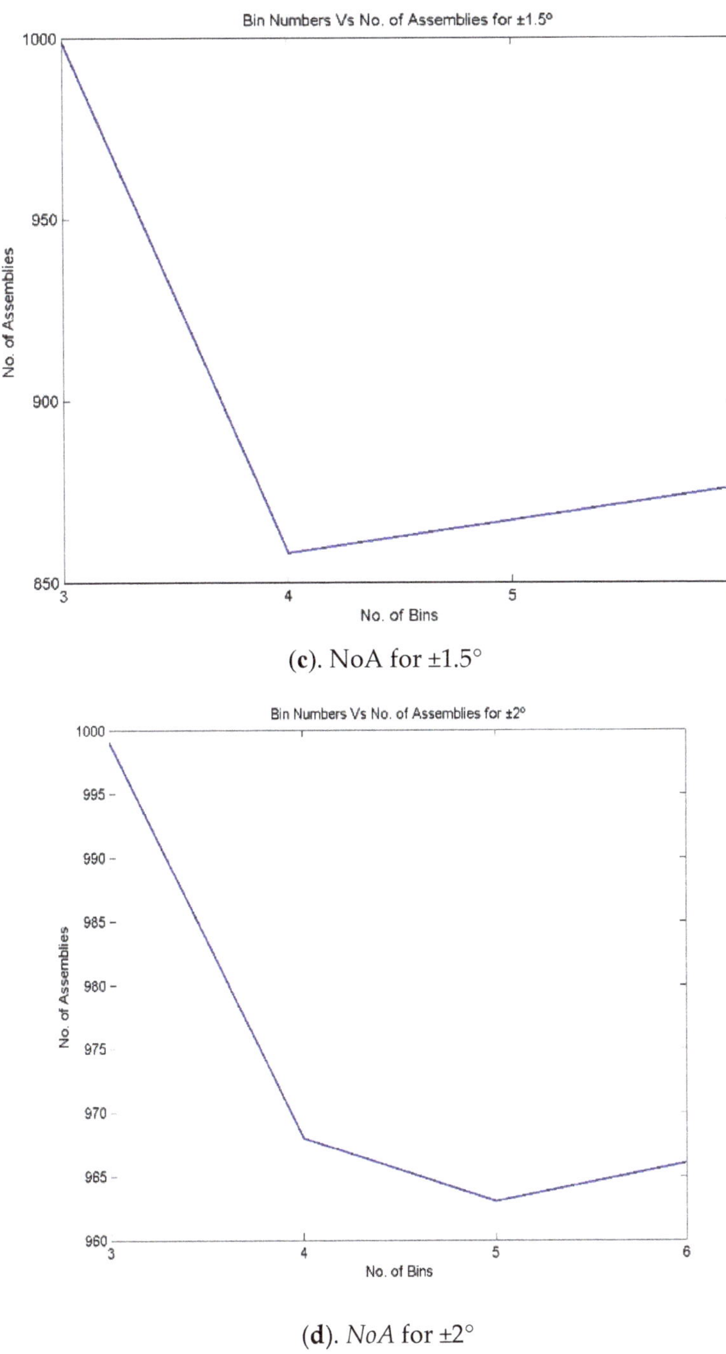

(**c**). NoA for ±1.5°

(**d**). *NoA* for ±2°

Figure 5. *NoA* for various bin numbers and assembly specifications for the 3213 process combination (**a**) *NoA* for ±0.5°, (**b**) *NoA* for ±1°, (**c**). *NoA* for ±1.5° and (**d**) *NoA* for ±2°.

6. Results and Discussion

An attempt has been made to make assemblies by considering the number of bins/ partitions, up to 9. Figure 6 reveals that in the equal area method for making non-linear assemblies, it is possible to make 996 assemblies out of 1000 components by partitioning them into 3 bins for the assembly specification of $\pm 2°$. It is also understood that while increasing the partition number, there may be a 5.8% (938 assemblies) drop in producing the number of assemblies for the same assembly specification. In the meantime, the number of assemblies is reduced for the same 1000 components of *X1*, *X2*, *X3*, and *X4* while reducing the assembly specification for the same partition number (equal to 3). The assemblies are reduced from 996 to 392 for the assembly specification of $\pm 2°$ to $\pm 0.5°$.

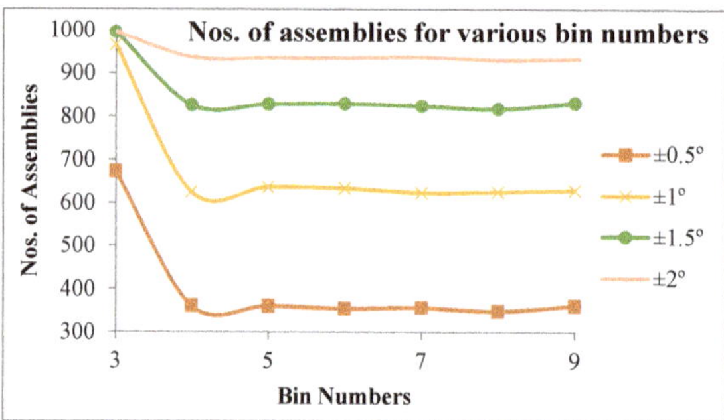

Figure 6. *NoA* vs. bin number for various assembly specifications for the 3213 process combination.

Figures 7–9 represent the number of assemblies produced for various assembly specifications while changing the partition number for the same set of 1000 components produced based on process combination 3213. A similarity can be observed from the above figures. Except in three bin partitions, all other partitions of components and matching components according to the best bin combinations obtained through the HSA produced almost a very close number of assemblies for various assembly specifications. Figure 10 indicates that maximum assemblies are produced for various assembly specifications and process combinations while partitioning the components into three bins. The assembly specification value after $\pm 1°$ in all the process combinations could produce an almost equal number of assemblies for bin number three. Table 15 represents the best bin combinations and their maximum number of assemblies for various process combinations.

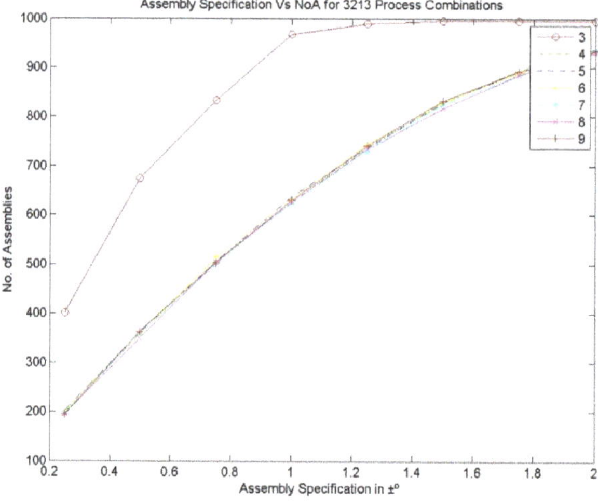

Figure 7. Assembly specification vs. *NoA* for various bin numbers—the 3213 process combination.

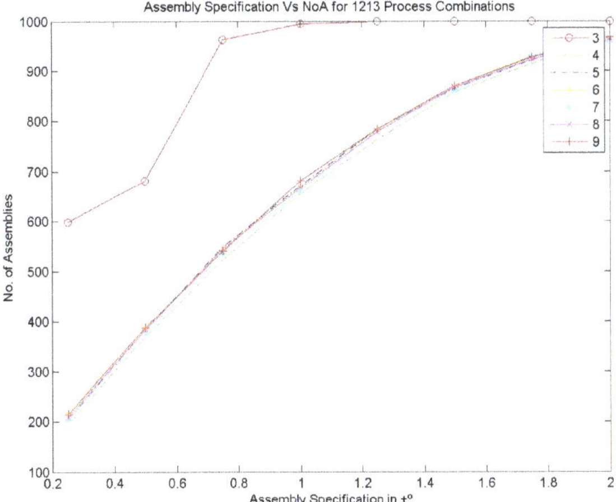

Figure 8. Assembly specification vs. *NoA* for various bin numbers—the 1213 process combination.

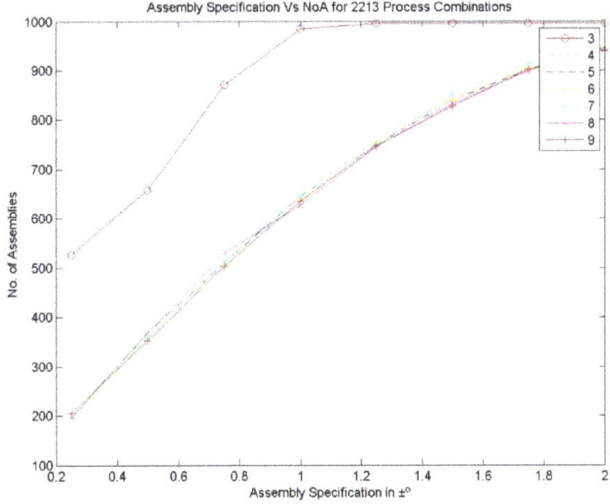

Figure 9. Assembly specification vs. *NoA* for various bin numbers—the 2213 process combination.

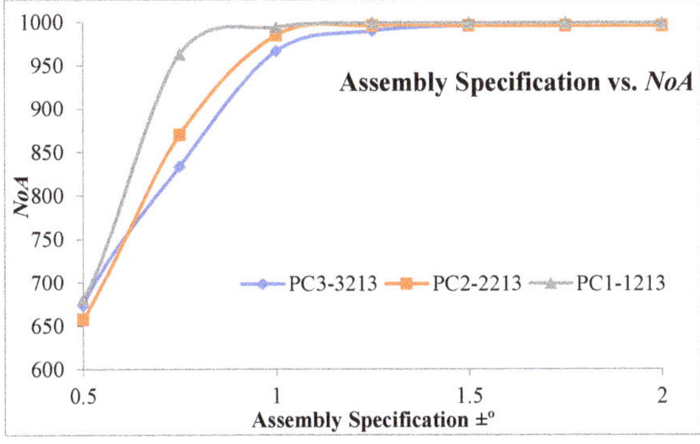

Figure 10. Maximum *NoA* produced for different assembly specifications and various process combinations.

Table 15. Best bin combinations for various assembly specifications and process combinations.

AS	PC3–3213					PC2–2213					PC1–1213				
	X1	*X2*	*X3*	*X4*	*NoA*	*X1*	*X2*	*X3*	*X4*	*NoA*	*X1*	*X2*	*X3*	*X4*	*NoA*
±0.5°	1 2 3	2 3 1	2 3 1	1 3 2	674	3 2 1	1 2 3	2 3 1	1 2 3	658	2 1 3	3 2 1	3 1 2	3 1 2	681
±1°	1 2 3	3 2 1	2 3 1	2 3 1	967	1 3 2	3 1 2	2 1 3	2 1 3	985	2 1 3	2 1 3	3 2 1	3 1 2	994
±1.5°	1 2 3	3 2 1	2 3 1	2 3 1	996	2 1 3	2 3 1	2 3 1	2 3 1	996	2 1 3	3 2 1	3 2 1	3 2 1	999
±2°	1 3 2	2 1 3	2 3 1	1 3 2	996	2 1 3	1 2 3	3 2 1	2 1 3	996	1 3 2	1 3 2	1 3 2	1 3 2	999

AS—assembly specification.

7. Conclusions

This paper addresses a novel methodology by combining the univariate search method and the harmony search algorithm in selective assembly for making non-linear assemblies for various assembly specifications. The best processes for the different components of the assembly are selected from the known alternative processes using the univariate search method, and these components are grouped into 3 to 9 bins. Further, the best bin combinations for making assemblies to reduce manufacturing cost are obtained through the harmony search algorithm. In this work, the component's dimensions are directly considered for making assemblies from the best bin combinations rather than considering tolerances, as in the existing method. The proposed method is demonstrated on a non-linear overrunning clutch assembly and has proved its efficiency by saving 24.9% of manufacturing cost compared with the existing method for the best process combination of 3213.

Author Contributions: Conceptualization and Methodology by L.N. and S.K.M.; Writing—Original Draft Preparation by L.N. and S.K.M.; Writing—Review and Editing by S.S., E.A.N., J.P.D. and H.M.A.H. All authors have read and agreed to the published version of the manuscript.

Funding: The authors extend their appreciation to King Saud University for funding this work through Researchers Supporting Project number (RSP-2021/164), King Saud University, Riyadh, Saudi Arabia.

Institutional Review Board Statement: Not Applicable.

Informed Consent Statement: Not Applicable.

Data Availability Statement: Not Applicable.

Conflicts of Interest: The authors declare that they have no conflict of interest.

Availability of Data and Material: Not applicable.

References

1. Kern, D.C. Forecasting Manufacturing Variation Using Historical Process Capability Data: Applications for Random Assembly, Selective Assembly, and Serial Processing. Ph.D. Thesis, Massachusetts Institute of Technology, Department of Mechanical Engineering, Cambridge, MA, USA, 15 May 2003.
2. Mease, D.; Nair, V.N.; Sudjianto, A. Selective assembly in manufacturing: Statistical issues and optimal binning strategies. *Technometrics* **2004**, *46*, 165–175. [CrossRef]
3. Matsuura, S.; Shinozaki, N. Selective assembly to minimize clearance variation by shifting the process mean when two components have dissimilar variances. *Quality-Tokyo* **2008**, *38*, 73.
4. Kwon, H.M.; Hong, S.H.; Lee, M.K. Effectiveness of the ordered selective assembly for two mating parts. In Proceedings of the 10th Asia Pacific Industrial Engineering & Management Systems, Kitakyushu, Japan, 6 September 2009; pp. 1402–1405. Available online: https://apiems.org/ (accessed on 6 September 2021).
5. Matsuura, S.; Shinozaki, N. Optimal process design in selective assembly when components with smaller variance are manufactured at three shifted means. *Int. J. Prod. Res.* **2011**, *49*, 869–882. [CrossRef]
6. Yue, X.; Wu, Z.; Tianze, H.; Julong, Y. A heuristic algorithm to minimize clearance variation in selective assembly. *Rev. Tech. Fac. Ing. Univ. Zulia* **2014**, *37*, 55–65.
7. Babu, J.R.; Asha, A. Tolerance modelling in selective assembly for minimizing linear assembly tolerance variation and assembly cost by using Taguchi and AIS algorithm. *Int. J. Adv. Manuf. Technol.* **2014**, *75*, 869–881. [CrossRef]
8. Ju, F.; Li, J. A Bernoulli model of selective assembly systems. *IFAC Proc. Vol.* **2014**, *47*, 1692–1697. [CrossRef]
9. Xu, H.-Y.; Kuo, S.-H.; Tsai, J.W.H.; Ying, J.F.; Lee, G.K.K. A selective assembly strategy to improve the components' utilization rate with an application to hard disk drives. *Int. J. Adv. Manuf. Technol.* **2014**, *75*, 247–255. [CrossRef]

10. Lu, C.; Fei, J.-F. An approach to minimizing surplus parts in selective assembly with genetic algorithm. *Proc. Inst. Mech. Eng. Part B J. Eng. Manuf.* **2014**, *229*, 508–520. [CrossRef]
11. Manickam, N.; De, T.N.; Laboratory, A.S. Optimum group size selection for launch vehicle sections linear assembly by selective assembly method. *Int. J. Eng. Res.* **2015**, *4*, 837–843. [CrossRef]
12. Babu, J.R.; Asha, A. Modelling in selective assembly with symmetrical interval-based Taguchi loss function for minimising assembly loss and clearance variation. *Int. J. Manuf. Technol. Manag.* **2015**, *29*, 288. [CrossRef]
13. Ju, F.; Li, J.; Deng, W. Selective assembly system with unreliable bernoulli machines and finite buffers. *IEEE Trans. Autom. Sci. Eng.* **2016**, *14*, 171–184. [CrossRef]
14. Malaichamy, T.; Sivasubramanian, R.; Sivakumar, M. Selective assem–A software approach. *Int. J. Adv. Eng. Technol.* **2016**, *7*, 729–735.
15. Malaichamy, T.; Sivasubramanian, R.; Kumar, S.M.; Sundaram, M.C. Simulated annealing algorithm for minimising the surplus parts in selective assembly-A software approach. *Asian J. Res. Soc. Sci. Humanit.* **2016**, *6*, 1567. [CrossRef]
16. Liu, S.; Liu, L. Determining the number of groups in selective assembly for remanufacturing engine. *Procedia Eng.* **2017**, *174*, 815–819. [CrossRef]
17. Chu, X.; Xu, H.; Wu, X.; Tao, J.; Shao, G. The method of selective assembly for the RV reducer based on genetic algorithm. *Proc. Inst. Mech. Eng. Part C J. Mech. Eng. Sci.* **2017**, *232*, 921–929. [CrossRef]
18. Geem, Z.W.; Kim, J.H.; Loganathan, G. A new heuristic optimization algorithm: Harmony search. *Simulation* **2001**, *76*, 60–68. [CrossRef]
19. Geem, Z.; Kim, J.; Loganathan, G. Harmony search optimization: Application to pipe network design. *Int. J. Model. Simul.* **2002**, *22*, 125–133. [CrossRef]
20. Lee, K.S.; Geem, Z.W. A new structural optimization method based on the harmony search algorithm. *Comput. Struct.* **2004**, *82*, 781–798. [CrossRef]
21. Lee, K.S.; Geem, Z.W. A new meta-heuristic algorithm for continuous engineering optimization: Harmony search theory and practice. *Comput. Methods Appl. Mech. Eng.* **2005**, *194*, 3902–3933. [CrossRef]
22. Wang, X.; Gao, X.Z.; Ovaska, S.J. Fusion of clonal selection algorithm and harmony search method in optimisation of fuzzy classification systems. *Int. J. Bio. Inspired Comput.* **2009**, *1*, 80. [CrossRef]
23. Ganesan, K.; Suresh, R.K.; Mohanram, P.V.; Vishnukumar, C.H. Optimum tolerance allocation and process selection using simulated annealing. In Proceedings of the 4th Int. Conference on Mechanical Engineering, Theni, Tamil Nadu, India, 26–28 December 2009; Volume VI, pp. 201–205.

applied
sciences

Article

A Hybrid Bat Algorithm for Solving the Three-Stage Distributed Assembly Permutation Flowshop Scheduling Problem

Jianguo Zheng and Yilin Wang *

Glorious Sun School of Business and Management, Donghua University, Shanghai 200051, China; zjg@dhu.edu.cn
* Correspondence: yilinwang0319@163.com

Abstract: In this paper, a hybrid bat optimization algorithm based on variable neighbourhood structure and two learning strategies is proposed to solve a three-stage distributed assembly permutation flowshop scheduling problem to minimize the makespan. The algorithm is firstly designed to increase the population diversity by classifying the populations, which solves the difficult trade-off between convergence and diversity of the bat algorithm. Secondly, a selection mechanism is used to update the bat's velocity and location, solving the difficulty of the algorithm to trade-off exploration and mining capacity. Finally, the Gaussian learning strategy and elite learning strategy assist the whole population to jump out of the local optimal frontier. The simulation results demonstrate that the algorithm proposed in this paper can well solve the DAPFSP. In addition, compared with other metaheuristic algorithms, IHBA has better performance and gives full play to its advantage of finding optimal solutions.

Keywords: hybrid bat algorithm; optimization problem; the distributed assembly permutation flowshop scheduling problem; variable neighborhood descent

Citation: Zheng, J.; Wang, Y. A Hybrid Bat Algorithm for Solving the Three-Stage Distributed Assembly Permutation Flowshop Scheduling Problem. *Appl. Sci.* **2021**, *11*, 10102. https://doi.org/10.3390/app112110102

Academic Editors: Vladimir Modrak and Zuzana Soltysova

Received: 17 September 2021
Accepted: 22 October 2021
Published: 28 October 2021

Publisher's Note: MDPI stays neutral with regard to jurisdictional claims in published maps and institutional affiliations.

1. Introduction

The distributed assembly scheduling problem is a derivative and extension of the assembly scheduling problems. It is mainly used to produce multiple products assembled from different parts. In a decentralized and globalized economy, the current economic trend of customization and intelligent manufacturing has led to the rapid development of assembly production. Internationally, international companies with multiple production centers or plants are even more common. This shows that distributed and intelligent production plays a pivotal and irreplaceable role in the modern manufacturing industry. Currently, this type of scheduling is widely used in the supply chain and manufacturing industry. In order to maintain a competitive position in a rapidly changing market, managers must quickly make the right decisions about how to allocate work to plants and how to schedule each plant efficiently. Therefore, the distributed scheduling problem has become a hot research topic among scholars and researchers.

The distributed assembly permutation flowshop scheduling problem (DAPFSP) is of great importance to both the industrial industry and the research community. As we all know, DAPFSP consists of two phases, including production and assembly. In detail, the production phase corresponds to a distributed displacement flowshop scheduling problem, where n jobs are assigned to be processed in f plants with m identical machines in a flowshop line. Each job belongs to a certain product. The processes contained in each product cannot be interrupted or inserted by processes of other products during the production process. The assembly phase corresponds to the assembly flow shop scheduling problem, where s products are made in an assembly plant with a single assembly machine, and each product must be assembled after all processes produced in the production phase are completed.

Total completion time has been identified as a more relevant and meaningful goal in today's dynamic manufacturing environment when a batch of work needs to be completed as quickly as possible. This allows the objective function to minimize the flow of work and efficiently improve resource utilization. DPFSP is an NP-hard with m greater than or equal to 2, and is a derivation and extension of the traditional permutation flow shop scheduling problem with this criterion. Similarly, DAPFSP can be seen as an extension of DPFSP, which is more complex than DPFSP. Therefore, the DAPFSP with completion time as the objective function is also a strong NP-hard problem. In recent years, as a simple yet efficient method, the bat algorithm has solved many scheduling problems in academia, including the job shop scheduling problem [1], the open shop scheduling problem [2], the task scheduling problem [3], and the production scheduling problem [4], by continuously performing local search in the neighborhood of the current solution to obtain the optimal solution.

In this paper, an improved hybrid bat algorithm, IHBA, is proposed for solving the DAPFSP with maximum makespan based on the original bat algorithm. The main contributions of this paper are:

- The algorithm firstly designs the population classification to increase the diversity of the population and solve the difficult trade-off between convergence and diversity of the bat algorithm.
- A selection mechanism is used to update the speed and location of bats, solving the difficulty of the algorithm to trade-off exploration and mining capacity.
- Gaussian learning strategy and elite learning strategy assist the whole population to jump out of the local optimal frontier.

This paper is organized as described below. Section 2 reviews the closely related literature. Section 3 presents the DAPFSP problem and its mathematical model. In Section 4, the hybrid bat algorithm proposed in this paper is described in detail. In addition, the parameter calibration and simulation calculation results are analyzed in Section 5. In the last section, the paper is summarized, and future work is provided.

2. Literature Review

It is well known that the distributed permutation flowshop scheduling problem (PFSP) is one of the most well-known production scheduling problems. It has a strong engineering background by having the same work order on all machines. The problem has been shown to be a strong NP-hard problem when more than two machines are involved. The problem of distributed assembly permutation flowshop scheduling problem (DAPFSP) has been gaining attention and popularity among scholars and researchers. The most leading exponent is Hatami et al. [5], whose work is very enlightening. He was the first to optimize and improve the DAPFSP for modeling and studying complex supply chains in 2013. They also introduced the mixed integer linear programming model, proposed three construction algorithms, tested the variable neighborhood descent (VND) algorithm, and demonstrated the good performance of the VND algorithm in solving this scheduling problem. Similarly, he went on to expand on his previous paper by adding time for sequence-related setups in both the production and assembly stages [6]. In 2015, Hatami et al. [7] considered a distributed assembly scheduling problem with sequence-dependent setup times and the goal of minimizing production time. The setup times of both phases are sequence-dependent, which also lays the theoretical foundation for the sequences in this paper. Heuristics and meta-heuristics are also proposed to solve it. Based on his literature and views, many of these ideas and opinions have been studied and explored in greater depth by many scholars. Based on these rather cutting-edge ideas, many scholars have built on and deepened their research, and Ying et al. [8] extended the DAPFSP for flexible assembly and sequence-independent setup times in supply chains. Using completion time as an optimization criterion, construction heuristic and custom meta-heuristic algorithms are proposed to solve this emerging scheduling extension. Gonzalez-Neira et al. [9] studied a stochastic version of the DAPFSP and proposed a hybrid algorithm to solve the stochastic problem, integrating biased randomization and simulation techniques. Wang [10] introduced a just-in-time constraint between the two phases of the traditional DAPFSP to form a just-in-time DAPFSP to minimize the maximum weighted tardiness cost. A variable neighborhood search based on memory algorithm is proposed. From the above reviewed studies, it can be concluded that these scholars and researchers have extended and improved this problem, and most of them have an objective function based on minimizing the makespan, which provides a theoretical basis for the objective function in this paper.

Most scheduling problems are NP-hard and exact methods simply cannot find the optimal solution in reasonable computational time. Therefore, heuristic and meta-heuristic algorithms have been used to find the optimal solution and near-optimal solutions in reasonable computation time. Zhang et al. [11] pointed out that the DAPFSP is a typical NP-hard combinatorial optimization problem, and proposed an innovative three-dimensional matrix cube-based distribution estimation algorithm to address the difficulties of the proposed DAPFSP-T model. Likewise, Zhang et al. [12] also assumed this idea, and their team integrated the problem with two machine environments, distributed production and flexible assembly, which can assemble job processing into customized products, and proposed a mixed integer linear programming model to characterize the problem nature and solve the small-scale problem. An efficient modulo algorithm is further proposed. They affirm that the modal algorithm is an efficient algorithm to solve the DAPFSP, while some scholars prefer to use the greedy algorithm to solve the problem. For example, Pan et al. [13] solved a novel DAPFSP by proposing a mixed integer linear model, three construction heuristics, two variable neighborhood search methods, and an iterative greedy algorithm to obtain important problem-specific knowledge to improve the effectiveness of the algorithm. Similarly, Ochi et al. [14] studied the DAPFSP with the objective of minimizing completion time and, proposed an iterative greedy based approach called bounded search iterative greedy algorithm. Similarly, in order to coordinate production and transportation scheduling, Yang [15] proposed a novel DAPFSP-FABD model. The objective is to minimize the total cost of delivery and delay, and four heuristic algorithms, a variable neighborhood descent algorithm, and two iterative greedy algorithms are proposed. Liu et al. [16] proposed a memory algorithm for DAPFSP based on variable neighborhood search with the objective function of, i.e., makespan. And, the initialization based on NEH heuristic is applied to the product ordering. The neighborhood structure is introduced into VNS and used to perturb the job assignment of the factory and adjust the job order of the factory. Huang et al. [17] considered the DAPFSP with a total flow time criterion, and proposed an improved iterative greedy algorithm based on groupthink to solve the problem. It can be seen that the exact algorithm is no longer able to solve the problem, and this has promoted the development of heuristic and meta-heuristic algorithms, which can very definitely find the optimal solution as a solution to this scheduling problem.

More and more scholars are focusing more on the improvement of algorithmic aspects. For the no-wait flow shop scheduling problem that depends on time series, Hu et al. [18] proposed an enhanced differential evolutionary algorithm. Subsequently, Seidgar [19] took minimizing the weighted sum of expected completion time and average completion time as the solution objective and proposed four metaheuristic algorithms; namely, genetic algorithm, imperialistic competition algorithm, cloud theory-based simulated annealing, and adaptive differential evolution algorithm. Moreover, Li et al. [20] argued that the imperialist competition algorithm can solve the fuzzy distributed assembly flow shop scheduling problem. Furthermore, Al-Behadili et al. [21] proposed a multi-objective optimization model and particle swarm optimization solution method for the robust dynamic scheduling of permutation flow shops with uncertainty. Likewise, Zhang [22] emphasized that DAPFSP is a new generalization of the distributed displacement flow shop scheduling problem and the assembly flow shop scheduling problem, proposed an enhanced population-based metaheuristic genetic algorithm, and designed an effective crossover strategy based on local search to accelerate convergence. Li et al. [23] studied the minimization problem of DAPFSP and proposed a genetic algorithm with an enhanced crossover strategy and three different local searches. It is no coincidence that Mao et al. [24] advocated an improved discrete artificial bee colony algorithm to solve the DAPFSP with preventive maintenance operator, optimized by the completion time criterion. Moreover, a local search method with insertion and exchange operators is used to generate adjacent solutions in the hiring bee phase and the bystander bee phase. Song and Lin [25] jointly proposed a genetic programming hyperheuristic algorithm to solve the DAPFSP with sequence-dependent setup times, minimizing the completion time as the objective. For the improvement of the algorithm, these scholars further study the genetic algorithm and particle swarm algorithm. This also provides a reference basis for our subsequent simulation experiments.

3. Problem Description

The traditional distributed assembly permutation flowshop scheduling problem (DAPFSP) can be divided into two phases; namely, the production phase and the assembly phase. In this section, the two-stage DAPFSP problem is extended to a three-stage DAPFSP, named the production stage, the transportation stage, and the assembly stage, and the problem is described as follows.

The job $j \in \{1, 2, \ldots, n\}$ consists of operations O_{ij}, O_{Tj} and O_{Aj}. In the production stage, there are n work processes and f identical plants, and job $j \in \{1, 2, \ldots, n\}$ needs to be assigned to any plant $l \in \{1, 2, \ldots, f\}$ for processing, and each plant is equipped with the same m machines, $M = \{M_1, M_2, \ldots, M_m\}$, which correspond to the same assembly line shop for part processing. Operations $O_{ij} \in \{O_{1j}, O_{2j}, \ldots, O_{mj}\}$ are processed on machine M_i, $i = 1, \ldots, m$ and require p_{ij} time units. Machine M_{ij} can process at most one job at a time. The production and transport operations O_{Tj} will be executed on machine M_T and take p_{Tj} time units. The rule is that in each assembly permutation flowshop, all jobs need to be processed on the same path in the order of machine i, which is equivalent to first on machine M_1, then on M_2. This goes on, until the end on machine M_m. Parts belonging to the same product must be processed continuously, and no partial parts are allowed to pass through. All jobs start timing at time $t = 0$ and on machine $i = 1$.

The subsequent phase is the transport stage, which means that after the completion of the first one in the first stage of the product operation, the transport machine collects all parts of the product and moves to the assembly stage.

In the assembly stage, there are s products and an assembly machine M_A. Each product $r \in \{1, 2, \ldots, s\}$ has a defined part of the operation. The assembly operation O_{Aj} will be executed on the machine M_A and uses p_{Aj} time units. The rule is that the assembly of a product can only start when the work contained in the product is completed and the assembly machine is idle. Continuous processing stages are part of the same product operation. The next product operation can only be performed after all jobs belonging to a product have been processed.

From this, it is clear that the objective function of the problem is to determine the allocation of products to plants, and the sequence of parts and products to each plant in order to minimize the completion time or maximum completion time for all products. Based on the above description of the problem, the minimization of the completion time is denoted as $F_m|nwt|C_{max}$. To satisfy the above constraint, a feasible schedule $\pi = \{\pi_1, \pi_2, \ldots, \pi_n\}$ is found for n jobs, such that the completion time $C_{max}(\pi)$ is minimized. This is the same where $C_{max}(\pi)$ is also equivalent to the time to complete the last job on the last machine. The Equation (1) is as follows.

$$C_{max}(\pi) = \sum_{k=2}^{n} D_{\pi_{k-1}, \pi_k} + \sum_{j=1}^{m} P_{\pi_n, j} \tag{1}$$

In the transportation phase, it is necessary to ensure that the completion time distance between two adjacent jobs is determined by the processing time of the two processes, independent of the other jobs in the arrangement. Therefore, the completion time distance is defined between each pair of jobs. The completion time distance between two adjacent job processes π_{k-1} and π_k is calculated as

$$C_{\pi_{k-1},\pi_k} = \max_{1 \le i \le m} \left\{ \sum_{j=1}^{i} P_{\pi_{k-1},j} - \sum_{j=2}^{i} P_{\pi_k,j-1} \right\}, k = 2,3,\ldots,n \tag{2}$$

In order to build a mathematical model based on the above description, the constraints, parameters and variables are as follows.

Subject to,

$$\sum_{j=0,j\ne k}^{n} X_{j,k} = 1, k \in \{1,2,\cdots,n\} \tag{3}$$

$$\sum_{k=0,k\ne j}^{n} X_{j,k} = 1, j \in \{1,2,\cdots,n\} \tag{4}$$

$$\sum_{k=1}^{n} X_{0,k} = f \tag{5}$$

$$\sum_{j=1}^{n} X_{j,0} = f \tag{6}$$

$$X_{j,k} + X_{k,j} \le 1, k \in \{1,2,\cdots,n-1\}, k > j \tag{7}$$

$$\sum_{j=0}^{z_{r-1}} \sum_{k=z_{r-1}+1}^{z_r} X_{j,k} + \sum_{j=z_r+1}^{z_r} \sum_{k=z_{r-1}+1}^{z_r} X_{j,k} = 1 \tag{8}$$

$$r \in \{1,2,\cdots,s\}, z_r = \sum_{t=0}^{r} n_t \tag{9}$$

$$C_{i,j} \ge C_{i-1,j} + p_{i,j}, i \in \{1,2,\cdots,m\}, j \in \{1,2,\cdots,n\} \tag{10}$$

$$C_{i,j} \ge C_{i-1,j} + p_{i,j} + \left(X_{j,k} - 1\right)g$$
$$i \in \{1,2,\cdots,m\}, j \in \{1,2,\cdots,n\}, k \in \{0,1,2,\cdots,n\}, j \ne k \tag{11}$$

$$\sum_{t=1,t\ne r}^{s} Y_{r,t} = 1, r \in \{1,2,\cdots,s\} \tag{12}$$

$$Y_{r,t} + Y_{t,r} \le 1, r \in \{1,2,\cdots,s-1\}, t > r \tag{13}$$

$$C_r \ge C_{m,j} + p_r$$
$$j \in \{z_{r-1}+1,\cdots,z_r\}, r \in \{1,2,\cdots,s\} \tag{14}$$

$$C_r \ge C_t + p_r + (Y_{r,t} - 1)g$$
$$r \in \{1,\cdots,s\}, t \in \{0,\cdots,s\}, r \ne t \tag{15}$$

$$TF \ge \sum_{r=1}^{s} C_r \tag{16}$$

$$X_{j,k} \in \{0,1\}, j \in \{0,\cdots,n\}, k \in \{0,\cdots,n\}, j \ne k \tag{17}$$

$$Y_{r,t} \in \{0,1\}, r \in \{0,\cdots,s\}, t \in \{0,\cdots,s\}, r \ne t \tag{18}$$

$$C_{i,j} \geqslant 0, i \in \{1, 2, \cdots, m\}, j \in \{1, 2, \cdots, n\} \tag{19}$$

$$C_r \geqslant 0, r \in \{1, 2, \cdots, s\} \tag{20}$$

where X_{jk} represents a binary variable. It is equal to 1 if job j is the predecessor of job k. Otherwise, $X_{jk} = 0$. The same is true for Y_{rt}. The product r is equal to 1 if it is the predecessor of the product t. Otherwise, $Y_{rt} = 0$. C_{ij} is the completion time of job j on machine i, and C_r is the completion time of product r. p_r is the assembly time of product r, and z_r is the total number of jobs contained in the first r products in the product sequence. In addition, g is a large enough positive number. r_t is the number of jobs contained in product t. It is worth noting that $C_{0,j} = C_{i,0} = C_0 = 0$.

Equation (1) shows that the goal of the mathematical model is to minimize the completion time. The constraint sets (3) and (4) indicate that each job must have only one preceding job and one following job. Constraint sets (5) and (6) show that the preceding and following jobs of virtual job 0 are forced to be executed f times. Constraint set (7) ensures that a job cannot be both a predecessor and a successor of another job. Constraint set (8) ensures that jobs belonging to the same product are processed consecutively. Constraint set (9) indicates that job j can be processed on machine $i - 1$ only after it is processed on machine i. Constraint set (10) indicates that job j can only be processed on machine i after completing the processing of job k if job j is a successor to job k, where g is positive and has a sufficiently large scale volume. The constraint sets (11) and (12) ensure that each product must have only one preorder and one subsequent job. Constraint set (13) ensures that a product cannot be both a preorder and a successor of another product. Constraint set (14) ensures that each product cannot enter the assembly stage before the production of the part of the job it contains is complete. Constraint set (15) shows that if product t is a preceding job of product r, then product r cannot start the assembly job until product t is completed. Constraint set (16) defines the minimum completion time. Constraint sets (17) to (20) define the domains of the decision variables.

The model uses a sequence-based variable and a set of $(\min\{f, s\} + 1)$ virtual parts. The sequence starts with a virtual part and ends with a virtual part. These virtual parts divide all parts into $\min\{f, s\}$ subsequences, each corresponding to a plant. Parts belonging to the same product are never separated. Thus, the product sequence is implicitly included in the part sequence. For example, $s = 4, n = 8, m = 2, f = 2$. Table 1 shows the machining times of parts and products.

Table 1. Machining time for parts and products.

Part	1	2	3	4	5	6	7	8
M_1	2	5	7	9	9	3	8	4
M_2	3	8	5	7	3	4	1	3

Product	1	2	3	4
M_A	9	8	7	6

One of the feasible solutions is $x_{04} = x_{43} = x_{35} = x_{50} = x_{01} = x_{12} = x_{28} = x_{86} = x_{67} = x_{70} = 1$. The sequence of parts is $\{0, 4, 3, 5, 0, 1, 2, 8, 6, 7, 0\}$, consisting of the subsequences $\{4, 3, 5\}$ and $\{1, 2, 8, 6, 7\}$.

4. Materials and Methods

To improve the local search capability of the algorithm, this section introduces the mathematical representation of feasible solutions, classification of populations, selection mechanism, domain structure and Gaussian learning strategy and elite learning strategy. A hybrid bat algorithm adapted to the three-stage DAPFSP problem is proposed.

4.1. Mathematical Expression of the Feasible Solution of the Three-Stage DAPFSP

The representation of feasible solutions is a crucial and important step in every meta-heuristic approach. The addition of a transportation process to the DAPFSP solved in this paper. The result is that the original expression for the process permutation of the PFSP is not adequate for this scheduling problem. Therefore, in this section, the feasible solution is divided into a sequence of products and a sequence of s processes based on the model of the three-stage DAPFSP model and the properties of the bat algorithm. Each sequence of these s processes corresponds to a product. The processes of a part can be viewed as the sequence of product assembly on the assembly flowshop.

In the production and transportation stages, parts need to be allocated one by one according to the product order. When allocating parts, their processes are sequentially arranged to the corresponding factories according to the process order of the product, and a minimum manufacturing cycle is obtained. Only after all the processes included in the current part are completed can the next part be produced. When a product has completed all processes, the product can proceed to the assembly stage for the installation and assembly stages. As an example, suppose a set of sequences f is used to represent a feasible solution, and each plant has one and only one solution. Each sequence is an ordering of all the parts assigned to that plant, indicating the order in which the parts enter the flow shop. Thus, the order of processed products in the assembly stage is also implied, and the feasible solution can be expressed as $\pi = \left\{ \pi_1, \pi_2, \ldots, \pi_f \right\}$, where $\pi_k = \left(\pi_{k,1}, \pi_{k,2}, \ldots, \pi_{k,\eta k} \right), k = 1, 2, \ldots, f$ is the sequence associated with sequence associated with plant k and η_k is the total number of parts assigned to plant k. For the example in Section 3, the representation corresponding to this feasible solution is $\pi = \{\pi_1, \pi_2\}$ and $\pi_1 = \{4, 3, 5\}, \quad \pi_2 = \{1, 2, 8, 6, 7\}$.

In the feasible solution of the original DAPFSP problem, the first stage represents the arrangement of all N processes, and the second stage represents the plants to which these N processes are assigned respectively. In this section, the concept of product priority is introduced to determine the order of product assembly, and a novel feasible solution of the three-stage DAPFSP is proposed. Specifically, the feasible solution in the first stage is expressed as the processing priority of the product, and the parts assembly of each product is equal to the number of processes that constitute that product. The specific steps are, firstly, calculating the total processing time of the product at each stage and randomly initializing the sequence of products. Second, the first two products in the initial product sequence are selected for evaluation, and whichever sequence is better is chosen as the current sequence. Again, this is done by placing it among the product sequences that are already scheduled properly. Finally, an optimal schedule can be obtained to select the current best sequence for the next iteration of the process. The second stage is represented as the processing priority of all processes, where the processes belonging to the same product are arranged in a random order. The smaller the number of its columns, the higher the priority of the element values; the third stage is the plant priority assignment vector, where each part is represented as the plant assigned to the corresponding process. The specific steps are to assign a part of the processes to the factories one by one in order, then select a process and find the minimum completion time by placing it after the last process in each factory. The best part assignment is selected for the next iteration.

Table 2 describes how the three-stage DAPFSP representation is decoded into a schedule. Suppose there are 3 products ($s = 3$) consisting of 9 jobs ($n = 9$) processed in 3 plants ($f = 3$). The mathematical representation is $P_1 = \{3, 4, 6\}, P_2 = \{1, 2, 8, 9\}, P_3 = \{5, 7\}$. According to the last two stages, called job sequence and factory assignment, it can be observed that jobs 5, 4, and 8 are assigned to the first factory and are processed in the order of 5, 4, and 8 according to the priority specified in layer 2. Then, it is instantly obtained that the partial scheduling of the factories can be decoded as $\pi_1 = \{5, 4, 8\}$. Similarly, the partial scheduling of the remaining two plants can be decoded as $\pi_2 = \{2, 7, 9\}$ and $\pi_3 = \{6, 3, 1\}$, respectively. According to the values of the first stage specifying the product assembly priority, the value of 1 in the table corresponds to jobs 5 and 7 belonging to product 3. It can be concluded that after the processing stage, product 3 should be assembled first. In a similar way, it can be concluded that product 1 is assembled for the second time immediately after product 2.

Table 2. Coding scheme.

Product Priority	2	1	2	3	3	2	1	3	3
Job	6	5	4	2	8	3	7	9	1
Factory Tasks	3	1	1	2	1	3	2	2	3

In the context of the traditional representation of feasible solutions, the order of assembly of all products follows a first-come, first-served rule. This results in products being ranked in the processing stage in ascending order of the total completion time of their component operations. In contrast, the priority of products in the assembly phase is determined by the order of arrival at the assembly, and there is no way to specify it. This observation is confirmed by a study of the two-stage assembly scheduling problem by Tozkapan [26]. The efficiency may be higher if the sequence of operations in the processing phase is different from the sequence of operations in the assembly phase under the total delay objective. With a very late delivery date, for example, the assembly of the product may be delayed, even if the entire processing phase of the product's component operations has been completed.

4.2. Population Classification

The optimization process is divided into two stages. The first stage is that when the individual bat is in a better search position and is close to the optimal solution, its loudness and pulse emission rate reach the best state. This process is called the search phase, its population is called the search-type population. The second stage is the population of bats in a disadvantaged search position; that is, the capture population, so two populations with different functions are obtained. After each iteration, by updating the search direction and step length, the loudness and pulse emission rate of the bat algorithm are improved to find the optimal solution. By adjusting the weights and deviations of the network, the difference between the average value and the standard deviation of the bat algorithm can be minimized.

4.2.1. Search-Type Population

The combination of the back propagation (BP) algorithm based on mean square error (MSE) and gradient descent method minimizes the mean square error [27]. The formula based on the mean square error measurement is shown in (21):

$$\mathrm{MSE}(d, y) = \frac{1}{N} \sum_{n=1}^{N} (d(n) - y(n))^2 \tag{21}$$

where $d(n)$ represents the n-th element of the required signal, and $y(n)$ represents the nth actual output. In the training process, the weight vector in the $(t+1)$ iteration is updated by (23). Where α is the learning rate or step length, W_t is the weight vector of the previous iteration and g_t is the gradient vector, which can be calculated by (22) and (23).

$$w_{t+1} = w_t - \alpha g_t \tag{22}$$

$$g_t = \left. \frac{\partial e}{\partial w} \right|_{w=w_t} = \left[\frac{\partial e}{\partial w_{11}} \cdots \frac{\partial e}{\partial w_{ij}} \cdots \frac{\partial e}{\partial w_{nn}} \right] \tag{23}$$

In (23), e is the MSE error output in the t-th step of the training process and $\partial e / \partial w$ is derived from the MSE error on each element of the w vector. The weight is expressed as

$w_{t+1} = w_t - \alpha_t g_t$, α_t will be adjusted to an appropriate value to achieve better convergence. At this time, suppose the probability pops of the search-type population, its loudness and pulse emission rate are updated to:

$$A_i^{t+1} = \alpha_t \times A_i^t \tag{24}$$

$$r_i^{t+1} = r_i^0 \left(1 - e^{-wt}\right) \tag{25}$$

4.2.2. Captive Population

As a branch of the gradient descent method, the conjugate gradient method (CG) is applied to nonlinear unconstrained optimization problems. This method has strong local and global convergence [26–28]. According to the cloud computing resource scheduling problem, the CG method uses (26) to generate a weight sequence:

$$w'_{t+1} = w'_t + \alpha'_t d \tag{26}$$

where α'_t is the result of the line search method, which can be an exact line search or an inaccurate line search, which is expressed as a step size here. In (26), d_t is in the descending direction, which is expressed as the search direction here, and its formula is shown in (27):

$$d_{t+1} = -g_{t+1} + \beta_t d_t \tag{27}$$

In (27), β_t is the conjugate parameter, g_{t+1} is the gradient of the objective function concerning the weight at step $t + 1$, and represents the direction of the last step, and the first step is $d_0 = -g_0$.

In the iterative process, the loudness A_i and the pulse emission rate r_i will also change. As the bat moves closer to its prey, its loudness will decrease, and the pulse emission rate will increase. At this time, the probability of the catching population is pop_h and $pop_s + pop_h = 1$. In summary, the loudness and pulse emission rate of the bat algorithm are updated to:

$$A_i^{t+1} = \alpha'_t \times A_i^t \tag{28}$$

$$r_i^{t+1} = r_i^0 \left(1 - e^{-w't}\right) \tag{29}$$

4.3. Selection Mechanism

When it comes to the bat algorithm as a learning algorithm, each individual in the population needs to learn from the individually optimal and globally optimal individuals in order to find the optimal solution. In the field of multi-objective optimization, there is no single optimal solution, and the optimal selection mechanism needs to be redesigned. Which strategy is used to update the individual optimal p_{best} and the global optimal g_{best} respectively, thus connecting the decision variable space and the objective space, is particularly important for weighing the exploration and exploitation capabilities of the algorithm evolution process. In this paper, we consider fully mining individuals with better diversity and convergence to assist populations to jump out of the local Pareto frontier. Three guides are selected from the searching and capturing populations to guide the entire individual bat flight, and how the guides are selected from the two populations is specified below.

Unlike the traditional bat algorithm, this algorithm first generates a set of θ with ω solutions and performs a local search process for the optimal solution in the set θ. Subsequently, IHBA enters the iterative phase. In the iteration, the solution φ is selected using the selection mechanism in the set θ. The solution φ is applied to the search phase and two partial sequences can be obtained, one for the solution φ_{Search} with the products and processes removed, and the other for the remaining part of the solution $\varphi_{Captive}$. The local search for the products is applied to $\varphi_{Captive}$, and the suboptimal solution $\varphi'_{Captive}$ is generated. IHBA again reinserts the removed products and jobs into $\varphi'_{Captive}$. This process is called the capture phase, and a complete solution φ' is regenerated. A local search is performed on φ' to generate φ''. Finally, acceptance criteria are used to determine whether the new solution φ'' updates the set.

Therefore, in scheduling problems, the selection mechanism is to select a solution from the solution set θ at the beginning of each iteration, and the selected solution is noted as φ, and proceeds to the subsequent stages such as search and capture. In this scheduling problem, the minimum completion time and the number of iterations are often used as selection factors. So, the proposed selection mechanism consists of two random options, both of which have a 50% probability of being selected. That is, two solutions are randomly selected in the set θ and the one with the lower minimum completion time is chosen as φ. Alternatively, two solutions are randomly selected in θ, and the one with the lower number of iterations is chosen.

The main focus of p_{best}, g_{best} and osd wizard selection mechanisms is on the target space, implying how to fully exploit the candidate solutions with good convergence and diversity in the target space. These wizards serve as a bridge between the target space and the decision variable space, and the new wizard selection mechanism can weigh the ability of exploration and exploitation of the decision variable space in an integrated way.

4.4. Three Neighborhood Structures Based on Insert Operator

To solve the three-stage DAPFSP problem, this paper introduces three neighborhood structures based on the INSERT operator, which constitutes the three-stage bat algorithm (3SBA). The insert operator is considered by many scholars as a superior performance operator, which is described as follows.

- Product-based Neighborhood (PBN)
  Regarding the order of product assembly, a randomly selected product is inserted into another random position.
- Factory-based Neighborhood (FBN)
  A process is randomly selected from the representation of the solution and assigned to another randomly selected plant. For this plant, all possible positions where jobs can be inserted are considered and the part of the job sequence with the earliest completion time is selected. It is worth noting that the sequence of jobs assigned to other plants remains unchanged during this process.
- Job Sequence-based Neighborhood (JSBN)
  A plant is randomly selected and its sequence of jobs is extracted from the solution representation. Next, a job is randomly selected from this partial sequence, and another location is randomly inserted. Finally, the jobs in this partial sequence along with their product priorities and plants are placed in an orderly manner in the distribution solution representation, which is partially the location belonging to that plant.

4.5. Local Search Method Based on Variable Neighborhood Structure

Based on the above three neighborhoods named PBN, FBN and JSBN, this section proposes a local search method with variable neighborhood search. Equation $n(s)$ denotes that the neighborhood structure and other parameters are consistent with the original bat algorithm, implying that the maximum number of iterations for each neighborhood at the same frequency is noted as T_{max} and the frequency f is initialized as f_0. In 3SBA, all three neighborhoods are iterated sequentially, avoiding premature convergence throughout the search process.

Variable neighborhood descent (VND) is the simplest variant of variable neighborhood search [29]. It has been shown to be effective in solving the flow shop [30] and distributed scheduling problems [31]. VND searches for different neighborhood structures from minimum to maximum. First, starting from the initial solution, VND searches the first neighborhood to find the local optimum. Then, while searching the second region, if a better local optimum solution is found, the algorithm returns to the first neighborhood.

Otherwise, it continues searching the third neighborhood and so on. The algorithm does not end until a local optimum is found with respect to all neighborhoods. Therefore, to facilitate the differentiation of the VNDs in this paper, two versions were developed based on the characteristics and features of the bat algorithm. The first one uses three neighborhood structures, named VND3BA, and the second one uses two neighborhood structures, named VND2BA, respectively. A VND that best fits the bat algorithm and the scheduling problem is selected by calibration through simulation experiments.

4.5.1. VND3BA Local Search Method Based on Three Neighborhood Structures

The first method is applied to each product contained in the plant. After removing each part of the product in succession, it is tested at all possible locations within the product. If a better manufacturing span is found, the part sequence is updated. This method runs until all parts contained in the plant have been taken into account and no further improvements are found; then, it can be stopped and ended. It can be named as Local Search Inside Product() and Part Local Search Inside Factory(). The former mainly searches individual products, while the latter searches the whole factory by calling the former, each in its own way.

The second method is a local search of the products within the factory. The principle of this method is to remove the products one by one from the beginning to the end and try to find the best position for them, equivalent to the optimal solution of this problem. If, in the process, a better manufacturing span is found, the sequence of parts in the plant is updated. The process is repeated until no products are found that can be improved.

The third method focuses on finding the most suitable location for a product that crosses plants. First, all plants are checked and the plant with the maximum completion time p is found. Then, the first product is removed and is tested in all possible locations in the remaining plants. If the generated best completion time is lower than the original completion time C_p, this product is inserted into the current location. The principle is that the algorithm iterates by finding the plant with the maximum completion time again, until a local optimum is found.

4.5.2. VND2BA Local Search Method Based on Two Neighborhood Structures

Since VND3BA considers the neighborhood of parts and the neighborhood of products separately, some available feasible solutions may be missed. Therefore, in this section, two local search algorithms are proposed to consider these two neighborhoods in an integrated manner.

The first approach explores the interior of the plant by extracting the range of products and finding the best location of the products within the plant. This step is to extend the scope of the local search. This method also adaptively finds the best part order for the product when inserting it.

The second approach exploits the local search capability of the bat algorithm itself to explore the optimal solution across plants. Therefore, it can be viewed as a combination of two processes named ProductLocalSearchBetweenFactories() and LocalSearchInsideProduct().

4.6. Gaussian Learning Strategy and Elite Learning Strategy

Gaussian learning strategy (GLS) [32] and elite learning strategy (ELS) [33] can assist individuals of bats to jump out of the local frontier and improve the local search ability of the algorithm. If the feasible solution is not updated many times, the whole bat population is most likely to fall into local optimum, and then only the GLS reset strategy can be executed, which is given by

$$x_i^{t+1} \sim N\left(\frac{x_{gbest} - x_{gbest,i}}{2}, \left| x_{gbest} - x_{gbest,i} \right|\right) \tag{30}$$

The collaborative learning mechanism of p_{best} and b_{best} ensures that the exploration and exploitation capability of the entire bat population is enhanced, allowing the algorithm to quickly jump out of the local frontier to approximate the true frontier.

Although the use of GLS can assist individual bats to jump out of the local frontier, this strategy alone still cannot meet the requirements of the bat algorithm for solving the DAPFSP. To further improve the performance of the bat algorithm, ELS is introduced to solve the scheduling problem with the following mathematical expressions.

$$E_i(j) \sim E_i(j) + (x_{ub}(j) - x_{lb}(j))N(0,1) \tag{31}$$

where $x_{ub}(j)$ and $x_{lb}(j)$ represent the upper and lower bounds of the j-th dimensional decision variable, respectively, and $N(0,1)$ represents the random number with a mean value of 0 and a variance of 1.

5. Simulation Results

5.1. Test Questions and Parameter Settings

In this section, the proposed IHBA is compared with the competitive memetic algorithm (CMA) [34], biogeography-based optimization algorithm (BBO) [35], estimation of distribution algorithm (EDA) [36], genetic algorithm [19,21], and particle swarm optimization algorithm [20], and only one of these algorithms was selected for comparison from the literature [37]. So far, the three DIWO algorithms have proved to be the state-of-the-art algorithms for the considered problem with the parameters shown in Table 3. All simulation experiments were implemented on the MATLAB r2020a platform. All programs were executed on an AMD A8-7100 Radeon R5@1.80 GHz and 12.0 GB RAM in Windows 10 operating system.

Table 3. Algorithms and Parameters.

Comparison	Parameters	Range
TDIWO	Initial population size	10
	Maximum population size	50
	permutation-based shift operator	5
	Max value	20
	Min value	1
	Control parameters	0.9
CMA	Selection probability of a subclass	0.5
	Selection probability of populations	0.5
	Number of populations	50
	Local search operator	100
BBO	Number of populations	40
	Elite retention number	10
	Max emigration rate	[0,1]
	Max migration rate	[0,1]
	Max mutation rate	[0,1]
EDA	Population size	50
	Probability vector	0.3
	Iterative Probability Sampling Parameters	0.4
GA	Number of populations	{40,50,60}
	Max number of iterations	300
	Crossover rate	[0.7, 0.9]
	Mutation rate	[0, 0.2]

Table 3. *Cont.*

Comparison	Parameters	Range
PSO	Number of populations	$2n$
	Decreasing coefficient	0.975
	Crossover rate	[0,1]
	Variation rate	[0,1]
	Uniform random number	[0,1]
	Inertia weights	[0.4,0.9]
BA	Frequency	[0,1]
	Loudness	[1,2]
	pulse emission rate	[−1,1]
IHBA	Frequency	[0,1]
	Loudness	[1,2]
	pulse emission rate	[0,1]
	Step size	[0,1]

To calibrate the proposed IHBA, a total of 810 instances were randomly generated according to the literature [5,34] written by Hatami et al. These instances can be found in http://soa.iti.es/problem-instances, accessed on 1 September 2021. These instances combine four factors, including the number of jobs (n), the number of machines (m), the number of products (s), and the number of plants (f). Each factor has three levels, $n \in \{100, 200, 500\}, m \in \{5, 10, 20\}, f \in \{4, 6, 8\}$ and $s \in \{30, 40, 50\}$. Thus, a total of $3^4 = 81$ factor combinations are studied. The processing time in the production stage is randomly generated in the range $[1, 99]$ with uniform distribution. The operation time of the transportation stage is obtained in the range $[1, 991]$ with uniform distribution. The assembly operation time for the assembly stage is obtained in the range between nl and $99nl$ with uniform distribution.

5.2. The Results and Discussion

Relative percentage deviations are the absolute deviations of a given measurement as a percentage of the mean, and can only be used to measure the deviation of a single measurement from the mean. In order to compare the efficiency between algorithms, the calculation of relative percentage deviations (RPD) is considered to compare and analyze these metaheuristic algorithms. Its formula is shown in (32).

$$RPD = \frac{C_{max}(\pi) - C_{max}(\pi)_{best}}{C_{max}(\pi)_{best}} \times 100\% \tag{32}$$

where $C_{max}(\pi)$ is the total completion time of the current algorithm, and $C_{max}(\pi)_{best}$ is the minimum value of $C_{max}(\pi)$ among all algorithms to be compared.

In this paper, the termination condition of the iteration is set to a maximum CPU runtime equal to $C \times m \times n$ milliseconds, and the results are calculated for $C = 20, 40$, and 60. Then, the results obtained from the calculation are compared with other algorithms. The methods used are the analysis of variance (ANOVA), which is widely used in parametric statistical procedures, and the Friedman test and Wilcoxon paired signed test, which are commonly used in nonparametric statistical procedures. These methods have been widely used and promoted in the recent scheduling literature.

5.2.1. Comparative Analysis at $C = 20$

When $C = 20$, the values of RPD for various heuristic algorithms are shown in Table 4. It can be seen from the bold figures that the value of PRD of the proposed algorithm IHBA in this paper is smaller than that of other algorithms in the vast majority of combinatorial solutions, including the recognized TDIWO algorithm and the original bat algorithm. Only in these 5 sets of data (1) $m = 5, n = 100, s = 30$ and $f = 4$, (2) $m = 5, n = 200, s = 40$

and $f = 4$, (3) $m = 10, n = 100, s = 50$ and $f = 8$, (4) $m = 15, n = 200, s = 50$ and $f = 4$, (5) $m = 15, n = 200, s = 40$ and $f = 8$, IHBA is 5–10% smaller than the other algorithms. Meanwhile, IHBA not only obtains the lowest relative percentage deviation value, but also the smallest average RPD value, which is 21.15%, with an error of about 0.83% from the optimal value. This is good proof that the IHBA algorithm has a strong advantage and superiority. It can also be seen that the GA and PSO performance is also very good, with average RPD values of 32.09% and 31.52%. Although TDIWO is not satisfactory in this comparison, it is significantly better than BBO and EDA. The most surprising is the original bat algorithm, with a result of 27.69, which is second only to the algorithm proposed in this paper. This also shows that the bat algorithm itself has better performance and has stronger search capability.

Table 4. The Average RPD Values at $C = 20$. It can be seen from the bold figures that the value of PRD of the proposed algorithm IHBA in this paper is smaller than that of other algorithms in the vast majority of combinatorial solutions, including the recognized TDIWO algorithm and the original bat algorithm.

Parameters				Algorithms for Comparison							
m	n	s	f	CMA	BBO	TDIWO	EDA	GA	PSO	BA	IHBA
5	100	30	4	11.85	14.07	25.59	28.36	12.28	**10.24**	53.91	18.83
5	100	40	4	41.58	49.38	46.39	48.22	29.16	49.85	46.87	**25.63**
5	100	50	4	26.73	31.74	38.40	41.56	47.26	43.52	**22.52**	23.37
5	200	30	4	48.36	57.43	40.74	47.44	35.20	56.69	35.68	**27.15**
5	200	40	4	8.87	10.53	7.14	11.10	**6.15**	10.70	10.16	17.69
5	200	50	4	47.76	56.71	55.76	59.54	53.07	37.67	43.21	**34.59**
5	500	30	4	30.86	36.64	33.26	41.60	33.05	25.58	19.19	**18.86**
5	500	40	4	25.67	30.49	36.49	34.20	37.50	11.28	37.71	**8.10**
5	500	50	4	35.44	42.09	43.98	55.36	21.82	45.52	58.99	**19.78**
5	100	30	6	30.36	36.05	33.53	42.51	34.00	23.69	20.28	**18.46**
5	100	40	6	16.92	18.22	20.91	26.45	**11.15**	22.00	18.97	15.03
5	100	50	6	14.64	17.38	27.64	29.01	16.31	17.51	51.16	**13.77**
5	200	30	6	42.98	51.04	46.84	46.54	46.33	35.33	28.28	**11.31**
5	200	40	6	12.65	13.15	19.69	13.68	14.52	14.22	14.55	**12.54**
5	200	50	6	23.56	27.98	26.25	35.09	17.27	27.49	25.69	**11.40**
5	500	30	6	46.72	55.48	60.63	60.97	57.59	31.18	58.46	**24.63**
5	500	40	6	51.32	60.94	56.66	55.71	49.76	47.74	41.95	**35.20**
5	500	50	6	40.78	48.42	41.75	42.01	44.55	32.92	16.54	**13.40**
5	100	30	8	20.84	24.75	25.22	21.40	38.58	21.02	16.60	**14.46**
5	100	40	8	**11.02**	13.09	15.01	19.47	21.27	19.67	28.64	14.66
5	100	50	8	22.67	26.92	33.08	34.28	24.49	18.57	45.03	**14.19**
5	200	30	8	31.21	37.06	30.75	46.38	38.70	50.60	33.70	**28.68**
5	200	40	8	53.90	54.00	49.87	62.06	49.33	43.08	32.29	**25.95**
5	200	50	8	35.10	41.69	43.78	45.73	35.73	40.96	55.08	**28.74**
5	500	30	8	**11.47**	13.62	19.68	24.89	20.14	21.65	19.63	15.58
5	500	40	8	27.18	32.28	24.95	24.47	26.94	24.71	23.42	**22.44**
5	500	50	8	30.93	36.73	31.26	37.29	21.58	47.19	33.29	**15.87**
10	100	30	4	29.22	34.70	25.77	36.59	26.01	39.50	18.83	**13.88**
10	100	40	4	44.89	53.31	39.65	44.46	32.99	52.31	21.54	**19.12**
10	100	50	4	47.60	56.52	44.56	47.68	28.89	21.54	17.43	**15.14**

Table 4. *Cont.*

Parameters				Algorithms for Comparison							
m	*n*	*s*	*f*	CMA	BBO	TDIWO	EDA	GA	PSO	BA	IHBA
10	200	30	4	42.81	50.84	45.18	52.77	23.63	27.72	**15.36**	16.31
10	200	40	4	24.49	29.09	24.96	30.24	22.21	24.33	24.06	**18.05**
10	200	50	4	22.92	25.34	35.40	46.44	33.68	30.87	29.50	**20.82**
10	500	30	4	32.87	39.03	42.94	48.66	33.78	28.67	28.91	**26.93**
10	500	40	4	50.92	60.46	59.45	67.94	30.49	26.25	**22.15**	22.22
10	500	50	4	25.84	30.68	29.11	33.74	20.88	28.21	17.38	**13.10**
10	100	30	6	23.73	26.31	25.34	36.70	23.18	22.91	18.00	**11.04**
10	100	40	6	38.41	39.99	36.67	44.55	20.91	25.07	**20.69**	23.65
10	100	50	6	29.15	34.62	38.62	41.89	27.51	27.88	27.73	**21.81**
10	200	30	6	39.89	37.37	31.57	36.12	26.45	29.34	29.94	**24.69**
10	200	40	6	36.05	34.69	42.13	34.93	23.31	24.19	23.62	**19.12**
10	200	50	6	30.22	35.89	49.16	32.23	26.75	**20.67**	24.09	21.13
10	500	30	6	42.37	50.32	48.31	50.49	48.61	31.90	32.61	**26.39**
10	500	40	6	29.31	**21.06**	21.56	33.01	21.97	25.72	25.27	23.94
10	500	50	6	59.07	70.15	66.60	64.97	24.49	27.74	26.04	**20.91**
10	100	30	8	51.95	61.69	49.99	52.54	29.46	29.25	26.85	**19.78**
10	100	40	8	34.00	36.63	32.67	40.62	22.16	24.44	28.81	**20.86**
10	100	50	8	56.42	67.00	50.65	54.65	55.99	51.22	**22.92**	28.90
10	200	30	8	41.14	48.85	49.44	46.66	56.14	22.03	21.86	**21.52**
10	200	40	8	49.89	59.25	51.00	48.80	52.47	42.32	**21.99**	22.23
10	200	50	8	41.96	49.82	50.83	50.00	45.85	33.87	45.36	**20.07**
10	500	30	8	40.46	48.05	39.82	46.89	40.35	36.53	24.49	**21.79**
10	500	40	8	33.12	39.33	29.86	43.19	21.71	21.22	23.33	**18.57**
10	500	50	8	22.82	25.22	29.75	30.31	22.57	21.79	20.48	**20.39**
15	100	30	4	41.00	48.68	48.43	33.07	27.57	30.33	27.09	**18.56**
15	100	40	4	32.39	38.46	41.61	50.28	25.80	25.73	29.15	**17.36**
15	100	50	4	24.60	29.21	27.17	25.18	21.45	25.28	22.54	**20.74**
15	200	30	4	39.11	46.44	39.60	46.76	33.83	40.48	**23.17**	27.04
15	200	40	4	32.44	38.52	44.61	46.40	30.73	30.90	28.91	**25.84**
15	200	50	4	35.44	36.46	27.70	28.21	28.96	**21.37**	26.87	27.48
15	500	30	4	29.43	34.94	29.08	37.33	38.61	27.30	**21.71**	24.60
15	500	40	4	39.05	30.74	32.58	35.84	34.42	32.77	27.07	**23.55**
15	500	50	4	33.60	39.90	29.44	36.31	27.27	36.57	25.20	**22.54**
15	100	30	6	48.79	57.94	44.76	54.99	45.33	47.37	29.07	**24.60**
15	100	40	6	35.70	42.39	34.74	41.65	32.14	45.68	24.84	**24.55**
15	100	50	6	30.59	34.45	31.89	33.88	34.72	34.40	29.02	**26.06**
15	200	30	6	48.61	57.73	34.31	62.97	28.05	24.32	21.15	**21.04**
15	200	40	6	36.57	43.43	32.65	41.21	37.58	31.90	29.96	**28.62**
15	200	50	6	**19.02**	22.59	25.25	32.25	25.43	20.72	23.85	20.05
15	500	30	6	34.44	39.02	36.17	39.27	43.01	33.42	33.74	**30.71**
15	500	40	6	38.32	45.50	38.05	42.53	32.75	40.05	20.60	**20.55**
15	500	50	6	34.04	40.43	40.17	45.40	40.42	34.26	29.71	**29.51**
15	100	30	8	28.09	33.36	24.98	29.91	30.08	53.29	25.50	**24.36**
15	100	40	8	31.40	37.29	37.76	38.73	33.05	46.61	28.17	**20.90**
15	100	50	8	44.26	52.56	49.95	51.98	34.90	49.19	28.10	**20.38**
15	200	30	8	32.96	39.14	33.75	48.87	25.10	27.53	24.32	**21.47**
15	200	40	8	26.57	31.55	34.71	42.07	**22.66**	27.82	24.17	28.35
15	200	50	8	23.99	28.49	28.64	32.36	25.55	20.04	15.67	**15.14**
15	500	30	8	44.80	53.20	42.29	43.54	52.56	32.56	**21.67**	22.26
15	500	40	8	53.71	63.78	45.43	54.28	44.49	57.55	**21.57**	22.33
15	500	50	8	39.05	46.38	42.36	45.63	56.73	27.47	29.36	**21.82**
	average			34.33	39.71	37.09	41.40	32.09	31.52	27.69	**21.15**

Then, the results of the descriptive statistics of the Friedman test were calculated for the case of $C = 20$, as shown in Table 5, where N is the number of test cases and takes the value of 4050. The minimum value of each item derived in the algorithm is marked in bold. The p-value calculated for the test statistic is 0 when the significance level $\alpha = 0.05$, which is less than or equal to 0.05. This indicates that there are significant differences between the algorithms used for comparison. The following conclusions can be clearly seen through Table 5. The rank of IHBA is only 2.60, which is close to one-half of CMA, BBO and TDIWO, while it is smaller than the ranks of GA, PSO and the original bat algorithm by 1.39, 1.09 and 0.78, respectively. Although IHBA does not take the best value in terms of maximum value, it means that standard deviation and minimum value are the smallest among all algorithms. The comparison between the groups also shows that the ranking of EDA is even more than three times that of IHBA, indicating the worst performance of the algorithm, which echoes the results and analysis in Table 4. In addition, it can also be seen by the mean and standard deviation that the three algorithms of DIWO are very close to the proposed algorithm.

Table 5. The Descriptive Statistics Achieved by the Friedman Test at $C=20$ and $\alpha=0.05$. The bold figures indicate the comparison between algorithms.

Algorithms	Rank	n	Mean	Std. Deviation	Min	Max
CMA	4.81	4050.00	1.17	0.92	0.00	7.05
BBO	4.92	4050.00	1.47	1.30	0.00	7.70
EDA	7.01	4050.00	16.89	12.80	0.00	50.37
TDIWO	3.36	4050.00	0.99	0.96	0.00	4.82
GA	3.99	4050.00	1.43	1.36	0.00	4.61
PSO	3.69	4050.00	1.31	1.25	0.00	5.35
BA	3.38	4050.00	1.00	0.95	0.00	4.47
IHBA	2.60	4050.00	0.74	0.71	0.00	5.51
p-value	0.00					

Next, the non-parametric test for paired samples is the Wilcoxon paired signed test method used. This method was developed based on the signed test for paired observations, and is more effective than the traditional test with positive and negative signs alone. The observed data are generally considered to be significantly different when p is less than 5%. When p is greater than or equal to 5%, the difference in the data is considered insignificant. The results of the Wilcoxon paired signed test for this paper are given in Table 6, assuming that there is no difference between each pair of algorithms ($\alpha = 0.05$). $R+$ in the table represents positive differences and $R-$ represents negative differences. In the results for each pair of algorithms, the sum of $R+$, $R-$ and bonding values is equal to n in Table 5. It can also be seen in Table 6 that the p-values in the last column are equal to 0 and all are less than 0.001, indicating that the differences between the algorithms compared are statistically significant with respect to 0. In other words, there is a significant difference between IHBA and the other algorithms.

Table 6. The Wilcoxon paired signed test result at $C = 20$ and $\alpha = 0.05$.

Algorithms for Comparison	R+	R−	Bonding Value	*p*-Value
IHBA VS CMA	756	3168	126	0.00
IHBA VS BBO	2341	274	1435	0.00
IHBA VS EDA	3840	100	110	0.00
IHBA VS TDIWO	1791	817	1442	0.00
IHBA VS GA	3113	783	154	0.00
IHBA VS PSO	2881	829	341	0.00
IHBA VS BA	2873	907	270	0.00

Finally, in order to make the observations more visual and relevant, this section analyzes and interprets the results of the ANOVA analysis for the eight algorithms. The methods used are the mean plot and the minimum difference method interval, and Figure 1 shows the plotted plots with 95% confidence level. It can be very clearly seen that the best performance is IHBA, which beats the other comparison algorithms by a more significant margin. The bat algorithm also has some advantages, while GA and PSO continue to perform consistently, and the EDA algorithm has the worst performance, followed by BBO.

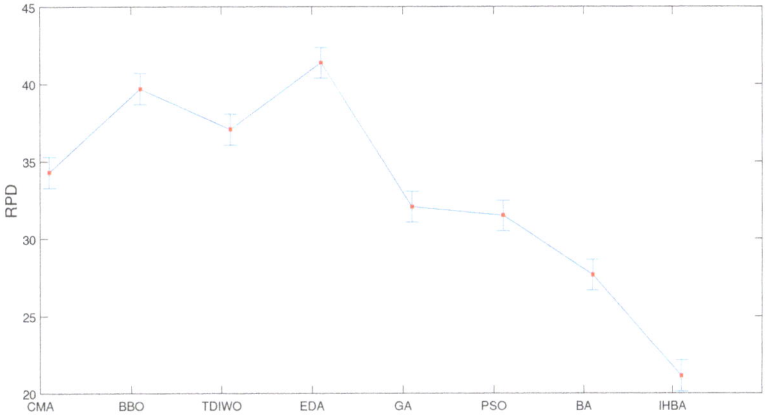

Figure 1. Mean plot and 95% LSD interval of the bat algorithm at $C = 20$.

5.2.2. Comparative Analysis at $C = 40$

The lower the RPD, the better the performance of the algorithms. When C takes the value of 40, the average RPD of these eight algorithms is shown in Table 7. As can be seen in the last row, the average value of IHBA is 22.73%, which is significantly lower than the average value of other algorithms including the original bat algorithm. It shows that the IHBA algorithm is more efficient than the other metaheuristic algorithms, while BA performs second, with an average RPD of 23.57%. Compared with Table 7, it can be concluded that the RPD value of IHBA changes from the original 21.15 to the current 22.73, as the value of C is taken to increase, which shows that the value of RPD of IHBA increases with the increase of the number of iterations and time. In addition, PSO with a mean value of 35.76% is significantly better than GA and the other four heuristics. Meanwhile, EDA has the largest value and its performance is the worst, which is also consistent with the results in Table 4.

Table 8 describes the results of the Friedman test calculations. As can be seen at a glance from the rankings, IHBA has a mean rank of 3.06, which is the smallest value among these algorithms. The next best performer is TDIWO, which also has a better performance, but its standard deviation is 0.89, which is larger than CMA, GA, and PSO. It can be seen from the mean and standard deviation that IHBA has values of 0.79 and 0.76, respectively, which are also the smallest values among all algorithms. The performance of EDA is still the worst, with its average rank, mean, standard deviation, and maximum value all being at a disadvantage. The performance of EDA is still the worst, with its ranking, mean, standard deviation and maximum value being at a disadvantage. In particular, the mean value is as high as 32 times that of IHBA. The p-value in the last row is 0, which is less than 0.05, indicating that the difference between the algorithms has statistical significance. This difference is not due to chance sampling, but rather, the difference between the two groups of algorithms is significant.

Table 7. The Average RPD Values at $C = 40$. It can be seen from the bold figures that the value of PRD of the proposed algorithm IHBA in this paper is smaller than that of other algorithms in the vast majority of combinatorial solutions, including the recognized TDIWO algorithm and the original bat algorithm.

Parameters				Algorithms for Comparison							
m	n	s	f	CMA	BBO	TDIWO	EDA	GA	PSO	BA	IHBA
5	100	30	4	**17.00**	21.80	22.45	25.09	18.96	**17.93**	18.42	17.76
5	100	40	4	45.33	46.15	47.52	54.11	49.00	44.16	30.39	**29.30**
5	100	50	4	31.97	35.80	36.86	49.68	42.45	40.10	**26.74**	25.79
5	200	30	4	48.36	46.67	48.06	56.50	52.94	47.03	29.99	**28.92**
5	200	40	4	8.76	9.22	9.50	10.91	**10.09**	**9.03**	7.37	7.49
5	200	50	4	52.88	55.13	56.77	70.34	57.50	52.03	35.04	**33.79**
5	500	30	4	33.25	35.74	36.80	45.17	37.22	33.69	21.18	**20.42**
5	500	40	4	30.58	32.43	33.39	43.34	32.52	29.70	20.40	**19.68**
5	500	50	4	40.10	45.33	46.68	51.02	45.22	41.34	30.30	**29.22**
5	100	30	6	32.98	35.93	36.99	45.65	37.06	33.62	21.29	**20.53**
5	100	40	6	18.50	21.02	21.64	23.98	**21.42**	19.55	14.14	**13.63**
5	100	50	6	19.69	23.73	24.43	28.23	22.68	21.36	19.19	**18.50**
5	200	30	6	**26.49**	26.29	27.66	39.61	29.83	**24.77**	26.50	25.55
5	200	40	6	15.01	14.91	15.35	19.08	16.28	14.90	11.01	**10.62**
5	200	50	6	25.67	28.63	29.48	33.31	29.19	26.55	17.68	**17.05**
5	500	30	6	53.74	56.76	58.44	73.33	57.88	52.64	36.23	**34.94**
5	500	40	6	55.75	55.55	57.20	69.71	59.65	53.63	35.43	**34.17**
5	500	50	6	43.22	42.37	43.62	55.23	46.38	41.51	23.60	**22.76**
5	100	30	8	23.37	22.88	23.55	34.36	28.11	25.93	16.95	**16.34**
5	100	40	8	**12.91**	15.25	15.70	21.51	18.43	17.53	14.66	**14.13**
5	100	50	8	27.28	30.22	31.12	37.12	29.63	27.20	20.94	**20.20**
5	200	30	8	32.68	36.60	37.69	47.78	43.46	40.30	28.25	**27.24**
5	200	40	8	52.07	53.18	54.76	67.27	57.82	51.16	32.42	**31.26**
5	200	50	8	39.79	42.05	43.30	52.17	45.00	41.17	30.87	**29.76**
5	500	30	8	**13.78**	18.65	19.20	24.48	20.64	19.80	15.01	**14.47**
5	500	40	8	27.86	26.19	26.96	33.95	29.73	26.41	18.14	**17.49**
5	500	50	8	32.65	33.74	34.75	39.64	37.96	34.47	23.02	**22.20**
10	100	30	4	29.60	31.11	32.03	38.46	35.52	32.19	19.82	**19.12**
10	100	40	4	45.50	44.04	45.35	53.25	49.56	44.10	25.93	**25.01**
10	100	50	4	49.07	47.68	49.10	55.52	45.70	39.44	21.63	**20.86**
10	200	30	4	45.82	47.69	49.11	53.88	44.99	39.63	**22.34**	21.54
10	200	40	4	25.92	27.02	27.82	33.28	28.76	25.91	17.76	**17.13**
10	200	50	4	27.61	34.35	35.37	44.02	36.05	34.01	24.29	**23.42**
10	500	30	4	37.90	41.87	43.11	51.38	41.84	38.23	25.91	**24.99**
10	500	40	4	56.38	60.21	62.00	68.23	54.72	48.43	**28.21**	27.20
10	500	50	4	28.26	29.98	30.87	35.75	31.20	28.24	17.58	**16.95**
10	100	30	6	24.88	28.32	29.16	34.85	29.29	26.62	16.93	**16.33**
10	100	40	6	37.98	38.85	40.00	44.41	38.07	33.11	**21.18**	20.42
10	100	50	6	33.79	36.90	38.00	44.58	36.98	33.77	22.89	**22.08**
10	200	30	6	35.92	33.67	**34.67**	41.10	37.17	31.85	21.99	**21.20**
10	200	40	6	37.25	35.82	36.88	42.21	36.17	31.53	**20.65**	19.92
10	200	50	6	38.04	37.59	38.71	45.01	36.10	**32.61**	21.49	20.72
10	500	30	6	46.53	47.79	49.21	61.79	50.37	45.47	29.42	**28.37**
10	500	40	6	23.74	**24.24**	24.96	30.50	28.26	24.42	18.70	**18.03**
10	500	50	6	64.63	64.65	66.57	70.69	57.97	50.29	28.49	**27.47**
10	100	30	8	54.00	52.63	54.20	60.53	50.90	44.14	25.66	**24.75**
10	100	40	8	34.09	35.23	36.28	41.28	35.28	30.99	20.93	**20.19**
10	100	50	8	57.45	55.22	56.86	71.34	62.21	55.35	**32.63**	31.47
10	200	30	8	46.02	46.46	47.84	62.84	48.94	44.18	26.87	**25.91**
10	200	40	8	52.85	50.98	52.49	66.10	56.25	50.27	**29.48**	28.43
10	200	50	8	47.07	48.29	49.72	61.41	50.43	45.62	30.37	**29.28**
10	500	30	8	42.35	43.19	44.48	54.72	46.69	41.91	25.91	**24.98**

Table 7. *Cont.*

Parameters				Algorithms for Comparison							
m	n	s	f	CMA	BBO	TDIWO	EDA	GA	PSO	BA	IHBA
10	500	40	8	33.77	36.02	37.09	41.90	34.89	30.75	19.49	**18.80**
10	500	50	8	25.67	27.33	28.15	33.70	28.23	25.67	17.94	**17.30**
15	100	30	4	45.58	41.72	42.96	49.30	42.42	37.24	22.85	**22.03**
15	100	40	4	37.12	41.78	43.02	48.80	39.68	36.02	23.45	**22.61**
15	100	50	4	26.73	26.14	26.92	32.19	28.31	25.40	17.58	**16.95**
15	200	30	4	41.30	42.56	43.83	52.07	45.60	41.01	**26.03**	25.10
15	200	40	4	38.14	41.52	42.75	50.08	41.41	37.85	25.60	**24.69**
15	200	50	4	32.87	29.61	30.49	37.92	32.99	**28.26**	19.83	19.12
15	500	30	4	30.84	32.48	33.45	43.74	36.42	33.12	**22.05**	21.27
15	500	40	4	33.79	31.78	32.73	41.74	38.04	32.94	22.99	**22.17**
15	500	50	4	33.97	33.86	34.87	41.54	37.61	33.56	21.89	**21.11**
15	100	30	6	50.00	50.54	52.04	63.44	55.40	49.58	30.39	**29.30**
15	100	40	6	37.24	38.07	39.20	47.16	43.02	38.93	25.14	**24.24**
15	100	50	6	31.99	32.12	33.08	42.17	37.02	33.53	23.45	**22.62**
15	200	30	6	46.42	49.68	51.16	57.21	47.41	41.07	23.68	**22.84**
15	200	40	6	37.18	37.59	38.71	48.40	41.36	36.98	24.93	**24.04**
15	200	50	6	**22.07**	25.67	26.43	32.98	26.90	25.00	18.22	17.57
15	500	30	6	36.18	36.69	37.78	49.21	41.73	37.80	26.71	**25.75**
15	500	40	6	40.22	40.41	41.61	49.63	43.93	39.38	24.02	**23.16**
15	500	50	6	37.83	40.38	41.58	52.01	43.47	39.74	27.10	**26.13**
15	100	30	8	28.52	28.29	29.13	36.98	36.98	33.98	23.22	**22.40**
15	100	40	8	35.13	36.47	37.55	45.88	41.64	38.30	25.34	**24.43**
15	100	50	8	48.44	49.52	50.99	59.18	52.38	47.24	28.95	**27.92**
15	200	30	8	34.93	39.03	40.18	45.89	38.40	34.53	22.35	**21.55**
15	200	40	8	30.64	34.72	35.75	40.93	**34.33**	31.45	22.20	21.40
15	200	50	8	26.77	28.68	29.53	35.95	29.46	26.75	16.96	**16.36**
15	500	30	8	46.30	44.56	45.88	59.87	49.81	44.39	**26.53**	25.58
15	500	40	8	53.77	52.40	53.96	64.99	59.12	52.58	**30.33**	29.24
15	500	50	8	42.17	43.07	44.35	59.72	47.71	43.28	27.58	**26.59**
average				36.42	37.64	38.76	46.72	39.78	35.76	23.57	**22.73**

Table 8. The Descriptive Statistics Achieved by the Friedman Test at $C = 40$ and $\alpha = 0.05$. The bold figures indicate the comparison between algorithms.

Algorithms	*Rank*	n	*Mean*	*Std. Deviation*	*Min*	*Max*
CMA	4.81	4050.00	1.17	0.92	0.00	7.05
BBO	4.92	4050.00	1.47	1.30	0.00	7.70
EDA	7.01	4050.00	16.89	12.80	0.00	50.37
TDIWO	3.36	4050.00	0.99	0.96	0.00	4.82
GA	3.43	4050.00	0.93	0.87	0.00	4.44
PSO	4.22	4050.00	1.11	0.96	0.00	5.98
BA	3.06	4050.00	0.79	0.76	0.00	4.67
IHBA	2.94	4050.00	0.77	0.75	0.00	4.71
p-value	0.00					

The results obtained from the Wilcoxon paired signed test are shown in Table 9. The *p*-values in the last column are all equal to 0, which is less than 0.05. It shows that the differences between these algorithms are statistically significant, and that there are significant differences between IHBA and the other algorithms. This also reconfirms the conclusion that $C = 20$. More definitely, it explains that IHBA is an effective group intelligence algorithm with outstanding performance.

Table 9. The Wilcoxon paired signed test result at $C = 40$ and $\alpha = 0.05$.

Algorithms for Comparison	R+	R−	Bonding Value	*p*-Value
IHBA VS CMA	671	3280	99	0.00
IHBA VS BBO	2434	220	1396	0.00
IHBA VS EDA	3879	100	71	0.00
IHBA VS TDIWO	1933	710	1407	0.00
IHBA VS GA	3198	804	48	0.00
IHBA VS PSO	2149	540	1361	0.00
IHBA VS BA	3050	948	52	0.00

Finally, in this section, EDA is excluded from the mean plot. The reason is that the algorithm is too intrusive and its performance is too poor. As can be seen in Figure 2, IHBA is once again stronger than the other metaheuristic algorithms by a margin. Once again, the variability between the algorithms is proven to be at a significant level. Since the LSD test has the highest sensitivity, Figure 2 verifies from the side that IHBA is an algorithm that performs well and can reach Pareto optimality.

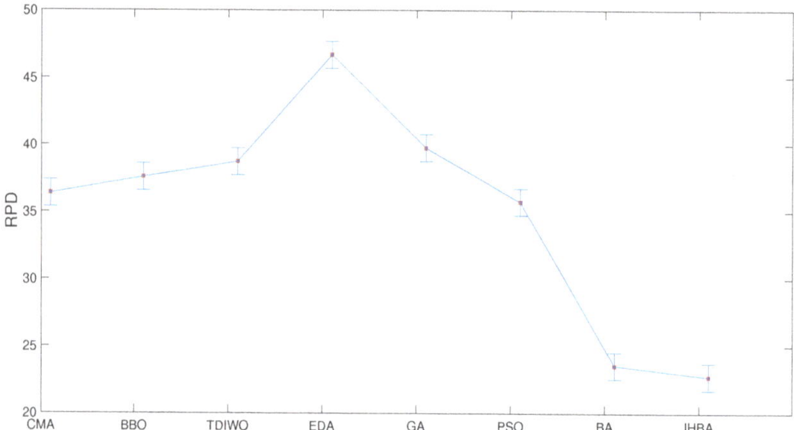

Figure 2. Mean plot and 95% LSD interval of the bat algorithm at $C = 40$.

5.2.3. Comparative Analysis at $C = 60$

Table 10 presents the results of the calculations at $C = 60$. Table 10 shows the validity of the IHBA, which yielded an overall mean RPD value of 31.55%. The results of the Friedman test in Table 11, the results of the Wilcoxon paired sign test in Table 12, and the mean plot in Figure 3 again show that the proposed IHBA performed best in the comparison.

Table 10. The Average RPD Values at $C = 60$. It can be seen from the bold figures that the value of PRD of the proposed algorithm IHBA in this paper is smaller than that of other algorithms in the vast majority of combinatorial solutions, including the recognized TDIWO algorithm and the original bat algorithm.

Parameters				Algorithms for Comparison							
m	*n*	*s*	*f*	**CMA**	**BBO**	**TDIWO**	**EDA**	**GA**	**PSO**	**BA**	**IHBA**
5	100	30	4	**21.03**	21.53	21.14	22.70	22.35	**19.26**	20.68	20.84
5	100	40	4	45.06	46.69	45.56	47.61	45.75	46.63	40.83	**40.08**
5	100	50	4	37.12	37.79	37.41	41.04	39.73	40.82	**34.18**	33.43
5	200	30	4	47.20	49.43	47.57	50.02	47.74	49.63	42.30	**41.33**
5	200	40	4	11.03	12.18	11.01	13.06	**11.75**	**9.04**	**9.69**	9.66
5	200	50	4	53.25	54.80	53.76	55.71	55.59	55.49	47.22	**46.29**
5	500	30	4	34.70	36.10	34.82	37.18	36.60	35.23	29.30	**28.58**
5	500	40	4	31.75	32.82	32.21	33.83	34.12	31.45	26.98	**26.47**
5	500	50	4	42.52	43.67	42.61	44.96	43.45	43.58	39.09	**38.53**
5	100	30	6	34.71	36.05	34.81	37.16	36.76	35.19	29.28	**28.57**
5	100	40	6	20.98	21.90	20.95	23.25	**21.89**	20.05	18.20	**17.91**
5	100	50	6	23.38	23.98	23.58	25.45	24.73	22.48	22.43	**22.41**
5	200	30	6	**31.96**	35.07	37.00	39.41	38.96	**40.95**	34.61	**35.04**
5	200	40	6	16.42	17.30	16.77	18.65	17.45	15.14	13.98	**13.78**
5	200	50	6	27.68	28.91	27.69	30.09	28.80	27.43	23.77	**23.27**
5	500	30	6	54.30	55.62	54.94	56.61	57.10	56.33	48.11	**47.22**
5	500	40	6	54.49	56.34	55.21	57.09	56.11	57.05	48.51	**47.50**
5	500	50	6	41.94	43.79	42.47	44.65	43.88	43.44	34.55	**33.48**
5	100	30	8	25.59	26.71	26.03	28.52	27.71	26.63	22.52	**22.01**
5	100	40	8	**17.68**	18.42	17.70	20.40	19.22	17.77	17.30	**17.20**
5	100	50	8	29.38	30.33	29.67	31.41	30.96	28.81	26.43	**26.14**
5	200	30	8	38.13	39.23	37.99	41.76	39.92	41.60	36.36	**35.72**
5	200	40	8	52.43	53.78	52.29	54.87	54.22	54.77	45.26	**44.12**
5	200	50	8	41.91	43.09	42.34	44.51	43.33	43.64	39.47	**38.96**
5	500	30	8	**19.50**	20.11	19.62	22.39	21.30	19.79	18.05	**17.81**
5	500	40	8	27.99	29.62	28.36	30.23	29.02	27.91	24.85	**24.40**
5	500	50	8	34.15	35.61	34.35	37.10	34.76	35.59	30.98	**30.35**
10	100	30	4	31.67	33.18	31.67	34.69	32.85	33.03	27.60	**26.88**
10	100	40	4	43.96	46.09	44.36	46.88	44.68	46.02	37.74	**36.66**
10	100	50	4	43.84	46.31	44.35	45.13	44.90	42.43	33.92	**32.84**
10	200	30	4	42.87	44.87	43.17	44.57	44.00	41.96	**33.59**	32.56
10	200	40	4	27.34	28.72	27.49	29.64	28.52	27.10	23.92	**23.48**
10	200	50	4	33.34	33.74	33.22	36.41	35.74	35.00	30.23	**29.70**
10	500	30	4	39.57	40.71	39.83	41.94	41.52	40.24	34.40	**33.72**
10	500	40	4	52.70	54.68	53.09	54.06	54.32	51.76	**41.54**	40.34
10	500	50	4	29.19	30.54	29.38	31.66	30.37	29.12	24.41	**23.79**
10	100	30	6	27.57	28.64	27.42	30.04	29.20	27.50	23.03	**22.46**
10	100	40	6	36.71	38.24	36.72	38.26	37.51	35.65	**30.23**	29.51
10	100	50	6	35.13	36.28	35.43	37.48	36.72	35.36	30.41	**29.81**
10	200	30	6	35.05	36.48	**25.00**	36.74	35.55	34.76	30.63	**30.04**
10	200	40	6	34.94	36.12	35.38	36.55	35.66	33.93	**28.12**	28.19
10	200	50	6	35.39	36.64	36.45	37.23	36.67	**34.44**	29.26	**28.63**
10	500	30	6	46.59	48.16	47.09	49.07	48.84	48.23	40.39	**39.48**
10	500	40	6	26.68	**27.32**	26.11	28.68	27.47	26.53	24.27	**23.91**
10	500	50	6	56.43	59.00	57.28	57.13	57.08	54.35	43.68	**42.38**
10	100	30	8	48.53	51.05	49.11	49.84	49.34	47.57	38.91	**37.82**
10	100	40	8	34.03	35.49	34.05	35.77	34.99	33.26	28.97	**28.39**
10	100	50	8	55.22	57.60	55.79	58.18	56.95	58.49	**47.45**	46.11
10	200	30	8	45.36	46.91	46.02	47.81	48.24	46.78	37.80	**36.78**
10	200	40	8	50.53	52.60	51.26	53.33	52.42	52.97	**42.70**	41.47
10	200	50	8	46.91	48.40	47.51	49.32	48.93	48.44	41.12	**40.28**
10	500	30	8	42.52	44.27	42.80	45.17	44.25	44.08	36.49	**35.56**
10	500	40	8	33.76	35.52	33.71	35.51	34.98	32.76	27.86	**27.22**

Table 10. *Cont.*

Parameters				Algorithms for Comparison							
m	n	s	f	CMA	BBO	TDIWO	EDA	GA	PSO	BA	IHBA
10	500	50	8	27.20	28.20	27.44	29.46	28.55	26.81	23.54	**23.11**
15	100	30	4	40.34	42.38	41.45	42.08	40.91	39.70	33.27	**32.44**
15	100	40	4	38.10	39.33	38.25	40.20	39.79	37.87	31.88	**31.17**
15	100	50	4	26.98	28.40	27.39	29.22	27.89	26.61	23.70	**23.28**
15	200	30	4	41.62	43.31	41.89	44.27	42.96	43.04	**36.18**	35.32
15	200	40	4	39.16	40.30	39.57	41.55	40.85	39.73	33.99	**33.32**
15	200	50	4	31.61	33.30	31.82	33.23	32.38	**30.95**	27.72	**27.25**
15	500	30	4	33.42	34.77	33.52	36.15	35.46	34.69	**29.66**	29.04
15	500	40	4	34.80	35.60	34.57	36.97	35.72	35.69	31.01	**30.37**
15	500	50	4	34.45	36.19	34.63	36.96	35.38	35.24	30.42	**29.76**
15	100	30	6	49.72	51.74	49.98	52.56	51.34	52.26	43.13	**42.01**
15	100	40	6	38.52	40.11	38.74	41.52	39.67	40.49	34.42	**33.64**
15	100	50	6	33.84	35.10	34.11	36.57	35.26	35.23	30.97	**30.42**
15	200	30	6	45.12	47.58	44.69	46.53	46.53	44.27	36.10	**35.07**
15	200	40	6	38.20	39.92	38.36	40.61	39.75	39.31	34.10	**33.44**
15	200	50	6	**25.76**	26.65	25.79	28.29	27.58	25.86	23.07	**22.72**
15	500	30	6	38.25	39.53	38.53	40.94	40.10	40.06	35.15	**34.55**
15	500	40	6	39.85	41.62	40.20	42.54	41.07	41.15	34.07	**33.17**
15	500	50	6	40.03	41.27	40.32	42.78	42.04	41.74	35.87	**35.17**
15	100	30	8	31.73	33.10	31.92	35.37	32.58	34.68	30.51	**29.92**
15	100	40	8	37.13	38.34	37.53	40.40	38.45	39.45	33.64	**32.91**
15	100	50	8	47.43	49.14	47.96	50.25	48.62	49.35	40.65	**39.58**
15	200	30	8	36.42	37.90	36.33	38.59	37.95	36.38	30.83	**30.14**
15	200	40	8	32.89	33.95	33.01	35.20	**34.33**	32.91	28.94	**28.45**
15	200	50	8	28.05	29.29	28.26	30.40	29.69	27.74	23.28	**22.72**
15	500	30	8	45.08	47.09	45.61	47.71	47.22	46.99	**38.27**	37.22
15	500	40	8	52.01	54.39	52.37	55.12	53.16	55.11	**44.53**	43.21
15	500	50	8	43.50	44.99	43.91	46.29	46.34	45.72	37.83	**36.93**
average				36.68	37.88	39.01	37.08	38.54	37.28	39.54	**38.60**

Table 11. The Descriptive Statistics Achieved by the Friedman Test at $C = 60$ and $\alpha = 0.05$. The bold figures indicate the comparison between algorithms.

Algorithms	*Rank*	n	*Mean*	*Std. Deviation*	*Min*	*Max*
CMA	4.81	4050.00	1.17	0.92	0.00	7.05
BBO	4.92	4050.00	1.47	1.30	0.00	7.70
EDA	7.01	4050.00	16.89	12.80	0.00	50.37
TDIWO	3.36	4050.00	0.99	0.96	0.00	4.82
GA	4.39	4050.00	0.98	0.85	0.00	4.46
PSO	4.44	4050.00	0.99	0.94	0.00	4.78
BA	4.33	4050.00	0.98	0.85	0.00	4.44
IHBA	3.53	4050.00	0.42	0.50	0.00	5.28
p-value	0.00					

Table 12. The Wilcoxon paired signed test result at $C = 60$ and $\alpha = 0.05$.

Algorithms for Comparison	R+	R−	Bonding Value	*p*-Value
IHBA VS CMA	639	3314	98	0.00
IHBA VS BBO	2428	205	1417	0.00
IHBA VS EDA	3840	100	110	0.00
IHBA VS TDIWO	1981	650	1419	0.00
IHBA VS GA	3222	755	72	0.00
IHBA VS PSO	2181	493	1376	0.00
IHBA VS BA	3099	879	72	0.00

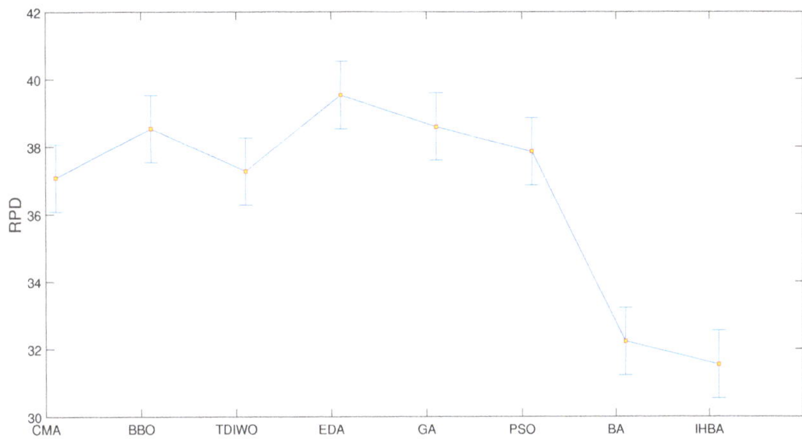

Figure 3. Mean plot and 95% LSD interval of the bat algorithm at $C = 60$.

5.3. VND Method

In this section, the proposed VND3BA and VND2BA are compared with VNDH12, VNDH22, and VNDH32, proposed by Hatami et al. These are 3 different versions of the VND method using H12, H22 and H32 to generate the initial solution. The VND method uses product local search to improve the product sequence and part local search to improve the part sequence for each product. Tables 13 and 14 show the computational results and CPU time for the 5 VND algorithms.

As the table shows, VND3BA and VND2BA perform significantly better than VNDH12, VNDH22 and VNDH12 in terms of solution quality and computational effort. The best overall average RPI value of 15.233% for all instances obtained by the existing VND algorithm is more than three times higher than the best overall average RPI value obtained by VND3BA and VND2BA. For all values of n, m, f and s, the proposed VND algorithm produces better results than the existing VNDH12, VNDH22, and VNDH32. The VND method shows stable advantages without considering the different n, m, f, and s involved. The proposed VND algorithm is also very fast, with an overall average CPU time of 0.009 and 0.057 s for VND3BA and VND2BA, respectively, compared with an overall average CPU time of more than 5 s for the existing algorithms.

Table 13. Average RPD value of VND method.

RPD		Algorithms				
		VNDH12	**VNDH22**	**VNDH32**	**VND3BA**	**VND2BA**
	100	19.177	14.919	16.791	4.656	4.365
n	200	16.752	16.180	19.497	4.957	4.423
	500	12.746	14.599	18.129	4.743	4.074
	5	17.606	15.753	18.614	4.181	3.977
m	10	16.616	15.345	18.731	5.015	4.520
	20	14.463	14.599	17.072	5.151	4.365
	4	13.095	11.824	13.706	4.462	3.919
f	6	16.859	15.627	18.449	4.821	4.317
	8	18.731	18.246	22.262	5.063	4.627
	30	14.608	14.928	18.246	5.160	4.695
s	40	15.850	15.423	18.449	4.860	4.326
	50	18.217	15.355	17.722	4.336	3.841
average		16.226	15.233	18.139	4.784	4.287

Table 14. CPU time of VND method.

CPU		Algorithms				
		VNDH12	VNDH22	VNDH32	VND3BA	VND2BA
n	100	0.459	0.508	0.525	0.004	0.009
	200	1.836	1.786	1.798	0.007	0.027
	500	13.424	13.885	13.417	0.016	0.135
m	5	4.765	4.591	4.326	0.008	0.028
	10	5.001	5.313	5.114	0.010	0.051
	20	5.953	6.274	6.300	0.011	0.091
f	4	3.662	3.522	3.668	0.012	0.083
	6	4.993	5.610	5.615	0.008	0.050
	8	7.065	7.045	6.457	0.008	0.037
s	30	5.321	5.836	5.763	0.007	0.049
	40	5.224	5.354	5.084	0.009	0.058
	50	5.173	4.989	4.894	0.012	0.064
average		5.240	5.393	5.247	0.009	0.057

6. Conclusions

To address the challenge that existing heuristic algorithms and meta-heuristic algorithms cannot fundamentally solve the three-stage distributed assembly permutation flow shop optimization problem, this paper proposes a hybrid bat algorithm optimization algorithm based on variable neighborhood structure and two learning strategies to minimize the completion time of this problem. The algorithm designs a search-based and capture-based two populations to solve the difficult trade-off between convergence and diversity of the bat algorithm in solving the scheduling optimization problem. Moreover, by fully mining the population information, a new selection mechanism and a new velocity and location update strategy are designed to solve the difficult trade-off between exploration and exploitation when the bat algorithm solves the scheduling optimization problem. The Gaussian learning strategy and elite learning strategy are used to assist the whole population to jump out of the local optimal frontier. Simulation results show that IHBA can solve the optimization problem of three-stage distributed assembly permutation flow shop scheduling well, and performs better than existing algorithms in the literature.

In future research, the algorithm is extended to more complex multi-objective optimization problems to further improve the efficiency of iterative search, and the proposed algorithm is applied to other scheduling problems.

Author Contributions: Conceptualization, Y.W.; methodology, Y.W.; software, Y.W.; validation, Y.W. and J.Z.; formal analysis, Y.W.; investigation, Y.W.; resources, Y.W. and J.Z.; data curation, Y.W.; writing—original draft preparation, Y.W.; visualization, Y.W. and J.Z.; supervision, Y.W. and J.Z.; project administration, J.Z.; funding acquisition, J.Z. All authors have read and agreed to the published version of the manuscript.

Funding: This research was funded by the Fundamental Research Funds for the Central Universities under Grant No. CUSF-DH-D-2018050 (18D310804).

Institutional Review Board Statement: Not applicable.

Informed Consent Statement: Not applicable.

Data Availability Statement: The benchmark of instances for the permutation flowshop scheduling problem can be found here: http://soa.iti.es/problem-instances, accessed on 1 September 2021.

Acknowledgments: We gratefully acknowledge the anonymous reviewers for their insightful comments on the manuscript.

Conflicts of Interest: The authors declare no conflict of interest.

References

1. Chen, X.; Zhang, B.; Gao, D. An Improved Bat Algorithm for Job Shop Scheduling Problem. In Proceedings of the 2019 IEEE International Conference on Mechatronics and Automation (ICMA), Tianjin, China, 4–7 August 2019; pp. 439–443.
2. Shareh, M.B.; Bargh, S.H.; Hosseinabadi, A.A.; Slowik, A. An improved bat optimization algorithm to solve the tasks scheduling problem in open shop. *Neural Comput. Appl.* **2021**, *33*, 1559–1573. [CrossRef]
3. Chen, P. S.; Tsai, C. C.; Dang, J. F.; Huang, W. T. Developing Three-phase Modified Bat Algorithms to Solve Medical Staff Scheduling Problems While Considering Minimal Violations of Preferences and Mean Workload. *Technol. Health Care* **2021**, 1–22. [CrossRef]
4. Tolouei, K.; Moosavi, E.; Hossein, A.; Tabrizi, B.; Afzal, P. Application of an improved Lagrangian relaxation approach in the constrained long-term production scheduling problem under grade uncertainty. *Eng. Optim.* **2021**, *53*, 735–753. [CrossRef]
5. Hatami, S.; Ruiz, R.; Andrés-Romano, C. The Distributed Assembly Permutation Flowshop Scheduling Problem. *Int. J. Prod. Res.* **2013**, *51*, 5292–5308. [CrossRef]
6. Hatami, S.; Ruiz, R.; Andrés-Romano, C. Simple constructive heuristics for the Distributed Assembly Permutation Flowshop Scheduling Problem with sequence dependent setup times. In Proceedings of the 2014 International Conference on Control, Decision and Information Technologies (CoDIT), Metz, France, 3–5 November 2014; pp. 19–23.
7. Hatami, S.; Ruiz, R.; Andrés-Romano, C. Heuristics and metaheuristics for the distributed assembly permutation flowshop scheduling problem with sequence dependent setup times. *Int. J. Prod. Econ.* **2015**, *169* , 76–88. [CrossRef]
8. Ying, K.C.; Pourhejazy, P.; Cheng, C.Y.; Syu, R.S. Supply chain-oriented permutation flowshop scheduling considering flexible assembly and setup times. *Int. J. Prod. Res.* **2020**, *58*, 1–24. [CrossRef]
9. Gonzalez-Neira, E.M.; Ferone, D.; Hatami, S.; Juan, A.A. A biased-randomized simheuristic for the distributed assembly permutation flowshop problem with stochastic processing times. *Simul. Model. Pract. Theory* **2017**, *79*, 23–36. [CrossRef]
10. Wang, K.; Li, Z.; Duan, W.; Feng, X.; Liu, B. Variable neighborhood based memetic algorithm for just-in-time distributed assembly permutation flowshop scheduling. In Proceedings of the 2017 IEEE International Conference on Systems, Man, and Cybernetics (SMC), Banff, AB, Canada, 5–8 October 2017; pp. 3700–3704.
11. Zhang, Z.Q.; Qian, B.; Jin, H.P.; Wang, L. A matrix-cube-based estimation of distribution algorithm for the distributed assembly permutation flow-shop scheduling problem. *Swarm Evol. Comput.* **2021**, *60*, 100785. [CrossRef]
12. Zhang, G.; Xing, K.; He, Z. Memetic Algorithm with Meta-Lamarckian Learning and Simplex Search for Distributed Flexible Assembly Permutation Flowshop Scheduling Problem. *IEEE Access* **2020**, *8*, 96115–96128. [CrossRef]
13. Pan, Q.K.; Gao, L.; Li, X.Y.; Jose, F.M. Effective constructive heuristics and meta-heuristics for the distributed assembly permutation flowshop scheduling problem. *Appl. Soft Comput.* **2019**, *81*, 105492. [CrossRef]
14. Ochi, H.; Driss, B. Scheduling the distributed assembly flowshop problem to minimize the makespan. *Procedia Comput. Sci.* **2019**, *164*, 471–477. [CrossRef]
15. Yang, S.L.; Xu, Z.G. The distributed assembly permutation flowshop scheduling problem with flexible assembly and batch delivery. *Int. J. Prod. Res.* **2021**, *59*, 4053–4071. [CrossRef]
16. Liu, B.; Wang, K.; Zhang, R. Variable neighborhood based memetic algorithm for distributed assembly permutation flowshop. In Proceedings of the 2016 IEEE Congress on Evolutionary Computation (CEC), Vancouver, BC, Canada, 24–29 July 2016; pp. 1682–1686.
17. Huang, Y.Y.; Pan, Q.K.; Huang, J.P.; Suganthan, P.N.; Gao, L. An improved iterated greedy algorithm for the distributed assembly permutation flowshop scheduling problem. *Comput. Ind. Eng.* **2021**, *152*, 107021. [CrossRef]
18. Hu, R.; Wu, X.; Qian, B.; Mao, J.L.; Jin, H.P. An Enhanced Differential Evolution Algorithm with Fast Evaluating Strategies for TWT-NFSP with SSTs and RTs. *Complexity* **2020**, *2020*, 8835359. [CrossRef]
19. Seidgar, H.; Fazlollahtabar, H.; Zieh, M. Scheduling two-stage assembly flow shop with random machines breakdowns: integrated new self-adapted differential evolutionary and simulation approach. *Soft Comput.* **2020**, *24*, 8377–8401 [CrossRef]
20. Li, M.; Su, B.; Lei, D. A Novel Imperialist Competitive Algorithm for Fuzzy Distributed Assembly Flow Shop Scheduling . *J. Intell. Fuzzy Syst.* **2021**, *1*, 4545–4561. [CrossRef]
21. Al-Behadili, M.; Ouelhadj, D.; Jones, D. Multi-objective Particle Swarm Optimization for Robust Dynamic Scheduling in a Permutation Flow Shop. *Intell. Syst. Des. Appl.* **2017**, *557*, 498–507.
22. Zhang, X.; Li, X.T.; Yin, M.H. An enhanced genetic algorithm for the distributed assembly permutation flowshop scheduling problem. *Int. J. Bio-Inspired Comput.* **2020**, *15*, 113–124. [CrossRef]
23. Li, X.; Zhang, X.; Yin, M.; Wang, J. A genetic algorithm for the distributed assembly permutation flowshop scheduling problem. In Proceedings of the IEEE Congress on Evolutionary Computation (CEC), Sendai, Japan, 25–28 May 2015; pp. 3096–3101.
24. Mao, J.; Hu, X.; Pan, Q.K.; Miao, Z.; He, C.; Tasgetiren, M.F. An improved discrete artificial bee colony algorithm for the distributed permutation flowshop scheduling problem with preventive maintenance. In Proceedings of the 2020 39th Chinese Control Conference (CCC), Shenyang, China, 27–29 July 2020; pp. 1679–1684.
25. Song, H.B.; Lin, J. A genetic programming hyper-heuristic for the distributed assembly permutation flow-shop scheduling problem with sequence dependent setup times. *Swarm Evol. Comput.* **2021**, *60*, 100807. [CrossRef]
26. Tozkapan, A.; Kırca, Ö; Chung, C.S. A branch and bound algorithm to minimize the total weighted flowtime for the two-stage assembly scheduling problem. *Comput. Oper. Res.* **2003**, *30*, 309–320. [CrossRef]

27. Luo, J.; Ren, R.; Guo, K. The deformation monitoring of foundation pit by back propagation neural network and genetic algorithm and its application in geotechnical engineering. *PLoS ONE* **2020**, *15*, e0233398. [CrossRef]
28. Cools, S.; Cornelis, J.; Vanroose, W. Numerically Stable Recurrence Relations for the Communication Hiding Pipelined Conjugate Gradient Method. *IEEE Trans. Parallel Distrib. Syst.* **2019**, *30*, 2507–2522. [CrossRef]
29. Hansen, P.; Mladenović, N. Variable neighborhood search: Principles and applications. *Eur. J. Oper. Res.* **2001**, *130*, 449–467. [CrossRef]
30. Peng, K.K.; Pan, Q.K.; Gao, L.; Li, X.Y.; Das, S.; Zhang, B. A multi-start variable neighbourhood descent algorithm for hybrid flowshop rescheduling. *Swarm Evol. Comput.* **2019**, *45*, 92–112. [CrossRef]
31. Zhao, F.Q.; Liu, Y.; Zhang, Y.; Ma, W.M.; Zhang, C. A hybrid harmony search algorithm with efficient job sequence scheme and variable neighborhood search for the permutation flow shop scheduling problems. *Eng. Appl. Artif. Intell.* **2017**, *65*, 178–199. [CrossRef]
32. Wang, Z.Q.; Broccardo, M. A novel active learning-based Gaussian process meta modelling strategy for estimating the full probability distribution in forward UQ analysis. *Struct. Saf.* **2020**, *84*, 101937. [CrossRef]
33. Mornell, A.; Osborne, M.S.; McPherson, G.E. Evaluating practice strategies, behavior and learning progress in elite performers: An exploratory study. *Music. Sci.* **2020**, *1*, 130–135. [CrossRef]
34. Deng, J.; Wang, L. A competitive memetic algorithm for multi-objective distributed permutation flow shop scheduling problem. *Swarm Evol. Comput.* **2017**, *32*, 121–131. [CrossRef]
35. Huang, J.L.; Gu, X.S. Distributed assembly permutation flow-shop scheduling problem with sequence-dependent set-up times using a novel biogeography-based optimization algorithm. *Eng. Optim.* **2021**, in press. [CrossRef]
36. Deng, C.; Hu, R.; Qian, B.; Jin, H.P. Hybrid Estimation of Distribution Algorithm for Solving Three-Stage Multiobjective Integrated Scheduling Problem. *Complexity* **2021**, *2021*, 5558949. [CrossRef]
37. Sang, H.Y.; Pan, Q.K.; Wang, P.; Han, YY.; Gao, K.; Duan, P. Effective invasive weed optimization algorithms for distributed assembly permutation flowshop problem with total flowtime criterion. *Swarm Evol. Comput.* **2019**, *44*, 64–73. [CrossRef]

applied sciences

MDPI

Article

A Multi-Criteria Assessment of Manufacturing Cell Performance Using the AHP Method

Zuzana Soltysova *, Vladimir Modrak and Julia Nazarejova

Faculty of Manufacturing Technologies, Technical University of Kosice, Bayerova 1, 080 01 Presov, Slovakia; vladimir.modrak@tuke.sk (V.M.); julia.nazarejova@tuke.sk (J.N.)
* Correspondence: zuzana.soltysova@tuke.sk

Abstract: Research of manufacturing cell design problems is still pertinent today, because new manufacturing strategies, such as mass customization, call for further improvement of the fundamental performance of cellular manufacturing systems. The main scope of this article is to find the optimal cell design(s) from alternative design(s) by multi-criteria evaluation. For this purpose, alternative design solutions are mutually compared by using the selected performance criteria, namely operational complexity, production line balancing rate, and makespan. Then, multi-criteria decision analysis based on the analytic hierarchy process method is used to show that two more-cell solutions better satisfy the determined criteria of manufacturing cell design performance than three less-cell solutions. The novelty of this research approach refers to the use of the modification of Saaty's scale for the comparison of alternatives in pairs based on the objective assessment of the designs. Its benefit lies in the exactly enumerated values of the selected criteria, according to which the points from the mentioned scale are assigned to the alternatives.

Keywords: multi-criteria assessment; cell manufacturing design; operational complexity; makespan; production line balancing rate

Citation: Soltysova, Z.; Modrak, V.; Nazarejova, J. A Multi-Criteria Assessment of Manufacturing Cell Performance Using the AHP Method. *Appl. Sci.* **2022**, *12*, 854. https://doi.org/10.3390/app12020854

Academic Editor: Aliaksandr Shaula

Received: 15 December 2021
Accepted: 13 January 2022
Published: 14 January 2022

Publisher's Note: MDPI stays neutral with regard to jurisdictional claims in published maps and institutional affiliations.

1. Introduction and Problem Description

The smart manufacturing concept, as an important part of the Industry 4.0 strategy, opens new opportunities for producers to implement new platform-based business models by embracing cutting-edge technologies. Cellular manufacturing (CM) systems belong among six fundamental manufacturing systems, for which Industry 4.0 has been conceived [1]. Their goal is to complete jobs as swiftly as possible, make a wide variety of similar products, and produce as little waste as possible. An important objective of CM systems is to be as easily reconfigurable as possible [2–4]. If the number of potential configuration design solutions is generated, the role of the user is to define constraints on design and performance to obtain solutions meeting their requirements. Subsequently, it is needful to explore design options and configurations to select an optimal solution.

Operation management research often reflects CM problems in order to make manufacturing operations more efficient and productive. The cellular manufacturing method brings scattered processes together with compact cells usually arranged in U-shapes and can significantly improve the operation of batch production [5]. Batch production with a wide range of product types presents a crucial problem for layout designers since the parts move in batches from one process to another, and ready parts must wait for the remaining parts to complete processing before they move to the next stage. Because operation times are distinct one from another, it causes unbalanced machine utilization, scheduling problems, and possible late deliveries. Unbalanced machine utilization is frequently considered one of the main criteria in CM design optimization since ineffective utilization of machines can result in unprofitable production [6,7]. Another approach to optimize CM design is based on using appropriate scheduling techniques that aim to determine the actual assignment of products or jobs to available operations in order to complete manufacturing orders on time [8]. These techniques are mostly developed using the minimum makespan criterion for CM design optimization [9]. Moreover, makespan minimization positively influences work-in-process inventories [10,11]. Taking into consideration the fact that CM systems have to be adaptable to changing market demands, their structural and operational complexity adequately increases as a consequence [12,13]. In this context, the cost of operational complexity is becoming a topical problem, and therefore, the trade-offs between the level of complexity and its added value are often discussed (see, e.g., [14–17]).

It is also worth mentioning that CM is not always the appropriate approach to take for certain scenarios [18]. This problem has been studied, e.g., by Flynn and Jacobs [19], who indicated that, through a well-organized job shop, it is possible to achieve at least as good a performance as the cellular layout with respect to several criteria such as work-in-process inventory levels and average flow times. For this reason, it is assumed in this study that a given manufacturing process satisfies basic criteria for a cellular layout, which include a high ratio of setup to process time, stable demand, unidirectional workflow within a cell, and a considerable level of material movement times between process departments [20].

The problem that is treated in the present work can be described as follows. The time required to complete all jobs, namely makespan, is one of the most frequent performance indicators for the job shop problem [21]. Considering that one-piece flow cellular manufacturing is the ultimate in lean production [22], CM is primarily aimed at reducing times within the production system. On the other hand, competitiveness through cost reduction in the design and implementation of production systems is an important and permanent task for process designers. In this order, identifying and monitoring cost items causes significant difficulties since there are various costs such as part holding cost at a facility, machine procurement cost, machine maintenance overhead cost, machine repair cost, production loss cost due to machine breakdown, machine operation cost, setup cost, tool consumption cost, inter-cell travel cost, intra-cell travel cost, etc. [23]. Therefore, from a practical viewpoint, it is reasonable to indicate the main portion of costs by using indirect indicators, namely production line balancing rate and minimization of operational complexity.

Taking into account the above-mentioned observations and factors influencing the efficiency of CM systems, our interest in this paper is to propose a multi-criteria decision-making approach for the selection of an optimal manufacturing cell from several alternative options. The three selected criteria, i.e., maximization of production line balancing rate, minimization of operational complexity of alternative CM designs, and makespan minimization, is employed to assess the layout design alternatives.

For the purpose of multi-criteria decision analysis, the analytic hierarchy process (AHP) method is applied. The AHP method is one of the most exploited multi-criteria decision-making tools and one of the most trusted decision-making methodologies. On the other hand, its disadvantage lies in possible flaws in the verbal scale often used in AHP pairwise comparisons. In the case of investigation of the given manufacturing problem, a numerical scale is used for the assessment of the criteria. Then, the results will not be influenced by personal opinions in considering and representing facts.

As testing examples, alternative CM designs along with input data from the existing representative case study is used.

2. Literature Background

The literature background conducts a review regarding the above-mentioned optimization criteria or factors influencing decision making in cellular manufacturing design.

Line balancing is considered an effective tool to optimize layout design and reduce product cycle times. The objectives of line balancing techniques in flow shop scheduling problems are usually focused on the reduction and/or redesign of workstations in order to minimize production costs, work in process in order to reduce storage space and bound capital, and minimization of makespan and flow times [24,25]. In general, line balancing techniques are divided into deterministic types, where all input parameters are known and not changed over time, and probabilistic types, which deal with parameter uncertainties [26]. An innovative stochastic line balancing method was proposed by Kottas-Lau et al. [27]. Their algorithm was developed for the purpose of achieving optimized total production costs and to allow a good level of line balancing. Among the other important deterministic line balancing techniques can be mentioned the largest candidate rule [28], the Kilbridge and Wester Column method [29], and the ranked positional weight method proposed by Hegelson and Bernie [30]. The latter method was developed to minimize idle times and the number of workstations. The specific type of these optimization tasks consists of large-scale line balancing problems that deal with uncertainty in line balancing. Hazr and Dolgui [31] proposed two optimization models that belong to these problems, by which it is possible to generate exact solutions of cellular manufacturing designs. Another optimization technique based on the identification of equilibrium between line efficiency and equipment cost was proposed by Gurevsky et al. [32]. It is also useful to note that their method supports decisions at the early design stage of production lines. As it is impossible to involve all pertinent works on the topic, comprehensive literature studies focused on the comparative evaluation of line balancing techniques overcome this drawback. Such review studies can be found, e.g., in [33–35].

Operational and structural complexity is another relevant factor affecting the efficiency of CM systems. According to Hon [36], the main reason for the investigation of manufacturing system complexity is to comprehend and control the behavior of such systems in order to make them more productive and predictive. Fredendall and Gabriel [37] argue that by measuring system complexity, managers can better identify problems in manufacturing systems that hinder production flow. Therefore, the determination of quantitative metrics of manufacturing system complexity, either static or dynamic, is one of the crucial elements in this effort [38]. There are several different approaches to the utilization of complex concepts in this application domain, but it has not yet been possible to find closed-form equations able to describe the dynamic behavior of manufacturing systems [39]. For this reason, available operational complexity measures reflect only selected facets of such systems. Manufacturing system complexity is often divided into structural and operational types. The second type of complexity measure, which is of interest in this study, is based on measuring uncertainties involved in manufacturing systems. This type of complexity is further divided into time independent and time dependent [40]. Zhang [41] analyzed the relationship between cellular manufacturing system complexity and utility in order to show that increasing complexity can be beneficial for manufacturers until it reaches a critical value. Beyond this critical value, the situation becomes the opposite. The mentioned system complexity consequences on design and managerial practice were originally introduced by Tainter [42] and are widely respected in many research communities. Therefore, managers might mitigate the negative aspects of complexity while managing its positive aspects, as complexity indirectly influences the performance of manufacturing systems [43]. These arguments inspire us to employ operational complexity as one of the criteria in the decision-making procedure.

Makespan is commonly used as a criterion of performance measure in the design of cellular manufacturing systems, because the advantages of cellular manufacturing also include simplified planning and scheduling [44]. The scheduling problem in a cellular manufacturing system assumes that intercellular moves can be eliminated by duplicating machines, but it is usually very costly and therefore infeasible [45]. If duplicating machines is not a viable solution, then a volume limit can enhance the choice of the optimal routing of jobs. One method to make this choice is to minimize the makespan since this performance measure is the most frequent objective in flow shop problems [46].

3. Methodological Framework

This section aims to describe in a nutshell the set of methods and overall procedure in chronological order and help the reader understand the context under which the research was conducted. In its first steps, the criteria or factors influencing the efficiency of CM systems have been mostly identified based on empirical findings and general knowledge of CM design, namely makespan and production line balancing rate. Similarly, measurement methods to quantify operational complexity have been chosen. Subsequently, the AHP method divided into three steps was used for the evaluation of individual CM alternatives. Summarily, the methodological framework of this research consists of five steps, which is illustrated in the figure below (Figure 1).

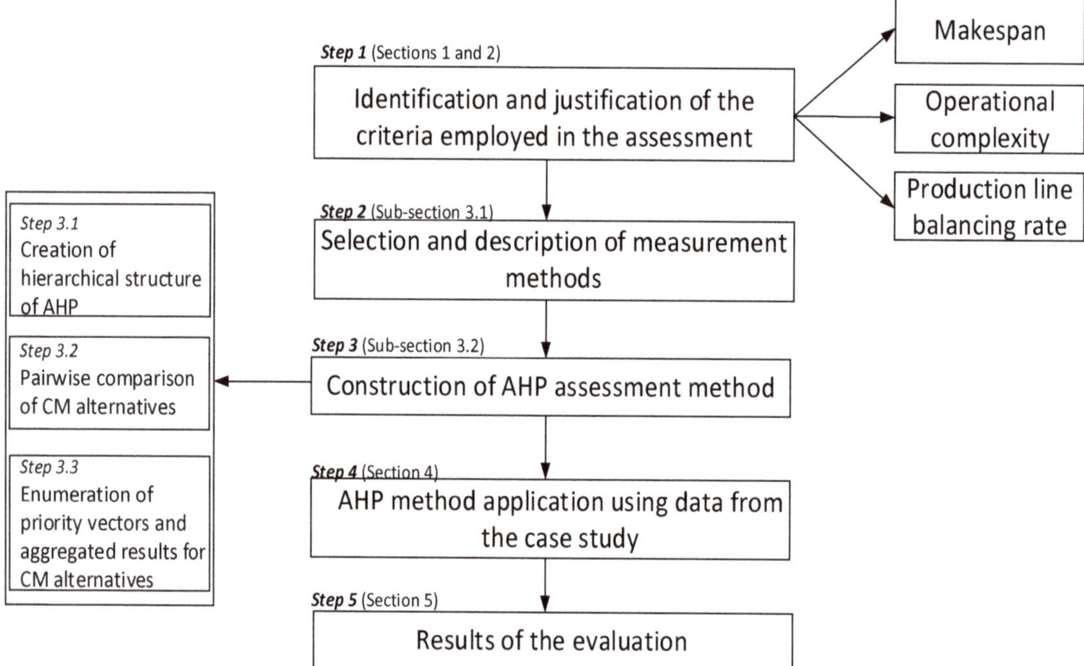

Figure 1. Overall research procedure in chronological order (Sections 1–5).

3.1. Indicators Used as Criteria to Assess Alternative CM Designs

3.1.1. Production Line Balancing Rate

The production line balancing rate (PLB) represents a measure of the average length of time of every cycle time in the working procedure on the processing line. It is equivalent to minimizing the number of workstations with a certain takt time [47]. It is expressed as [48]:

$$PLB = \frac{\sum_{j=1}^{n} t_j}{m * \max(T_i)} \ [\%],\tag{1}$$

where

t_j stands for standard work time of the j-th job elements;
n represents the number of the work elements;
m represents the number of total lines in the production system;
T_i represents the work time in the production line(s) (PL(s));
$\max(T_i)$ represents the biggest line operating time.

3.1.2. Operational Complexity Indicators

There are several potential operational complexity indicators that can be applied to measure the operational complexity properties of manufacturing systems from different or similar perspectives (see, e.g., [49–54]). Two of them that suit the available data characterizing the benchmarked alternatives are further introduced.

Process Complexity Indicator

This indicator, similar to the other concurrent complexity measures, is derived from Shannon's information theory [55]. The process complexity indicator (PCI) is specified for the quantification of manufacturing process complexity, taking into account the operational complexities of individual machines. The PCI indicator is enumerated by using equation [54]:

$$PCI = -\sum_{i=1}^{M}\sum_{j=1}^{P}\sum_{k=1}^{O} p_{ijk} \cdot \log_2 p_{ijk} \ [\text{bits}],\tag{2}$$

where

p_{ijk} stands for the probability that part j is processed due to operation k by individual machine i according to scheduling order;
O is the number of operations according to parts production;
P is the number of parts produced in the manufacturing process;
M is the number of all machines of all types in the manufacturing process.

Balanced Complexity Indicator

The balanced complexity indicator (BCI) takes into account the rate of mutual differences between the individual complexities of machines. This indicator expresses the variability of the partial complexities of workstations/machines and calculates the deviation of partial machine complexities from their mean value. It is calculated using the following formula [54]:

$$BCI = \frac{\sum_{i=1}^{N} MCI_{i(\max)} - \sum_{i=1}^{N} MCI_{i(\min)}}{N} \ [\text{bits}],\tag{3}$$

where

$MCI_{i(\max)}$ represents the first N-max complexity values;
$MCI_{i(\min)}$ represents the first N-min complexity values;
N represents the number of max and min machine complexity values.

3.1.3. Makespan Indicators

For the purpose of calculating makespan, the scheduling algorithm to minimize the completion of n-jobs of m-machines is used. As known, there are many different algorithms for the given purpose. In this work, the freely available online software is utilized [56], and the following input data for this algorithm are collected:

Processing times in minutes for each job, which are included in matrix $m \times n$ (Figure 2a);
Number of transport batch (Figure 2b);
Transport batch sizes for each job (Figure 2c);
Sequences of individual jobs numbered by order (Figure 2d).

Its application in the first step requires that input data are presented in the form of a Microsoft Office Excel table and pasted into the input data window. A flowchart of how to generate makespan is shown in Figure 3.

As can be seen from Figure 3, makespans are enumerated assuming two scenarios. According to the first scenario, makespan is calculated for determined batch sizes, and for the second scenario, the one-piece flow (OPF) principle is applied, i.e., when the transport batch size for each job equals 1.

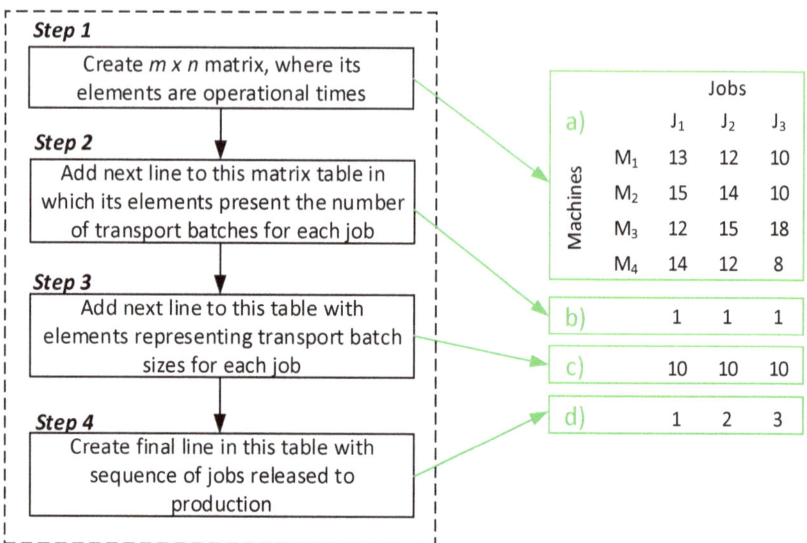

Figure 2. Table format for insertion of input data, (**a**) processing times in minutes for each job, which are included in matrix $m \times n$, (**b**) number of transport batch, (**c**) transport batch sizes for each job, (**d**) sequences of individual jobs numbered by order.

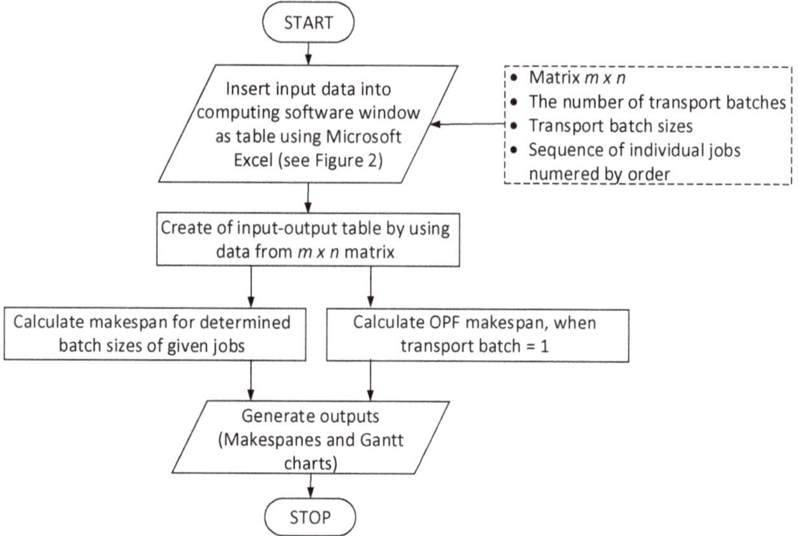

Figure 3. Software flowcharts with acquired data.

3.2. Description of AHP Method

Prior to describing the AHP method with the modified Saaty scale for a comparison of design alternatives in pairs using a multi-criteria decision-making approach, the five selected related AHP approaches are reported in Table 1 in order to point out the differences in the research addressed in our paper.

Table 1. Comparison of existing studies based on usage of AHP method.

Publication Title	Publication Characteristics
The use of AHP method for selection of supplier [57]	This article presents the general design of the model for the selection of a suitable supplier from three potential suppliers by the AHP method using Saaty's point scale. The proposed model is applied in a manufacturing company.
A multi-Criteria decision support concept for selecting the optimal contractor [58]	This paper presents a decision support concept for selecting the optimal contractor. This concept increases the transparency of decision-making and the consistency of the decision-making process, and it has potential for application in similar decision-making problems.
Fuzzy AHP group decision analysis and its application for the evaluation of energy sources [59]	The evaluation of a multi-criteria decision problem by use of fuzzy logic is the main concern of this research. It considers the specific problem of the searching of energy alternatives and a proper evaluation of these alternatives in comparison with existing ones.
Modeling procedure for the selection of steel pipes supplier by applying fuzzy AHP method [60]	This work is focused on the evaluation and selection of suppliers by applying fuzzy multi-criteria analysis using the AHP method to choose the optimal supplier from five suppliers for the production of pre-insulated pipes. These suppliers are compared based on nine criteria, e.g., material cost, delivery time, transport distance, etc.
An application of analytic hierarchy process (AHP) in a real-world problem of store location selection [61]	This study presents a new store location selection problem of Carglass Turkey, which includes tangible and intangible criteria, and the analytic hierarchy process (AHP) was applied. The hierarchical model established for this problem may provide insight regarding location selection problems.
Simultaneous customers and supplier's prioritization: an AHP based fuzzy inference decision support system (AHP-FIDSS) [62]	This research paper introduced a novel analytical hierarchical process-based fuzzy inference decision support system (AHP-FIDSS), which involves factor screening, hierarchical structure modeling, quantification of qualitative factors, and their conversion to quantitative values.

All of these AHP method applications are based on a subjective approach in mutual comparison of alternatives. On the contrary, the proposed procedure to find the most suitable cell design alternative(s) uses an objective approach of the pairwise comparison of cell design alternatives. In addition, it is necessary to select one of the possible methods in the application of the AHP method [63,64]. For the given purpose, the most accurate procedure seems to be that which consists of the following steps:

Step 1. Creation of hierarchical structure of AHP method. The hierarchical structure of the AHP method is created in the form of a diagram, where the criteria, sub-criteria, and CM alternatives are specified.

Step 2. Pairwise comparison of CM alternatives. All CM alternatives are pairwise benchmarked with respect to the criteria and sub-criteria. The pairwise comparison is provided in matrix form by comparing one CM alternative to another to determine the weights of importance. For this purpose, the proposed modified scale of relative importance is applied (see Section 4.3).

Step 3. Enumeration of priority vectors and aggregated results. The priorities are derived using the values of the principal right eigenvectors of the compared matrices. These priorities are expressed as absolute numbers bounded between 0 and 1, without units, and are calculated according to the so-called additive normalization method using the following simple procedure:

Sum each column values separately for each matrix, divide each element of the column with the sum of that column for each matrix, and compute the average of all elements in each row of all matrices to obtain the priority vector.

To obtain aggregated results, it is needed to summarize the determined priorities of all the individual indicators for each CM alternative. Then, the aggregated priorities are compared by ranking them in order from most to least important. Finally, the optimal CM alternative is selected.

4. A Practical Example

4.1. Description of Manufacturing Cell Designs Alternatives

This section aims to introduce five manufacturing cell alternatives for their mutual comparison in order to determine the optimal design. The CM alternative designs along with basic input data are taken from a case study by Yan and Irani [65]. The essential data from this case study include routing of parts (P) through machines (M), sequencing of parts by way of recording, operational times in minutes, and batch sizes for individual parts. The parts routings and operational times for all the parts are shown in Figure 4 (e.g., part P1 is firstly processed for 96 min on machine M1. It is subsequently machined for 36 min on machine M4, processed for 36 min on machine M8, and finally, it is machined for 72 min on machine M9).

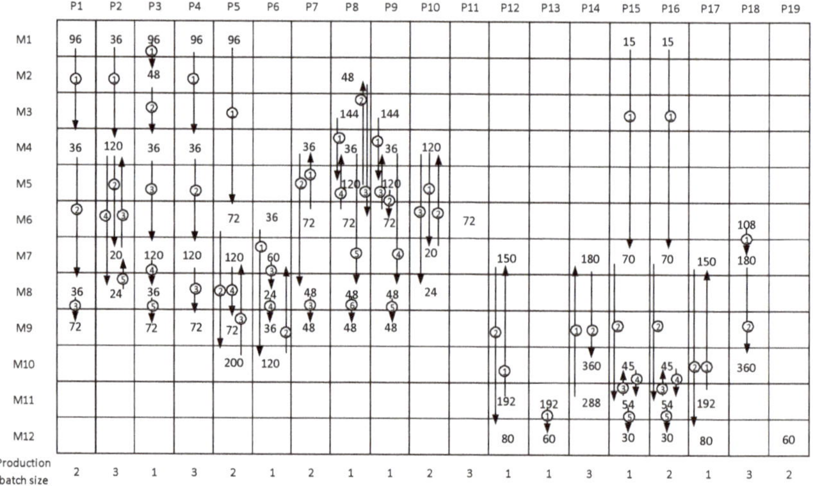

Figure 4. Sequences of machines for each job/part.

The alternative designs that are depicted in Figure 5 can be divided into two groups: two-cell solutions (three design alternatives, i.e., Cell Designs 1–3) and three-cell solutions (two design alternatives, i.e., Cell Designs 4–5). The sequence of machines for each job is not violated in all the alternatives.

As can be seen in this figure, Cell Design 1 consists of two cells: 11 machines are located in the first cell, and 12 machines are located in the second cell. Parts P1–P4 and P7–P11 are processed in the first cell, and the rest of the parts marked as P12–P19 are processed only in the second cell. Parts P5 and P6 are partially processed in the second cell and finalized in the first cell.

Cell Design 2 contains 25 machines. There are 16 machines in the first cell and 9 machines in the second cell. Parts P1–P11 and P18 are machined in the first cell, and the rest of the parts are produced in the second cell.

Cell Design 3 is quite similar to Cell Design 2, except that machine M1 is eliminated in the second cell, and parts P15 and P16 are first produced in the first cell and subsequently finished in the second cell.

Cell Design 4 includes three cells with 8 machines in the first cell, 10 machines in the second cell, and 8 machines in the third cell. Parts marked as P1, P3, P7–P9, and P11 are machined in the first cell, while P3 is completed in the second cell. Parts P2, P4–P6, P10, P15, P16, and P18 are produced in the second cell, but P15 and P16 are finished in the third cell. Parts P12–P14, P17, and P19 are machined in the third cell.

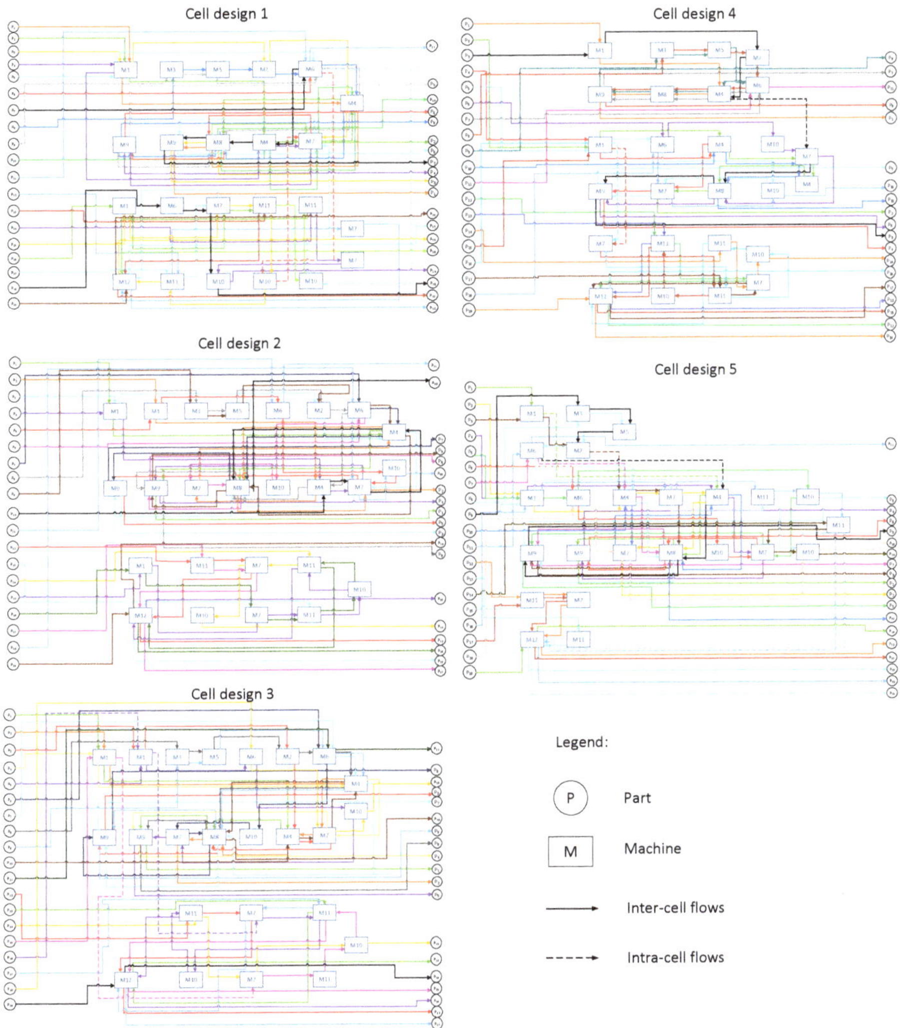

Figure 5. Cell design alternatives and their material flows.

Cell Design 5 is also divided into three cells. The first cell consists of 5 machines, 15 machines are located in the second cell, and 4 machines are placed in the third cell. P1, P3, P7–P9, and P11 are machined in the first cell, while P1, P3, and P7–P9 are completed in the second cell. P2, P4–6, P10, P14–P16, and P18 are produced in the second cell, while P15 and P16 are finalized in the third cell. P12, P13, P17, and P19 are machined only in the last cell.

4.2. Application of the Performance Indicators on Cell Designs

In this sub-section, the above-mentioned indicators are applied to the five CM alternatives. Obtained operational complexity values, makespans, and production line balancing rates are summarily shown in Table 2.

The obtained values are further used as input data for the purpose of multi-criteria comparison applying the AHP method.

Table 2. Enumerated results of all the criteria and sub-criteria.

Cell Designs	Makespan (min)	OPF Makespan (min)	PCI (bits)	BCI (bits)	PLB (%)
1	5926	3610	47.8	2.1	92.4 *
2	5014	3108	50.7	2.24	78.7
3	5049	3079	53.9	2.45	78.2
4	5438	3194	44.8 *	1.62 *	85.6
5	4650 *	2927 *	47.1	2.14	50.7

* Best obtained value.

4.3. Assessment of Manufacturing Cell Designs Using AHP Method

Firstly, the hierarchical structure of the AHP method is created. The overall focus is aimed at the selection of the optimal manufacturing cell design(s) from the five cell design alternatives. For this purpose, these five alternatives are compared using the criteria shown in Figure 6.

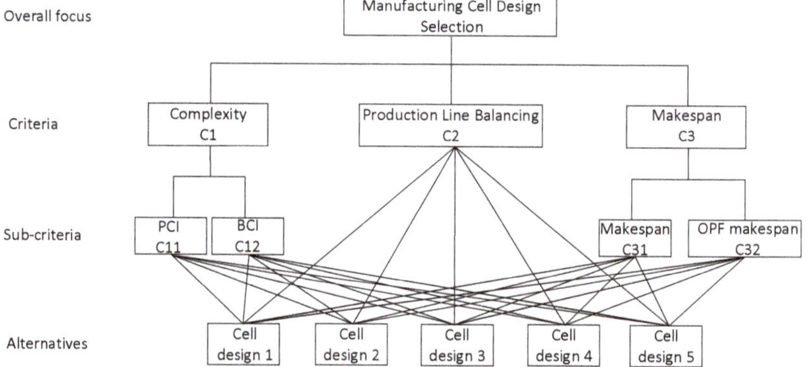

Figure 6. Hierarchical structure of the AHP method for selection of an optimal manufacturing cell.

Once the hierarchy is constructed, the alternatives are pairwise compared for each of the criteria based on the preferences using the scale of relative importance (see Table 3).

Table 3. Scale of relative importance.

Scale	Numerical Rating	Explanation
Equal importance	1	Two alternatives contribute equally to the objective (0% difference)
Moderate importance	3	One alternative is slightly favored over another (no more than 25% difference)
Strong importance	5	One alternative is strongly favored over another (25–50% difference)
Very strong importance	7	One alternative is very strongly favored over another (50–75% difference)
Absolute importance	9	One alternative is absolutely favored over another (more than 75% difference)
Intermediate values	2, 4, 6, 8	When compromise is needed between two alternatives

Note: When comparing the best alternative with the worst alternative, the difference between them is maximally 100%. Based on this difference value, the percentages 25%, 50%, and/or 75% are proposed.

The performance of each cell design (CD) with regard to each criterion is indicated by the following pairwise comparison matrices, and at the same time, it is assumed that all criteria are equal to each other (see Figure 7).

Subsequently, the priority vectors are calculated according to the additive normalization method for all the criteria shown in Table 4.

For PCI:

	CD 1	CD 2	CD 3	CD 4	CD 5
CD 1	1	5	7	1/5	1/3
CD 2	1/5	1	5	1/7	1/5
CD 3	1/7	1/5	1	1/9	1/7
CD 4	5	7	9	1	5
CD 5	3	5	7	1/5	1

For BCI:

	CD 1	CD 2	CD 3	CD 4	CD 5
CD 1	1	3	5	1/7	3
CD 2	1/3	1	5	1/7	1/3
CD 3	1/5	1/5	1	1/9	1/5
CD 4	7	7	9	1	7
CD 5	1/3	3	5	1/7	1

For Makespan:

	CD 1	CD 2	CD 3	CD 4	CD 5
CD 1	1	1/7	1/7	1/5	1/9
CD 2	7	1	3	5	1/5
CD 3	7	1/3	1	5	1/5
CD 4	5	1/5	1/5	1	1/7
CD 5	9	5	5	7	1

For OPF makespan:

	CD 1	CD 2	CD 3	CD 4	CD 5
CD 1	1	1/7	1/9	1/7	1/9
CD 2	7	1	1/3	3	1/5
CD 3	9	3	1	3	1/3
CD 4	7	1/3	1/3	1	1/5
CD 5	9	5	3	5	1

For PLB:

	CD 1	CD 2	CD 3	CD 4	CD 5
CD 1	1	5	5	3	9
CD 2	1/5	1	3	1/3	7
CD 3	1/5	1/3	1	1/3	7
CD 4	1/3	3	3	1	9
CD 5	1/9	1/7	1/7	1/9	1

Note: Due to the fact that precisely calculated quantitative values were used for the purpose of the pairwise comparison, it is not meaningful to assess them using the consistency coefficient.

Figure 7. Obtained values of pairwise comparison matrices for Cell Designs 1–5.

Table 4. Obtained priority vectors values for all the criteria.

Cell Designs	PCI	BCI	Makespan	OPF Makespan	PLB
1	0.15877167	0.175165699	0.029623447	0.027892418	0.476996214
2	0.072721123	0.085558568	0.221688177	0.148542569	0.148542569
3	0.029623447	0.03317987	0.15877167	0.245125107	0.101443692
4	0.517195582	0.580746424	0.072721123	0.101443692	0.245125107
5	0.221688177	0.125349438	0.517195582	0.476996214	0.027892418

Finally, the aggregated relative priorities from Table 4 are enumerated, which allows ranking CM alternatives, as shown in Figure 8.

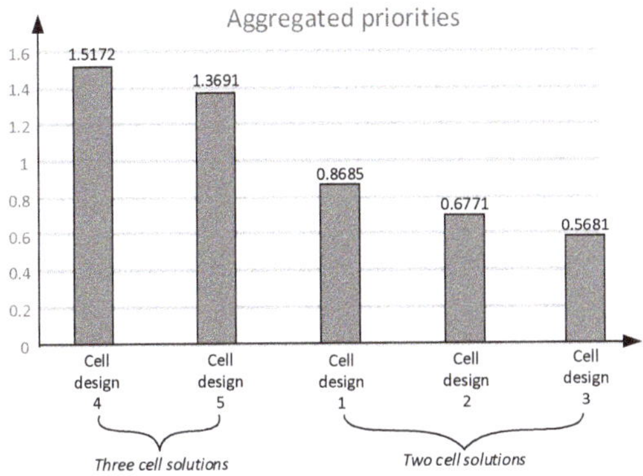

Figure 8. Final comparison of cell designs.

As can be seen from Figure 8, Cell Design 4 is considered the optimal design.

5. Conclusions

Summarily, it can be stated that both of the three-cell solutions better satisfied the determined criteria of manufacturing cell design performance than all the three two-cell solutions. It can be empirically explained by this that cells practically represent modules, and a modular manufacturing layout design is better than an integral design.

Moreover, from the obtained results of the computational experiments, it can be noted that:

According to both complexity indicators, the three-cell solutions are less complex than the two-cell solutions. The lower complexity of the three-cell designs against the two-cell designs can be comprehended in a way that the scheduling of cell designs with a higher number of cells is less complicated than in the case with a smaller number of cells. This statement comes from the fact the probability that parts are produced on given machines is higher than in the case of the CM design with a smaller number of cells;

Based on the makespan results, the three-cell solutions better satisfied the minimization of the total time needed to finish all the jobs than the two-cell solutions;

From the viewpoint of the PLB indicator, the two-cell solutions offer better balancing of machines than the three-cell solutions.

As mentioned, the case study was taken from the work of Yan and Irani [65], who compared two-cell solutions with three-cell solutions based on selected criteria such as the number of intra-cell flows and inter-cell flows, scheduling, etc., to point out the advantages and disadvantages of two-cell solutions and three-cell solutions. Our aim was to identify the optimal cell design solution(s) from the alternatives based on a multi-criteria decision-making approach, where five selected criteria were used. The main benefit of the used approach lies in the objective approach of the pairwise comparison of cell design alternatives in the decision-making process.

Related future research could be oriented to employ other criteria to assess manufacturing cells design performance in order to bring new findings for practitioners and researchers.

Author Contributions: Conceptualization, V.M.; methodology, V.M. and Z.S.; validation, Z.S.; formal analysis, V.M. and Z.S.; writing—original draft preparation, V.M. and Z.S.; writing—review and editing, V.M.; visualization, Z.S. and J.N.; supervision, V.M.; project administration, V.M. All authors have read and agreed to the published version of the manuscript.

Funding: This research was funded by the KEGA Project No. 025TUKE-4/2020 granted by the Ministry of Education of the Slovak Republic.

Institutional Review Board Statement: Not applicable.

Informed Consent Statement: Not applicable.

Notations

PLB	Production line balancing rate, in %
PCI	Process complexity indicator, in bits
BCI	Balanced complexity indicator, in bits
t_j	Standard work time of the j-th job elements
n	Number of the work elements
m	Number of total lines in production system
T_i	Work time in the production line(s) (PL(s))
$\max(T_i)$	Biggest line operating time
p_{ijk}	Probability that part j is processed due to operation k by individual machine i according to scheduling order
O	Number of operations according to parts production
P	Number of parts produced in manufacturing process
M	Number of all machines of all types in manufacturing process
$MCI_{i(max)}$	First N-max complexity values
$MCI_{i(min)}$	First N-min complexity values
N	Number of max and min machine complexity values

References

1. Qin, J.; Liu, Y.; Grosvenor, R. A Categorical Framework of Manufacturing for Industry 4.0 and Beyond. *Procedia CIRP* **2016**, *52*, 173–178. [CrossRef]
2. Tzafestas, E.S. A Study of Self-Organization in a Production Environment. In *Advances in Manufacturing*; Springer: London, UK, 1999; pp. 37–44.
3. Bortolini, M.; Galizia, F.G.; Mora, C.; Pilati, F. Reconfigurability in cellular manufacturing systems: A design model and mul-ti-scenario analysis. *Int. J. Adv. Manuf. Technol.* **2019**, *104*, 4387–4397. [CrossRef]
4. Modrak, V.; Pandian, R.S. *Operations Management Research and Cellular Manufacturing Systems: Innovative Methods and Approaches*; IGI Global: Hershey, PA, USA, 2012.
5. Huber, V.L.; Hyer, N.L. The human factor in cellular manufacturing. *J. Oper. Manag.* **1985**, *5*, 213–228. [CrossRef]
6. Kia, R.; Baboli, A.; Javadian, N.; Tavakkoli-Moghaddam, R.; Kazemi, M.; Khorrami, J. Solving a group layout design model of a dynamic cellular manufacturing system with alternative process routings, lot splitting and flexible reconfiguration by simulated annealing. *Comput. Oper. Res.* **2012**, *39*, 2642–2658. [CrossRef]
7. Renna, P.; Ambrico, M. Design and reconfiguration models for dynamic cellular manufacturing to handle market changes. *Int. J. Comput. Integr. Manuf.* **2015**, *28*, 170–186. [CrossRef]
8. Herrmann, J.W. A History of Production Scheduling. In *Handbook of Production Scheduling*; Springer: Boston, MA, USA, 2006; pp. 1–22.
9. Srimongkolkul, O.; Chongstitvatana, P. Minimizing Makespan Using Node Based Coincidence Algorithm in the Permutation Flowshop Scheduling Problem. In *Industrial Engineering, Management Science and Applications*; Springer: Berlin/Heidelberg, Germany, 2015; pp. 303–311.
10. Paul, S.K.; Azeem, A. Minimization of work in process inventory in hybrid flow shop scheduling using fuzzy logic. *Int. J. Ind. Eng. Theory Appl. Pract.* **2010**, *17*, 115–127.
11. Modrak, V.; Soltysova, Z. Batch Size Optimization of Multi-Stage Flow Lines in Terms of Mass Customization. *Int. J. Simul. Model.* **2020**, *19*, 219–230. [CrossRef]
12. Kuzgunkaya, O.; ElMaraghy, H.A. Assessing the structural complexity of manufacturing systems configurations. *Int. J. Flex. Manuf. Syst.* **2007**, *18*, 145–171. [CrossRef]
13. Modrak, V.; Marton, D. Configuration complexity assessment of convergent supply chain systems. *Int. J. Gen. Syst.* **2014**, *43*, 508–520. [CrossRef]
14. Trattner, A.; Hvam, L.; Forza, C.; Herbert-Hansen, Z.N.L. Product complexity and operational performance: A systematic literature review. *CIRP J. Manuf. Sci. Technol.* **2019**, *25*, 69–83. [CrossRef]
15. Calinescu, A. Manufacturing Complexity: An Integrative Information-Theoretic Approach. Ph.D. Dissertation, University of Oxford, Oxford, UK, 2002.
16. Park, K.; Kremer, G. The impact of complexity on manufacturing performance: A case study of the screwdriver product family. In Proceedings of the 2013 International Conference on Engineering Design, Seoul, Korea, 19–22 August 2013; The Design Society: Copenhagen, Denmark, 2013; Volume 3, pp. 141–150.
17. Bucki, R.; Suchánek, P. Comparative Simulation Analysis of the Performance of the Logistics Manufacturing System at the Operative Level. *Complexity* **2019**, *2019*, 7237585. [CrossRef]
18. Rogers, D.F.; Shafer, S.M. Measuring cellular manufacturing performance. *Manuf. Res. Technol.* **1995**, *24*, 147–165.
19. Flynn, B.B.; Jacobs, F.R. Applications and Implementation: An Experimental Comparison of Cellular (Group Technology) Layout with Process Layout. *Decis. Sci.* **1987**, *18*, 562–581. [CrossRef]
20. Morris, J.S.; Tersine, R.J. A Simulation Analysis of Factors Influencing the Attractiveness of Group Technology Cellular Layouts. *Manag. Sci.* **1990**, *36*, 1567–1578. [CrossRef]
21. Yuan, Y.; Xu, H.; Yang, J. A hybrid harmony search algorithm for the flexible job shop scheduling problem. *Appl. Soft Comput.* **2013**, *13*, 3259–3272. [CrossRef]
22. Liker, J.K. *The Toyota Way*; McGraw Hill: New York, NY, USA, 2004.
23. Kant, R.; Pattanaik, L.N.; Pandey, V. Framework for strategic implementation of cellular manufacturing in lean production environment. *J. Manuf. Technol. Res.* **2014**, *6*, 177.
24. Saraçoğlu, I.; Süer, G.A.; Gannon, P. Minimizing makespan and flowtime in a parallel multi-stage cellular manufacturing company. *Robot. Comput. Manuf.* **2021**, *72*, 102182. [CrossRef]
25. Belfiore, G.; Falcone, D.; Silvestri, L. Assembly line balancing techniques: Literature review of deterministic and stochastic methodologies. In Proceedings of the 17th International Conference on Modeling and Applied Simulation (MAS 2018), Budapest, Hungary, 17–19 September 2018; pp. 185–190.
26. Shinde, A.; More, D. Production Line Balancing: Is it a Balanced Act? *Vikalpa* **2015**, *40*, 242–247. [CrossRef]
27. Kottas, J.F.; Lau, H.-S. A cost oriented approach to stochastic line balancing. *AIIE Trans.* **1973**, *5*, 164–171. [CrossRef]
28. Moodie, C.L.; Young, H.H. A heuristic method of assembly line balancing assumptions of constant or variable work element times. *J. Ind. Eng.* **1965**, *16*, 23–29.
29. Kilbridge, M.D.; Wester, L. A heuristic method of assembly line balancing. *J. Ind. Eng.* **1961**, *12*, 292–298.
30. Helgeson, W.B.; Birnie, D.P. Assembly Line Balancing Using the Ranked Positional Weighting Technique. *J. Ind. Eng.* **1961**, *12*, 394–398.
31. Hazır, Ö.; Dolgui, A. Assembly line balancing under uncertainty Robust optimization models and exact solution method. *Comput. Ind. Eng.* **2013**, *65*, 261–267. [CrossRef]

32. Gurevsky, E.; Hazır, Ö.; Battaïa, O.; Dolgui, A. Robust balancing of straight assembly lines with interval task times. *J. Oper. Res. Soc.* **2013**, *64*, 1607–1613. [CrossRef]
33. Ponnambalam, S.G.; Aravindan, P.; Naidu, G.M. A comparative evaluation of assembly line balancing Heuristics. *Int. J. Adv. Manuf. Technol.* **1999**, *15*, 577–586. [CrossRef]
34. Özceylan, E.; Kalayci, C.B.; Güngör, A.; Gupta, S.M. Disassembly line balancing problem: A review of the state of the art and future directions. *Int. J. Prod. Res.* **2018**, *57*, 4805–4827. [CrossRef]
35. Hazır, Ö.; Dolgui, A. A Review on Robust Assembly Line Balancing Approaches. *IFAC-Pap.* **2019**, *52*, 987–991. [CrossRef]
36. Hon, K.K.B. Performance and Evaluation of Manufacturing Systems. *CIRP Ann.* **2005**, *54*, 139–154. [CrossRef]
37. Fredendall, L.D.; Gabriel, T.J. Manufacturing complexity: A quantitative measure. In Proceedings of the POMS Conference, Savannah, GA, USA, 4–7 April 2013.
38. Wiendahl, H.-P.; Scholtissek, P. Management and Control of Complexity in Manufacturing. *CIRP Ann.* **1994**, *43*, 533–540. [CrossRef]
39. Efthymiou, K.; Pagoropoulos, A.; Papakostas, N.; Mourtzis, D.; Chryssolouris, G. Manufacturing Systems Complexity Review: Challenges and Outlook. *Procedia CIRP* **2012**, *3*, 644–649. [CrossRef]
40. Suh, N.P. *Complexity: Theory and Applications*; Oxford University Press: New York, NY, USA, 2005.
41. Zhang, Z. Modeling complexity of cellular manufacturing systems. *Appl. Math. Model.* **2011**, *35*, 4189–4195. [CrossRef]
42. Tainter, J.A. Social complexity and sustainability. *Ecol. Complex.* **2006**, *3*, 91–103. [CrossRef]
43. ElMaraghy, W.; ElMaraghy, H.; Tomiyama, T.; Monostori, L. Complexity in engineering design and manufacturing. *CIRP Ann.* **2012**, *61*, 793–814. [CrossRef]
44. Delgoshaei, A.; Gomes, C. A multi-layer perceptron for scheduling cellular manufacturing systems in the presence of unreliable machines and uncertain cost. *Appl. Soft Comput.* **2016**, *49*, 27–55. [CrossRef]
45. Solimanpur, M.; Elmi, A. A tabu search approach for cell scheduling problem with makespan criterion. *Int. J. Prod. Econ.* **2013**, *141*, 639–645. [CrossRef]
46. Saeed, K.; Dvorský, J. *Computer Information Systems and Industrial Management, Proceedings of the 19th International Conference, CISIM 2020, Bialystok, Poland, 16–18 October 2020*; Springer Nature Switzerland AG: Cham, Switzerland, 2020.
47. Hu, X.; Zhang, Y.; Zeng, N.; Wang, D. A Novel Assembly Line Balancing Method Based on PSO Algorithm. *Math. Probl. Eng.* **2014**, *2014*, 743695. [CrossRef]
48. Bhattacharjee, T.K.; Sahu, S. Complexity of single model assembly line balancing problems. *Eng. Costs Prod. Econ.* **1990**, *18*, 203–214. [CrossRef]
49. ElMaraghy, H.A.; Kuzgunkaya, O.; Urbanic, R.J. Manufacturing Systems Configuration Complexity. *CIRP Ann.* **2005**, *54*, 445–450. [CrossRef]
50. Dolgui, A.; Kovalev, S. Scenario based robust line balancing: Computational complexity. *Discret. Appl. Math.* **2012**, *160*, 1955–1963. [CrossRef]
51. Sivadasan, S.; Efstathiou, J.; Calinescu, A.; Huatuco, L.H. Advances on measuring the operational complexity of supplier–customer systems. *Eur. J. Oper. Res.* **2006**, *171*, 208–226. [CrossRef]
52. Isik, F. An entropy-based approach for measuring complexity in supply chains. *Int. J. Prod. Res.* **2009**, *48*, 3681–3696. [CrossRef]
53. Zhang, Z. Manufacturing complexity and its measurement based on entropy models. *Int. J. Adv. Manuf. Technol.* **2012**, *62*, 867–873. [CrossRef]
54. Modrak, V.; Soltysova, Z. Development of operational complexity measure for selection of optimal layout design alternative. *Int. J. Prod. Res.* **2018**, *56*, 7280–7295. [CrossRef]
55. Shannon, C.E. A Mathematical Theory of Communication. *Bell Syst. Tech. J.* **1948**, *27*, 623–656. [CrossRef]
56. Online Computing Software. Available online: http://paseka.epizy.com/ (accessed on 5 December 2021).
57. Hruška, R.; Průša, P.; Babić, D. The use of AHP method for selection of supplier. *Transport* **2014**, *29*, 195–203. [CrossRef]
58. Marović, I.; Perić, M.; Hanak, T. A Multi-Criteria Decision Support Concept for Selecting the Optimal Contractor. *Appl. Sci.* **2021**, *11*, 1660. [CrossRef]
59. Meixner, O. Fuzzy AHP group decision analysis and its application for the evaluation of energy sources. In Proceedings of the 10th International Symposium on the Analytic Hierarchy/Network Process, Pittsburgh, PA, USA, 29 July–1 August 2009; Volume 29, pp. 2–16.
60. Zavadskas, E.K.; Turskis, Z.; Stević, Ž.; Mardani, A. Modelling Procedure for the Selection of Steel Pipes Supplier by Applying Fuzzy AHP Method. *Oper. Res. Eng. Sci. Theory Appl.* **2020**, *3*, 39–53. [CrossRef]
61. Koç, E.; Burhan, H.A. An application of analytic hierarchy process (AHP) in a real world problem of store location selection. *Adv. Manag. Appl. Econ.* **2015**, *5*, 41–50.
62. Imran, M.; Agha, M.H.; Ahmed, W.; Sarkar, B.; Ramzan, M.B. Simultaneous Customers and Supplier's Prioritization: An AHP-Based Fuzzy Inference Decision Support System (AHP-FIDSS). *Int. J. Fuzzy Syst.* **2020**, *22*, 2625–2651. [CrossRef]
63. Saaty, T.L. A scaling method for priorities in hierarchical structures. *J. Math. Psychol.* **1977**, *15*, 234–281. [CrossRef]
64. Saaty, T.L. *The Analytic Hierarchy Process*; McGraw-Hill: New York, NY, USA, 1980.
65. Yan, L.; Irani, S.A. Classroom Tutorial on the Design of a Cellular Manufacturing System. In *Handbook of Cellular Manufacturing Systems*; Irani, S.A., Ed.; John Wiley & Sons, Inc.: Hoboken, NJ, USA, 1999.

applied sciences

Article

Meta-Heuristic Technique-Based Parametric Optimization for Electrochemical Machining of Monel 400 Alloys to Investigate the Material Removal Rate and the Sludge

Vengatajalapathi Nagarajan [1], Ayyappan Solaiyappan [1,*], Siva Kumar Mahalingam [2], Lenin Nagarajan [2], Sachin Salunkhe [2], Emad Abouel Nasr [3], Ragavanantham Shanmugam [4] and Hussein Mohammed Abdel Moneam Hussein [5,6]

[1] Department of Mechanical Engineering, Government College of Technology, Coimbatore 641 013, Tamilnadu, India; vengat92vengat@gmail.com
[2] Department of Mechanical Engineering, Vel Tech Rangarajan Dr. Sagunthala R&D Institute of Science and Technology, Chennai 600 062, Tamilnadu, India; drmsivakumar@veltech.edu.in (S.K.M.); drleninn@veltech.edu.in (L.N.); drsalunkhesachin@veltech.edu.in (S.S.)
[3] Industrial Engineering Department, College of Engineering, King Saud University, Riyadh 11421, Saudi Arabia; eabdelghany@ksu.edu.sa
[4] School of Engineering, Math and Technology, Navajo Technical University, Crownpoint, NM 87313, USA; rags@navajotech.edu
[5] Mechanical Engineering Department, Faculty of Engineering, Helwan University, Cairo 11732, Egypt; hussein@h-eng.helwan.edu.eg
[6] Mechanical Engineering Department, Faculty of Engineering, Ahram Canadian University, 6th of October City 19228, Egypt
* Correspondence: ayyappan.s.mech@gct.ac.in

Citation: Nagarajan, V.; Solaiyappan, A.; Mahalingam, S.K.; Nagarajan, L.; Salunkhe, S.; Nasr, E.A.; Shanmugam, R.; Hussein, H.M.A.M. Meta-Heuristic Technique-Based Parametric Optimization for Electrochemical Machining of Monel 400 Alloys to Investigate the Material Removal Rate and the Sludge. *Appl. Sci.* **2022**, *12*, 2793. https://doi.org/10.3390/app12062793

Academic Editors: Vladimir Modrak and Zuzana Soltysova

Received: 23 January 2022
Accepted: 1 March 2022
Published: 9 March 2022

Publisher's Note: MDPI stays neutral with regard to jurisdictional claims in published maps and institutional affiliations.

Abstract: Electrochemical machining (ECM) is a preferred advanced machining process for machining Monel 400 alloys. During the machining, the toxic nickel hydroxides in the sludge are formed. Therefore, it becomes necessary to determine the optimum ECM process parameters that minimize the nickel presence (NP) emission in the sludge while maximizing the material removal rate (MRR). In this investigation, the predominant ECM process parameters, such as the applied voltage, flow rate, and electrolyte concentration, were controlled to study their effect on the performance measures (i.e., MRR and NP). A meta-heuristic algorithm, the grey wolf optimizer (GWO), was used for the multi-objective optimization of the process parameters for ECM, and its results were compared with the moth-flame optimization (MFO) and particle swarm optimization (PSO) algorithms. It was observed from the surface, main, and interaction plots of this experimentation that all the process variables influenced the objectives significantly. The TOPSIS algorithm was employed to convert multiple objectives into a single objective used in meta-heuristic algorithms. In the convergence plot for the MRR model, the PSO algorithm converged very quickly in 10 iterations, while GWO and MFO took 14 and 64 iterations, respectively. In the case of the NP model, the PSO tool took only 6 iterations to converge, whereas MFO and GWO took 48 and 88 iterations, respectively. However, both MFO and GWO obtained the same solutions of EC = 132.014 g/L, V = 2406 V, and FR = 2.8455 L/min with the best conflicting performances (i.e., MRR = 0.242 g/min and NP = 57.7202 PPM). Hence, it is confirmed that these metaheuristic algorithms of MFO and GWO are more suitable for finding the optimum process parameters for machining Monel 400 alloys with ECM. This work explores a greater scope for the ECM process with better machining performance.

Keywords: electrochemical machining (ECM); material removal rate (MRR); nickel presence (NP); grey wolf optimizer (GWO); moth-flame optimization algorithm (MFO); Monel 400 alloys

1. Introduction

For the last few decades, electrochemical machining (ECM) has been used for machining the macro- and micro-components of tool steel, carbides, superalloys, and titanium alloys. These materials are heavily used in the automotive and aerospace industries, as well as electronics, optics, medical devices, and communications. ECM is used as a high-priority machining tool because of its specific advantages, such as minor tool wear, mechanical forceless machining, no heat-affected zones, high surface quality, low roughness, and stress-free surface products. ECM is the reverse process of electroplating and can machine any hard materials or complicated shapes [1]. In ECM, the workpiece and tool are connected with the anodic pole and the cathodic pole of the electrochemical circuit, respectively, and the electrolyte is pumped through the inter-electrode gap (IEG) between the tool and the workpiece. A high-current DC power supply is allowed to pass through this electrochemical circuit to dissolve the metal from the workpiece in the form of metal hydroxides (i.e., sludge). ECM is an efficient and low-cost machining method for Ni-based alloys [2]. Several works were carried out to study the machinability of Ni alloys by ECM [3–5]. Ni alloys are primarily used in aircrafts, power generation turbines, rocket engines, chemical processing plants, and nuclear power plants. Poor surface traits, a short insert life, high manufacturing costs, and low productivity are associated with the machining of nickel alloys.

Monel 400 alloy is one of Ni alloys that exhibits excellent characteristics of corrosive resistance, strength, and toughness in challenging conditions [6]. When machining Ni-based alloys, ECM generates a large quantity of toxic sludge in the form of Ni and chromium hydroxides aside from gaseous byproducts, acids, sulphates, nitrates, oils, and metal ions. Environmentally sustainable machining of Ni-based alloys in ECM is needed while maintaining productivity and quality [7]. However, the discharge of nickel ions into the electrolyte slurry is significant. Therefore, it is necessary to use ECM with proper cutting conditions so that it produces sludge with fewer toxic byproducts. This hazardous emission may be retained in the electrolyte by repeated use of the electrolyte. However, such a process might decrease the machining performance. Thus, it is proposed to have optimized ECM process parameters to produce a better MRR and fewer hazardous material emissions. To the best of the authors' knowledge, investigating the amount of nickel present in the sludge during electrochemical machining of Monel 400 alloys has not been attempted before. Optimizing the process parameters is an essential task in the ECM process, as the optimum process parameters improve the performance and economics of machining. The selection of ECM process parameters is carried out based on the experience and expertise of the machinist or machining handbooks. The process parameters selected based on the operator's experience rarely assure high efficiency and quality machining. The machining handbooks can be a handy choice for a few applications only. In most cases, the selected optimum process parameters are far from the best, and this hampers the best utilization of ECM. Selecting the optimum values for the process parameters without proper optimization requires elaborate, costly, time-consuming, and tedious experimentation. Therefore, researchers have used different soft computing techniques to optimize the ECM process parameters in various operations [8]. Mukherjee and Chakraborty implemented a biogeography-based optimization (BBO) algorithm in ECM to determine the optimum machining conditions [9]. The BBO algorithm was inspired by the migration behavior of species among the habitats in nature. A genetic algorithm (GA) was integrated with a desirability function (DF) for simultaneous optimization of multiple objectives of the ECM process [10]. The particle swarm optimization (PSO) algorithm was adopted as an optimization strategy for determining the optimum parameters [11]. The PSO algorithm mimics the concept of the social interaction of birds in a flock. To perform the multi-objective optimization, the cuckoo optimization algorithm (COA) was used with the objective function, which was derived from the adaptive neuro-fuzzy inference system (ANFIS) model [12]. The COA was developed based on the obligate brood parasitism of some cuckoo species. A multi-objective Jaya (MO-Jaya) algorithm was implemented in the ECM process to obtain multiple optimal solutions [13]. The differential evolution (DE) algorithm has been used for both the single- and multi-objective optimization of process parameters [14]. Singh and Shukla (2017) considered the ECM process to evaluate the performance of the black hole algorithm (BHA). The performance measures of the MRR and overcut (OC) were taken to obtain the optimum machining parameters using the BHA [15]. The harmony search (HS) algorithm was combined with the DF so that an HS-DF optimizer was proposed to optimize the ECM process parameters [16]. Diyaley and Chakraborty conducted a comparative analysis with meta-heuristics such as the firefly algorithm (FA), DE algorithm, ant colony optimization (ACO) algorithm, and teaching-learning-based optimization (TLBO) algorithm for the optimization of various control parameters of an ECM process [17]. Apart from these meta-heuristics, artificial neural networks (ANNs) and fuzzy logic (FL) were also employed to model the ECM experiments and subsequent optimization process [18,19]. Several optimization algorithms have been reported in the literature, and their efficiency was tested with many benchmark functions and real-time applications. The grey wolf optimizer (GWO), a recently developed meta-heuristic, was proposed by Mirjalili et.al [20]. The GWO is inspired by the hunting behavior of the grey wolves in nature, and it ensures a better trade-off between the exploration and exploitation abilities of the algorithm. It was applied in many engineering applications, emphasizing its efficiency [21]. The significant characteristic of this algorithm is that it does not converge toward some local optima and helps store the best possible solutions obtained so far by its social hierarchy nature [22]. Kharwar and Verma implemented GWO to minimize milling performances (i.e., MRR, cutting force, and surface roughness) and proved the application potential in a manufacturing environment [23]. Omkar and Shalaka optimized the wire electrical discharge machining (WEDM) parameters with the GWO algorithm [24]. Shankar and Ankan experimented with the efficiency of the GWO algorithm for parametric optimization of the abrasive water jet machining (AWJM) process and found it to be successful [25].

A performance comparison of the response surface methodology (RSM), genetic algorithm (GA), and GWO algorithm was conducted for prediction of the surface roughness in ball-end milling of hardened steel, and GWO was found to be superior [26]. The grey relational analysis (GRA)-based GWO was implemented for milling experiments to determine the machining conditions of the spindle speed, feed rate, and depth of cut for the optimum surface roughness, cutting force, and MRR [27]. Mirjalili proposed another meta-heuristic, the moth-flame optimization (MFO) algorithm, which was inspired by the navigation method of moths in nature [28]. It gained attention for solving engineering optimization problems immediately. The improved MFO algorithm was proposed by Li et al. to maintain a balance between global and local searching [29]. A milling optimization problem was solved to emphasize its effectiveness in solving manufacturing problems [30]. The MFO algorithm was implemented to identify the optimal set of turning parameters to minimize the machinability indices. Its effectiveness was compared against the genetic algorithm (GA), grasshopper algorithm, GWO, and PSO algorithms and to be found superior to the other algorithms [31]. The optimization of the surface roughness in deep hole drilling was performed with the MFO algorithm [32]. The MFO algorithm was employed to optimize the process parameters of plasma arc cutting (PAC) of Inconel 718 superalloy [33]. Based on the literature on GWO, this algorithm is not utilized much in manufacturing optimization problems. These reviews conclude that the recently developed intelligent techniques are more advanced, and their scope of implementation in broader manufacturing applications can be explored. It is also noted that the efficiency of these new evolutionary algorithms (i.e., GWO and MFO algorithms) were not explored in the ECM process. The famous PSO algorithm is still being used to optimize machining parameters [34,35]. Hence, PSO is considered a benchmark algorithm to compare GWO and MFO algorithms in terms of convergence, computational time, and the actual Pareto optimal front.

2. Objectives

To determine the optimum ECM parameters for machining Monel 400 alloys for the maximum MRR and minimum nickel toxic emission in the electrolyte, meta-heuristics such as GWO, MFO, and PSO algorithms were employed.

3. Materials and Experimentation

The commercially available nickel-based alloy Monel 400 was used as a test specimen in this investigation. The chemical composition of the Monel 400 alloys is shown in Table 1 [36]. The machinability of Monel 400 alloy is very difficult, as it work hardens during machining. Therefore, it is a tough task to machine these alloys using conventional machining techniques.

Table 1. Chemical composition of Monel 400 alloys.

Monel 400 Alloys								
C	Si	Mn	P	S	Cr	Mo	Fe	V
0.047	0.172	1.03	0.012	0.01	0.1	0.1	1.66	0.029
W	Cu	Al	Co	Nb	Ti	Mg	Ni	
0.1	29.24	0.01	0.103	0.1	0.047	0.031	67.4	

Composition and weight (%)

The experiments were carried out on the eco-friendly electrochemical machining (EECM) tool. The schematic diagram of the EECM tool is shown in Figure 1. EECM comprises a power supply system, electrolyte supply system, filtering system, tool feed mechanism, work holding and position system, control panel, frame, and housing. The electrolyte (e.g., NaCl aqueous solution) is allowed to flow at the rate between 1 L/min and 10 L/min through the inter-electrode gap (IEG) of 0.1–0.6 mm. The direct current (DC) potential (11–15 V) with a current density of 20–110 A/cm^2 is applied across IEG between a cathodic copper tool and an anodic workpiece. At low current densities, the MRR is low [37]. The higher current density of 110 A/cm^2 was used to attain high current efficiency. According to Faraday's laws of electrolysis, the anodic dissolution rate of workpiece depends on the electrochemical properties of the workpiece metal, the electrolyte properties, and the power supply conditions. The tool is fed against the anodic workpiece, which is firmly fixed on the vice. The machining conditions of IEG, voltage (V), and current (C) are set in the control panel. The filtration system consists of layers of bio-absorbents compartments and membrane filter. A photograph of the indigenously developed EECM is shown in Figure 2. The tool's diameter and electrolyte exit hole were 8 mm and 2 mm, respectively. The operating conditions of EECM are summarized in Table 2.

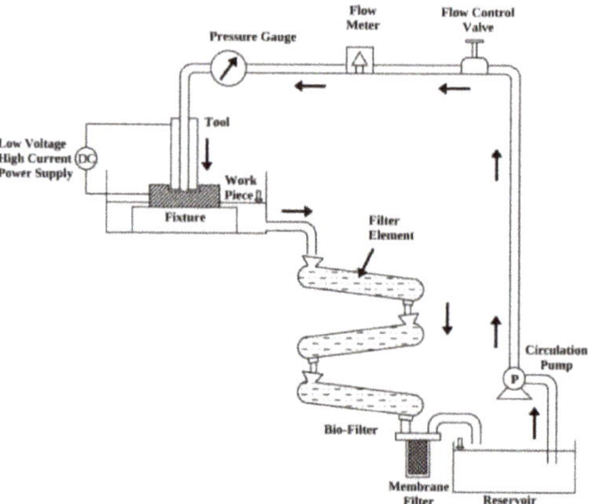

Figure 1. Schematic diagram of EECM.

Figure 2. EECM experimental set-up.

Table 2. Factors and conditions of ECM experiments.

Factors	Type	Condition/size
Work piece	Monel 400 alloys	Hardened material
Electrolyte	NaCl	130–190 g/L
Tool	Copper	C101
Voltage	DC	11, 13, 15 V
Tool feed rate	Horizontal feed	0.1 mm/min
IEG		0.1 mm
Current	DC	50 A
Flow rate		1–3 L/min
Machining time		5 min

Experimental Design and Measurements

The MRR is directly proportionate to the feed rate of the tool [36]. A high feed rate leads to electrolyte boiling or choking in the tool and workpiece gap [37]. Therefore, the tool feed rate was set to the possible minimum of 0.1 mm/min to have stable movement of the tool through the workpiece. When the applied voltage is high, the current machining increases to a high MRR. A low voltage results in poor machining performance [10,11]. The voltage range of 11–16 V was considered in many such investigations in ECM to have better performance in terms of the MRR and surface finish [38,39]. Therefore, the voltage range (10–15 V) available in the set-up was used for this investigation. When the voltage is raised above a particular level, it increases the hydrogen gas bubble generation at the tool electrode, resulting in high resistivity of the electrolyte and a decrease in the current density at the workpiece [40–43]. A high concentration decreases the mobility of ions, which results in poor anodic dissolution. The concentration levels in the range of around 100–200 g/L influence the MRR significantly [10]. According to the literature, the main process parameters governing the ECM are the voltage (V), flow rate, and EC. Therefore, these parameters were considered in this investigation, and their levels are shown in Table 3.

Table 3. Process parameters and levels.

S.No.	Process Parameters	Levels				
		−2	−1	0	1	2
1	Voltage (V)	11	12	13	14	15
2	Electrolyte concentration (g/L)	130	145	160	175	190
3	Flow rate (L/min)	1.0	1.5	2.0	2.5	3.0

The IEG was set to 0.1 mm, and a current of 50 A was set in the DC rectifier. The machining was performed for 5 min. The central composite design (CCD) of the response surface methodology (RSM) was adopted with the help of Minitab-17 software for the process parameters, and its levels are shown in Table 3. In this work, 20 experimental tests were carried out with different parameter combinations as shown in Figure 3. According to the concept of CCD, the center point combination was repeated 5 more times during experimentation to reduce the pure error.

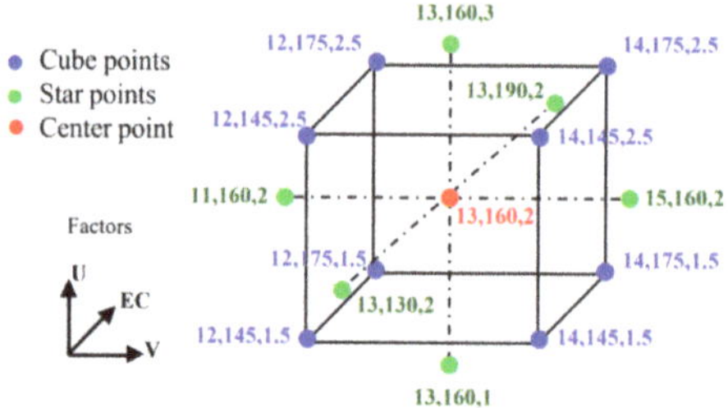

Figure 3. Central composite design for the experimentation.

A HCl concentration of 1% was added to the NaCl electrolyte to minimize the sludge formation in the IEG. The electrolyte was pneumatically pumped from a stainless steel reservoir. A 191E water analysis kit (Environmental & Scientific Instruments Co, Haryana, India.) was used to observe the electrolyte's pH, conductance, and temperature.

The sludge discharged from ECM was tested with atomic absorption spectroscopy (Agilent Technologies, Bangalore, India), which is shown in Figure 4. The results show the nickel content of 250 mg/L in the sludge. The precipitate of the electrolyte was tested via energy dispersive X-ray spectroscopy (EDAX, JEOL-JSM-6390, Chennai, India) and confirmed the presence of nickel. Figure 5 represents the EDAX results of the electrolyte sludge. The peaks in the figure confirm the presence of more nickel, chloride, and sodium particles. Therefore, in this invetigation, the experimental results for MRR and Nickel Presence (NP) were observed and tabulated in Table 4.

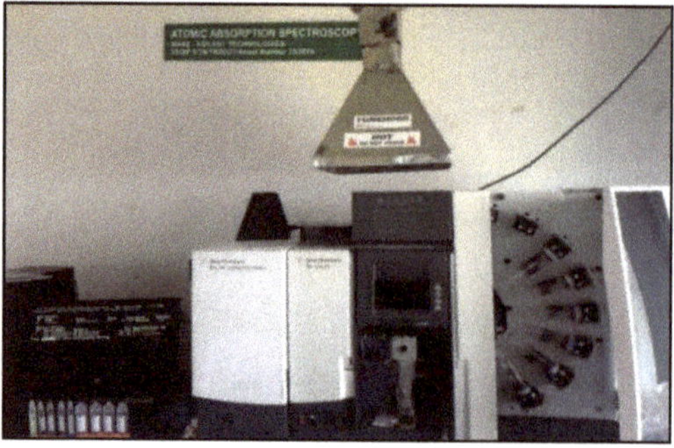

Figure 4. Atomic absorption spectroscopy.

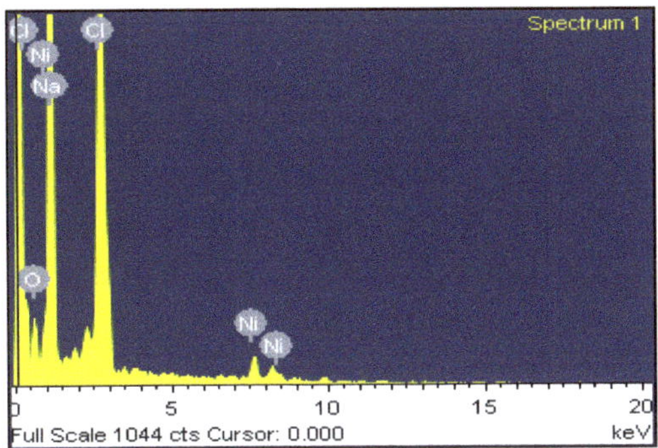

Figure 5. EDAX spectrum for nickel in precipitate.

Table 4. Experimental design and performance results.

Exp.No.	EC (g/L)	V (V)	FR (L/min)	Experimental Value		Calculated Values	
				MRR (g/min)	NP (PPM)	MRR (g/min)	NP (PPM)
1	190	13	2.0	0.225	65.08	0.20327	64.44446
2	160	13	2.0	0.219	57.26	0.20438	56.57423
3	145	12	2.5	0.204	58.06	0.19676	56.35493
4	175	14	2.5	0.215	67.45	0.22844	68.97187
5	160	13	2.0	0.208	55.05	0.20438	56.57423
6	130	13	2.0	0.129	50.65	0.15130	51.46866
7	160	13	1.0	0.062	55.48	0.07344	55.19648
8	160	13	2.0	0.216	56.45	0.20438	56.57423
9	175	12	1.5	0.157	58.63	0.16119	58.03033
10	160	11	2.0	0.142	57.29	0.14094	59.88398
11	160	13	2.0	0.171	56.45	0.20438	56.57423
12	160	13	3.0	0.216	66.75	0.20555	67.22471
13	145	12	1.5	0.114	55.24	0.09985	53.53437
14	145	14	2.5	0.243	60.08	0.23782	60.49203
15	160	13	2.0	0.209	55.40	0.20438	56.57423
16	175	14	1.5	0.187	58.24	0.19324	59.76420
17	160	15	2.0	0.212	68.16	0.21405	65.75495
18	145	14	1.5	0.132	54.03	0.11226	55.20372
19	160	13	2.0	0.203	58.65	0.20438	56.57423
20	175	12	2.5	0.149	66.13	0.16774	64.77025

4. Optimization Techniques and Procedures

The following details describe the development of mathematical models representing EECM characteristics for the MRR and the amount of nickel present, as well as the GWO and MFO techniques for optimizing the process parameters.

4.1. Mathematical Modeling of Experimentation

To study the effect of the EECM process parameters on the MRR and the discharge of nickel elements into the electrolyte, a regression model was developed to calculate the response value (as the output) in terms of different parameters (as the input). In this work, the quadratic form of the regression equation, shown in Equation (1), was established using the experimental data. The regression coefficients obtained for both the MRR and NP are shown in Table 5:

$$CV_{ij} = a_j + b_j * X_{1i} + c_j * X_{2i} + d_j * X_{3i} + e_j * X_{1i} * X_{2i} + f_j * X_{1i} * X_{3i} + g_j * X_{1i} * X_{3i} + h_j * X_{1i}^2 + i_j * X_{2i}^2 + j_j * X_{3i}^2 \qquad (1)$$

where $CVij$ is the predicted value of the ith trial for the jth response, $a, b, \ldots, j$ are the coefficients of the variables, and $X1, X2,$ and $X3$ are independent variables.

Table 5. Regression coefficients for MRR and NP models.

Response	a	b	c	d	e	f	g	h	i	j
MRR	-2.381	0.0123	0.112	0.621	0.0003	-0.003	0.0143	0.00003	-0.007	-0.065
NP	389	-0.551	-41.8	-49.5	0.0011	0.131	1.23	0.00154	1.561	4.64

The regression models for the given objectives can be formulated as in Equations (2) and (3):

$$
\begin{aligned}
MRR = -2.381 \quad &+0.0123 * EC + 0.112 * V + 0.621 * FR + 0.0003 * EC * V \\
&-0.003 * EC * FR + 0.0143 * V * FR - 0.00003 * EC^2 \\
&-0.007 * V^2 - 0.065 * FR^2
\end{aligned}
\tag{2}
$$

$$
\begin{aligned}
NP = \quad &389 - 0.551 * EC - 41.8 * V - 49.5 * FR + 0.0011 * EC * V + 0.131 \\
&* EC * FR + 1.23 * V * FR + 0.00154 * EC^2 + 1.561 * V^2 \\
&+4.64 * FR^2
\end{aligned}
\tag{3}
$$

4.2. TOPSIS Method for Multi-Objective Optimization

The Technique for Order of Preference by Similarity to Ideal Solution (TOPSIS) is a method used for converting multiple objectives into a single objective, which is consequently applied in optimization techniques for decision making [44,45]. TOPSIS is based on choosing the alternative with the shortest geometric distance from the positive ideal solution and the longest geometric distance from the negative ideal solution. By implementing the TOPSIS method, multiple objectives are converted into a single objective value. Algorithm 1 shows the the pseudo-code for TOPSIS method.

Algorithm 1 Pseudo-code for TOPSIS.

1: Read alternate and objectives matrix—O_{ij} with weights (W_j) and types of objectives (ot_j)
2: For each alternative (i = 1, 2, 3, $\ldots$, m) and objective (j = 1, 2, 3, $\ldots$, n)
3: Compute Normalized value of O_{ij} using $N_{ij} = \dfrac{O_{ij}}{\sqrt{\sum_{i=1}^{m} O_{ij}^2}}$
4: Calculate Performance matrix (A_{ij}) using $A_{ij} = N_{ij} * W_j$
5: End
6: For each objective (j = 1, 2, 3, $\ldots$, n)
7: Determine positive ideal (P_j) and negative ideal solution (M_j)
8: For minimization objective—$P_j = \min\limits_{1 < i < m} \left(A_{ij} \right)$ and $M_j = \max\limits_{1 < i < m} \left(A_{ij} \right)$
9: For maximization objective—$P_j = \max\limits_{1 < i < m} \left(A_{ij} \right)$ and $M_j = \min\limits_{1 < i < m} \left(A_{ij} \right)$
10: End
11: For each alternative (i = 1, 2, 3, $\ldots$, m)
12: Calculated Ideal (SP_i) and negative ideal separation (SM_i)
$SP_i = \sqrt{\sum_{j=1}^{n} \left(A_{ij} - P_j \right)^2}$ $SM_i = \sqrt{\sum_{j=1}^{n} \left(A_{ij} - M_j \right)^2}$
13: Determine relative closeness (R_i)
14: End
15: Rank alternatives w.r.t. R_i in descending order

4.3. Grey Wolf Optimizer

The grey wolf optimizer (GWO) was inspired by the hunting behavior of grey wolves in nature. Grey wolves usually live in a pack of 5–12 members with their dominant social hierarchy [20]. The alpha (α) wolf, the pack leader, makes decisions about hunting. The beta (β) wolf helps the alpha wolf in making decisions. The omega (ω) wolves, the lowest in ranking, are responsible for submitting information to all the wolves. Others, called delta (δ)

wolves, should respect alpha and beta wolves and dominate the omega wolves. The three steps in the hunting process of grey wolves are (1) tracking the prey, (2) encircling the prey, and (3) attacking the prey. GWO is simulated with the mathematical representation of the hunting process of the grey wolves to solve complex engineering problems. The optimal solution to the problem is considered the prey. To mathematically model the hunting process of these grey wolves when designing GWO, the alpha (α) wolf is considered the fittest solution. Consequently, the second- and third-best solutions are represented by beta (β) and delta (δ) wolves, respectively. The remaining candidate solutions are considered to be omega (ω) wolves.

The first step in the hunting process is encircling the prey. The mathematical formulation to mimic the encircling process for the prey and wolf are represented in the Equations (4) and (5):

$$\vec{D} = \left| \vec{C} * \vec{X_p}\,(t) - \vec{X}\,(t) \right| \tag{4}$$

$$\vec{X}\,(t+1) = \vec{X_p}\,(t) - \vec{A} * \vec{D} \tag{5}$$

where $\vec{D}$ is a distance vector between the position of the prey and grey wolf. $\vec{C}$ and $\vec{A}$ are coefficient vectors, t is the current iteration, and $\vec{X_p}$ and $\vec{X}$ are the positions of prey and a randomly chosen grey wolf, respectively [20]. The coefficients $\vec{C}$ and $\vec{A}$ are determined using the following formulas:

$$\vec{A} = 2\vec{a} * \vec{r_1} - \vec{a} \tag{6}$$

$$\vec{C} = 2 * \vec{r_2} \tag{7}$$

where $\vec{a}$ is a decrease from 2 to 0 linearly over the course of iterations and $\vec{r_1}$ and $\vec{r_2}$ are random vectors (0, 1).

A grey wolf is in the position of $\vec{X}$, updating its position $\vec{X}\,(t+1)$ according to the position of the prey $\vec{X_p}$. The position can be updated with respect to the current position by adjusting the values of the $\vec{A}$ and $\vec{C}$ vectors. This process enables the GWO to search the n-dimensional solution space of the given problem efficiently. The alpha (α), beta (β), and delta (δ) wolves guide the hunting process, and their positions are represented mathematically as in Equations (8)–(14), while the other wolves (ω) update their positions randomly:

$$\vec{D_\alpha} = \left| \vec{C_1} * \vec{X_\alpha} - \vec{X} \right| \tag{8}$$

$$\vec{D_\beta} = \left| \vec{C_1} * \vec{X_\beta} - \vec{X} \right| \tag{9}$$

$$\vec{D_\delta} = \left| \vec{C_1} * \vec{X_\delta} - \vec{X} \right| \tag{10}$$

$$\vec{X_1} = \vec{X_\alpha} - \vec{A_1} * \vec{D_\alpha} \tag{11}$$

$$\vec{X_2} = \vec{X_\beta} - \vec{A_2} * \vec{D_\beta} \tag{12}$$

$$\vec{X_3} = \vec{X_\delta} - \vec{A_3} * \vec{D_\delta} \tag{13}$$

$$\vec{X}\,(t+1) = \frac{\vec{X_1} + \vec{X_2} + \vec{X_3}}{3} \tag{14}$$

As the value of $\vec{}$ decreases, the search radius of the grey wolves reduces, and they get closer to the prey and attack over the last iterations of the GWO. This ensures the proper trade-off between exploration and exploitation by focusing on exploration initially and exploitation in the last iterations.

4.4. Moth-Flame Optimization Algorithm (MFO)

The moth-flame optimization (MFO) algorithm takes inspiration from the navigational mechanism (i.e., transverse orientation) of moths during night flights. The moth flies by maintaining a fixed angle relative to the moon. Since the moon is so far away from the moth, this mechanism assures the linear flight of moths. Although a lateral orientation is practical, moths are often trapped to repeatedly circle many close artificial or natural point light sources until they are exhausted. This behavior of moths is shown in Figure 6.

Figure 6. Spiral flight path of moths near artificial or natural point light sources [20].

Mathematical Formulation of the MFO Algorithm

According to the MFO algorithm, the moth is considered the candidate solution to the problem. The variables are represented by the positions of the moths in the solution space. Moths can fly in the solution space by changing the position vectors. Since this algorithm is a swarm-based intelligence optimization algorithm, the position vectors of the moth population can be represented in the following matrix (Equation (15)):

$$M = \begin{bmatrix} m_{1,1} & \cdots & m_{1,d} \\ \vdots & \ddots & \vdots \\ m_{n,1} & \cdots & m_{n,d} \end{bmatrix} \tag{15}$$

where n is the number of moths and d represents the number of dimension variables to be solved. The following array represents the list of fitness value vectors corresponding to all moths:

$$OM = \begin{vmatrix} OM_1 \\ \vdots \\ OM_n \end{vmatrix} \tag{16}$$

Each moth updates its position with the corresponding unique flame to avoid the algorithm being trapped in the optimal local value, which significantly supports the algorithm's exploration ability. Hence, the flame positions in the solution space corresponding to its moths are represented in the following array (Equation (6)):

$$F = \begin{bmatrix} f_{1,1} & \cdots & f_{1,d} \\ \vdots & \ddots & \vdots \\ f_{n,1} & \cdots & f_{n,d} \end{bmatrix} \tag{17}$$

For these flames, the corresponding column of the fitness value vectors is represented as follows:

$$OF = \begin{vmatrix} OF_1 \\ \vdots \\ OF_n \end{vmatrix} \tag{18}$$

The position update strategies for both the moth and flame matrices are different during the iteration process. The moths are the solution points that move within the solution space, and the flames are the best solution points obtained by moths iteratively so far. The best solutions of the moths are updated to the positions of the flames in the next iteration. With this approach, the MFO algorithm can find the optimal global solution. The position update behavior of each moth relative to a flame is expressed mathematically by the following equation:

$$M_i = S(M_i, F_j) \tag{19}$$

where M_i is the ith moth, F_j is the jth flame, and S is the spiral function.

The spiral function's initial point should start from the position of the moth, and the endpoint should be at the position of the flame. The range fluctuation of the spiral function should not exceed the boundaries of the search space. According to the above conditions, the logarithmic spiral function of the moth's flight path is defined as in Equation (20):

$$S(M_i, F_j) = D_i * e^{bt} * \cos(2\pi t) + F_j \tag{20}$$

where D_i is the linear distance of the ith moth for the jth flame and b is an index of the shape of the logarithmic spiral. The magnitude of t is represented by Equation (21), where a is represented by Equation (22):

$$t = (a - 1) * rand + 1 \tag{21}$$

The path coefficient $t \in [r, 1]$ represents the distance between the moth and the flame's position in the next optimization iteration. The variable r decreases linearly from -1 to -2 as the number of iterations in the optimization iteration increases. The coefficient from $t = -1$ to -2 represents that the moths' position is close to the flame, and $t = 1$ shows that the moths are farther from the flame:

$$a = -1 + Iteration * \left(-\frac{1}{T_{max}}\right) \tag{22}$$

Di is expressed as follows:

$$D_i = |F_j - M_i| \tag{23}$$

The above equations ensure the algorithm's balanced global and local search capabilities. When the value of t is smaller, the moth converges to the flame. As the moth gets closer to the flame, its position around the flame is updated more quickly. After each iteration, the flames are updated based on fitness values. In the next iteration, the moth updates its position according to the updated sequence of the flame. The first moth updates its position with the flame's best fitness value, and the last moth updates its position with the worst fitness value in the list. Hence, the flame position matrix F contains the optimal solutions of the current iteration. The number of flames can be reduced adaptively in the iterative process to balance the algorithm's global search and local search capability in the solution space as follows:

$$flame_{no} = round\left(N - l * \frac{N-1}{T}\right) \tag{24}$$

where l is the current iteration number, N is the initial maximum number of flames, and T represents the maximum number of iterations.

5. ANOVA and Parametric Influence on Performances

Analysis of variance (ANOVA) on the process parameters and the effects of the main process parameters on the performance measures was performed. This and Fisher's test (F-ratio) were performed to justify the adequacy of the developed regression models as in Equations (2) and (3). Values of "Prob > F" less than 0.05 indicate the model's significance and terms. Values greater than 0.1 indicate that the model terms are insignificant.

The model F-values in Table 6 indicate that the developed models were very significant. The two-way interaction terms such as EC*V, EC*EC, and V*FR for the MRR and NP models were insignificant when the values of "Prob > F" were greater than 0.05. The terms V*V and EC*FR for the MRR and NP models, respectively, were not significant, but the coefficient of determination values for both models were good, as shown in Table 7. The normal probability curves in Figure 7 confirm the adequacy of the model equations. The high coefficient of determination values (R2 = 90.34% for the MRR and R2 = 92.29% for the NP) for both responses confirm its adequacy for optimization study.

Table 6. Analysis of variance for quadratic models of MRR and NP.

Source	DF		Adj SS		Adj MS		F-Value		*p*-Value	
	MRR	NP	MRR	NP	MRR	NP	MRR	NP	MRR	NP
Model	9	9	0.037278	437.944	0.004142	48.660	10.39	13.29	0.001	0.000
Linear	3	3	0.025497	347.518	0.008499	115.839	21.32	31.64	0.000	0.000
EC	1	1	0.002700	168.372	0.002700	168.372	6.77	45.99	0.026	0.000
V	1	1	0.005345	34.468	0.005345	34.468	13.41	9.42	0.004	0.012
FR	1	1	0.017452	144.678	0.017452	144.678	43.78	39.52	0.000	0.000
Square	3	3	0.007095	79.698	0.002365	26.566	5.93	7.26	0.014	0.007
EC*EC	1	1	0.001154	3.003	0.001154	3.003	2.89	0.82	0.120	0.386
V*V	1	1	0.001136	61.290	0.001136	61.290	2.85	16.74	0.122	0.002
FR*FR	1	1	0.006615	33.779	0.006615	33.779	16.59	9.23	0.002	0.013
2-Way Interaction	3	3	0.004685	10.728	0.001562	3.576	3.92	0.98	0.044	0.442
EC*V	1	1	0.000193	0.002	0.000193	0.002	0.48	0.00	0.502	0.981
EC*FR	1	1	0.004082	7.681	0.004082	7.681	10.24	2.10	0.009	0.178
V*FR	1	1	0.000410	3.045	0.000410	3.045	1.03	0.83	0.334	0.383
Error	10	10	0.003987	36.607	0.000399	3.661				
Lack of Fit	5	5	0.002498	28.127	0.000500	5.625	1.68	3.32	0.292	0.107
Pure Error	5	5	0.001489	8.480	0.000298	1.696				
Total	19	19	0.041264	474.551						

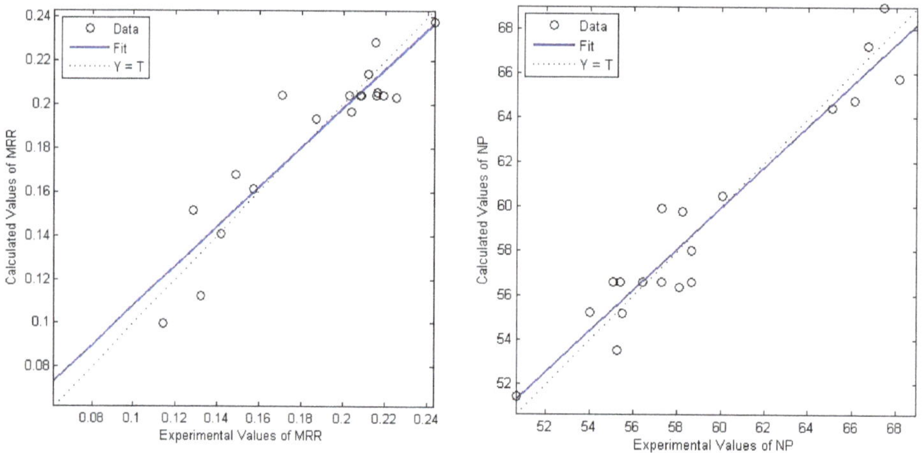

Figure 7. Normal probability plots for MRR and NP.

Table 7. Model summary.

Model	S	R-sq	R-sq (adj)	R-sq (pred)
MRR	0.01997	90.34%	81.64%	44.22%
NP	1.91331	92.29%	85.34%	47.85%

Parametric Influence on the Performance Measures

Figure 8 depicts the surface plots of the MRR and NP responses for electrochemical machining of Monel 400 alloys. In each surface plot, one parameter was kept constant, and other parameters were varied with defined levels. The surface plots show that the machining parameters V, FR, and EC influenced the material removal and the discharge of nickel elements in the sludge significantly. It can be observed that the change in electrolyte concentration from 150 g/L to 190 g/L gradually increased the sludge removal. A higher concentration helped increase the speed of the electrochemical reactions and resulted in more removal (Zhang, 2010; Ayyappan and Sivakumar, 2015). A lower concentration led to a lower ion content, which reduced the rate of the electrochemical reaction (Trimmer, Hudson, Kock, and Schuster, 2003). A flow rate above 1.5 L/min removed the sludge from the gap between the tool and the workpiece better and exposed the specimen's new surface for different electrochemical reactions (Ayyappan and Sivakumar, 2014). High flow rates increased the nickel ions in the electrolyte discharge and the MRR (Bhattacharyya and Sorkhel, 1999).

Figure 9 represents the main effect plots of the MRR and NP responses using MiniTab17 software. It was noticed that the electrolyte flow rate and electrolyte concentration were the influencing factors for the MRR and NP, respectively. Variations in the voltage also produced significant results in the responses.

Figure 10 shows the interaction plots of the ECM parameters for different process attributes. As shown, the interaction terms such as EC*V and V*FR had a lean influence on the responses. The MRR and NP models were insignificant when the "Prob > F" values were more significant than 0.05. The term EC*FR influenced the NP at a high flow rate of 3 L/min.

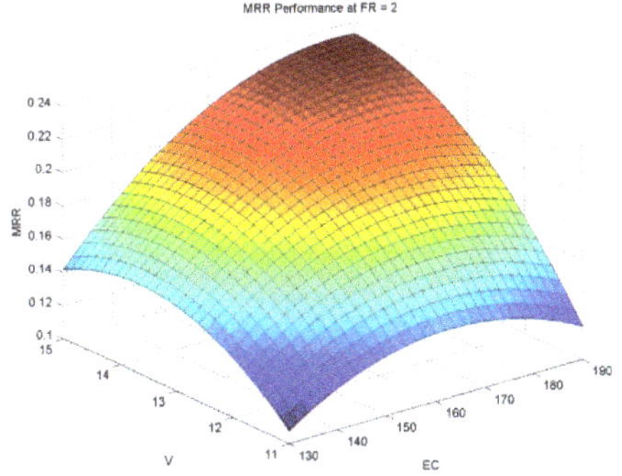

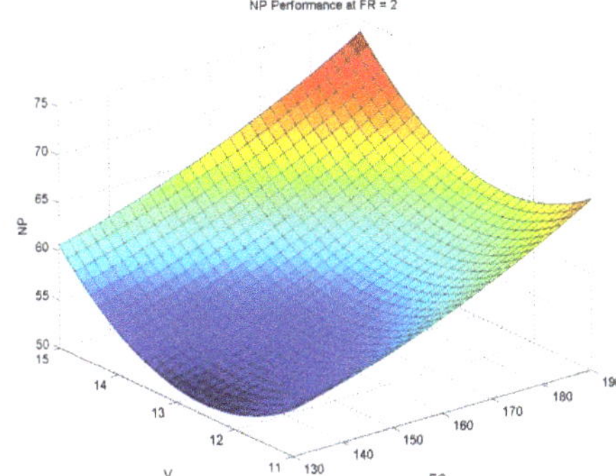

Figure 8. *Cont.*

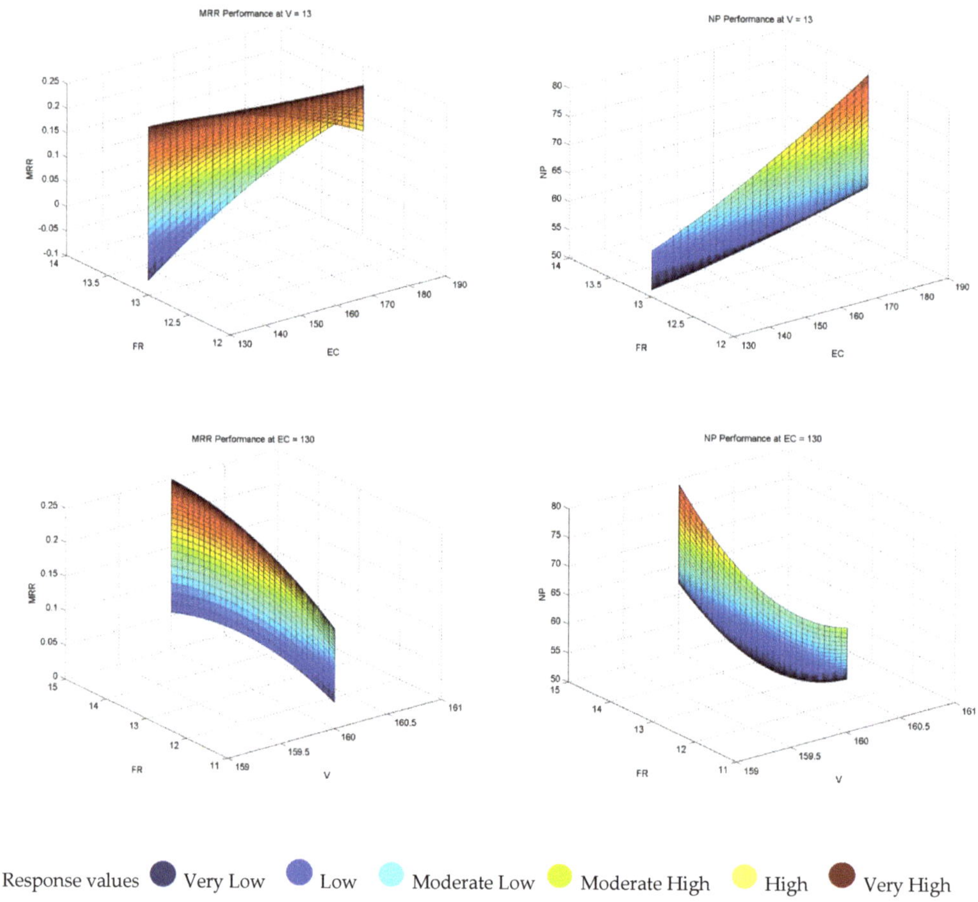

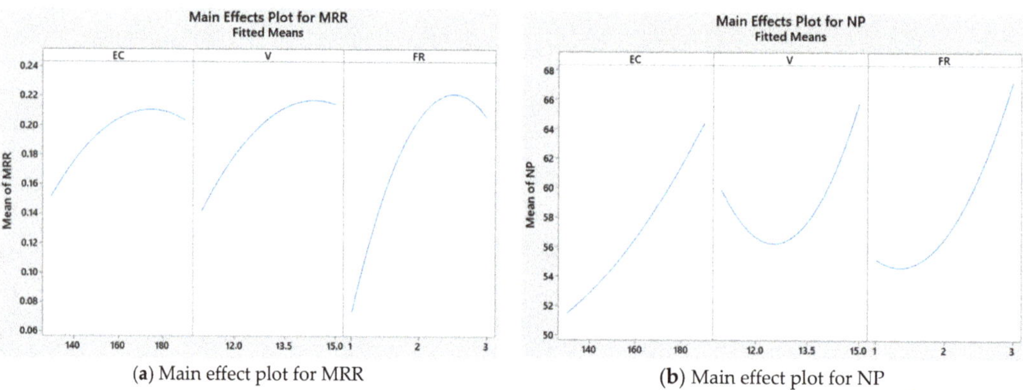

Figure 8. Surface plot of responses for various combinations of parameters.

(**a**) Main effect plot for MRR

(**b**) Main effect plot for NP

Figure 9. Main effects plots of MRR and NP.

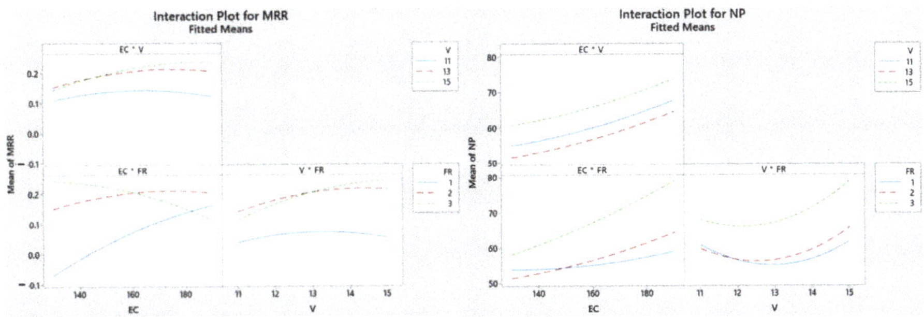

Figure 10. Interaction plots for MRR and NP.

6. Multi-Objective Optimization of ECM Process Parameters

6.1. RSM Optimization Tool

Since the RSM as a DOE strategy was employed in this work, the optimum process parameters were obtained with its built-in optimization tool. The RSM optimization tool used the steepest ascent (SA) method combined with the desirability function (DF) for finding the optimum parameters. The variables and response constraints were set as shown in Table 8.

Table 8. Limits of process and performance parameters.

Response	Goal	Lower	Target	Upper	Weight	Importance
NP	Minimum		50.6452	68.1613	1	1
MRR	Maximum	0.062	0.2429		1	1

The optimization results are shown in Table 9, and the optimization plot is shown in Figure 11.

Table 9. Multiple response prediction results using RSM.

Solution	EC	V	FR	NP Fit	MRR Fit	Composite Desirability
1	130	12.8586	2.57576	53.9394	0.217350	0.835117

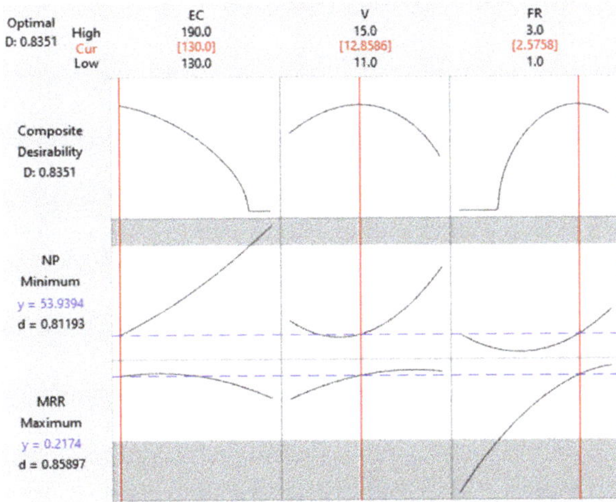

Figure 11. Response optimization plot.

6.2. Applications of Meta-Heuristics for Optimization of the Process Parameters

To apply the various meta-heuristic algorithms such as MFO, GWO, and PSO in this work, the TOPSIS method was adopted to convert the multiple objectives into a single objective. A performance comparison of these algorithms was carried out. The Algorithms 2–4 show the pseudo-codes of the MFO, GWO, and PSO respectively.

Algorithm 2 Grey Wolf Optimization Algorithm

1: Initialize no. of grey wolf (X_{ij}—i =1,2,..nw and j = 1,2,..nd)
2: While ($it < nitr$)
3: Determine the fitness function F_{ik}
4: Calculate Pareto optimal distance f_i
5: Sort f_i in descending order and set as sf_i
6: Store the first wolf's data as X_{it}. and F_{it}.
7: Using the sorted data, assign X_a. = X_1, X_b. = X_2. and X_d. = X_3.
8: Compute $a = 2\text{-}it*(2/nitr)$
9: For each wolf, Update the position using $A1 = 2*a*rand()\text{-}a$
10: $C1 = 2*rand()$
11: D_a. = $abs(C1*X_a..X_i.)$
12: X_1. = $X_a.\text{-}A1*D_a$.
13: $A2 = 2*a*rand()\text{-}a$
14: $C2 = 2*rand()$
15: D_b. = $abs(C1*X_b..X_i.)$
16: X_2. = $X_b.\text{-}A2*D_b$.
17: $A3 = 2*a*rand()\text{-}a$
18: $C3 = 2*rand()$
19: D_d. = $abs(C1*X_d..X_i.)$
20: X_3. = $X_d.\text{-}A3*D_d$.
21: X_i. = $(X_1. + X_2. + X_3.)/3$
22: Check X_i. within bounds
23: End
24: End
25: Using TOPSIS method convert F_{it}. into f_i
27: Sort fi in descending order and display the first wolf's data (optimum data)
28: Print the best solution

Algorithm 3 Moth-Flame Optimization Algorithm

1: Initialize the parameters for Moth-flame
2: Initialize Moth position Mi randomly
3: For each $i = 1$:n Calculate the fitness valute fi
4: End
5: While ($i \leq i_{max}$)
6: Update the position of Mi
7: Calculate the no. of flames
8: Compute the fitness value fi
9: If ($i = 1$) then $F = sort\ (M)\ OF = sort\ (OM)$
10: Else $F = sort\ (Mt\text{-}1,\ Mt)\ OF = sort\ (Mt\text{-}1,\ Mt)$
11: End
12: For each $i = 1$:n
13: For each $j = 1$:d Update the values of r and t
14: Calculate the value of D w.r.t. corresponding Moth
15: Update $M(i,j)$ w.r.t. corresponding Moth
16: End
17: End
18: End
19: Print the best solution

Algorithm 4 Particle Swarm Optimization Algorithm

1: Initialize Particle Position P
2: For $i = 1$ to itr_{max}
3: For each particle p in P
4: Evaluate $fp = f(p)$
5: If fp is better than $f(pB)$; $pB = p$;
6: End
7: End
8: $gB =$ best p in P
9: For each particle p in P
10: Compute $v = v + c_1 *rand()*(pB - p) + c_2 *rand()*(gB - p)$
11: Update $p = p + v$
12: End
13: End
14: Print the best solution

The terms in GWO and their equivalent meanings in the optimization problem are listed in Table 10.

Table 10. Terms in GWO and optimization problems.

Optimization Problem	Grey Wolf Optimization (GWO) Algorithm
Number of solutions	Number of grey wolves (i = 1, 2, … nw)
Combination of parameters within their bounds	Position of grey wolf (Xij)
Number of parameters, factors, and independent variables	Number of dimensions involved in defining the position of a wolf (j = 1, 2, … nd)
Value of best parameters	Position of prey (Xbest)
Response value, output, and dependant variable	Fitness of grey wolf (Fik)
First best three solution's parameters	Position of alpha, beta, and delta grey wolves (Xa., Xb., and Xd.)
First best three solution's fitness values	Fitness of alpha, beta, and delta grey wolves (Fa., Fb., and Fd.)
Except for best first three solutions	Omega grey wolf

Similarly, the equivalent terms for the MFO and PSO algorithms for an optimization problem are shown in Table 11. The algorithm parameters are shown in Table 12.

Table 11. Terms in MFO and PSO optimization problems.

Optimization Problem	Moth Flame Optimization (MFO) Algorithm	Particle Swarm Optimization (GWO) Algorithm
Number of solutions	Number of moths ($i = 1, 2, … nm$)	Number of particles (i = 1, 2, … np)
Combination of parameters within their bounds	Position of moth (M_{ij})	Position of particle (P_{ij})
Number of parameters, factors, andindependent variables	Number of dimensions involved in defining the position of moth ($j = 1, 2, … nd$)	Number of dimensions involved in defining the position of particle ($j = 1, 2, … nd$)
Number of better solutions	Number of flames ($i1 = 1, 2, … nf$)	Fitness of particle (f_i)
Combination of parameters within their bounds	Position of flames (F_{i1j})	
Response value, output, and dependant variable	Fitness of moth (f_i)	Global best (gB)

Table 12. Algorithm parameters for GWO, MFO, and PSO algorithms.

GWO Algorithm		MFO Algorithm		PSO Algorithm	
Parameter	Value	Parameter	Value	Parameter	Value
Maximum number of grey wolves	100	Maximum number of moths	100	Maximum number of particles	100
Constant a	2 to 0	Position of moth close to the flame (t)	From -1 to -2	Learning factors (C_1 and C_2)	2 and 2
Coefficient vectors A	$-2a$ to $2a$	Update mechanism	Logarithmic spiral	Inertia weight (ω)	From 0.4 to 0.9
Coefficient vectors C	$2*rand(0,1)$	Adaptive number of flames	round((mf-(itr*(mf-1)/max_itr)))	Number of particles	30
No. of iterations (nitr)	100	No. of iterations (nitr)	100	No. of iterations (nitr)	100

The steps in implementing the GWO algorithm in this experimentation are shown in Figure 12. The flowchart for implementing the MFO algorithm in the present work is shown in Figure 13.

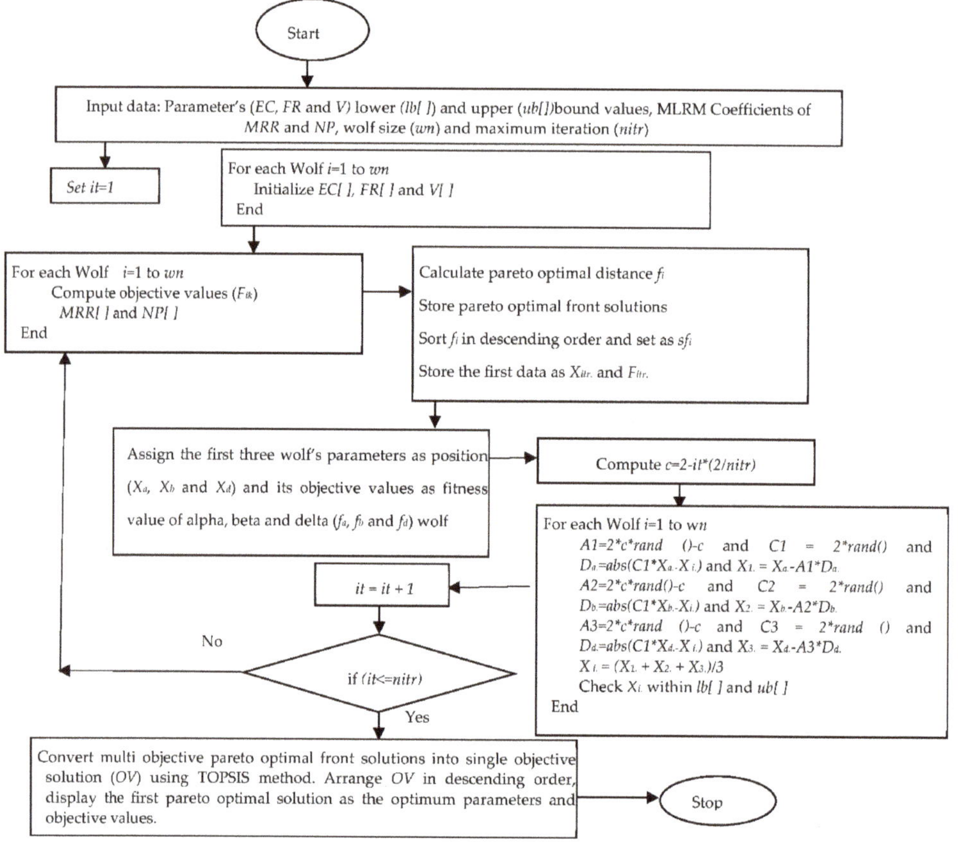

Figure 12. Implementation of GWO algorithm.

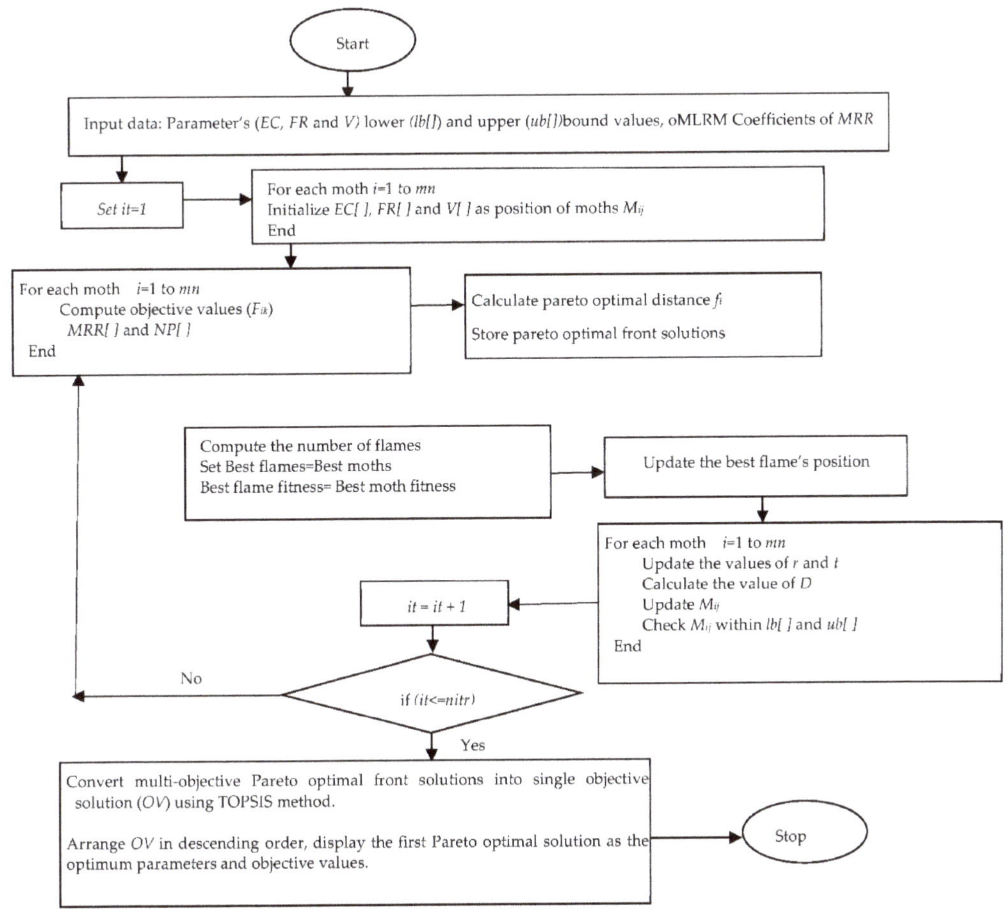

Figure 13. Implementation of MFO algorithm.

The PSO implementation is shown in Figure 14. The efficiency of these algorithms is compared in the convergence plot of Figure 15. In maximization of the MRR, the MFO convergence plot was attained after the 60th iteration, while GWO and PSO performed better. In minimization of the NP, PSO converged well before MFO and GWO.

The optimum process parameters obtained by different meta-heuristics and the RSM are tabulated in Table 13. The convergence plot of the Pareto optimal solutions is shown in Figure 16, and the Pareto optimal set of process parameters out of 12 runs is presented in Table 14.

Table 13. Optimum process parameters and the response values.

Algorithms	EC	V	FR	MRR	NP
RSM	130.000	12.8586	2.5758	0.217	53.9400
PSO	130.401	12.7735	2.0378	0.156	51.3507
GWO	132.014	13.2406	2.8455	0.242	57.7202
MFO	132.014	13.2406	2.8455	0.242	57.7202

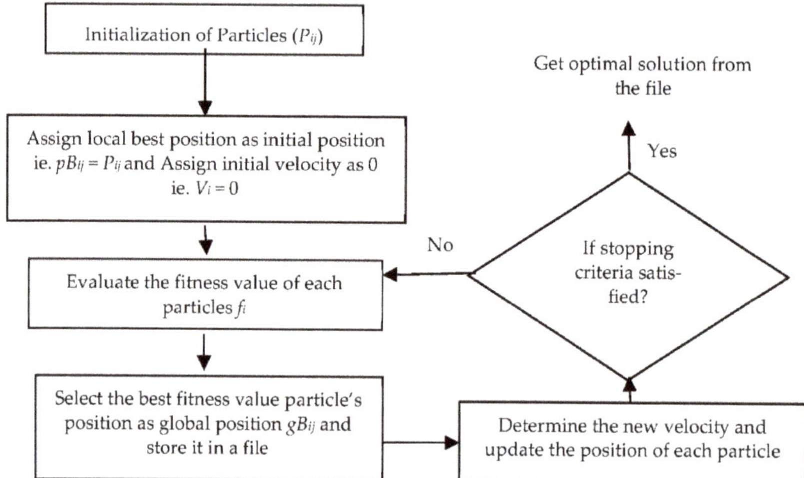

Figure 14. Implementation of PSO algorithm.

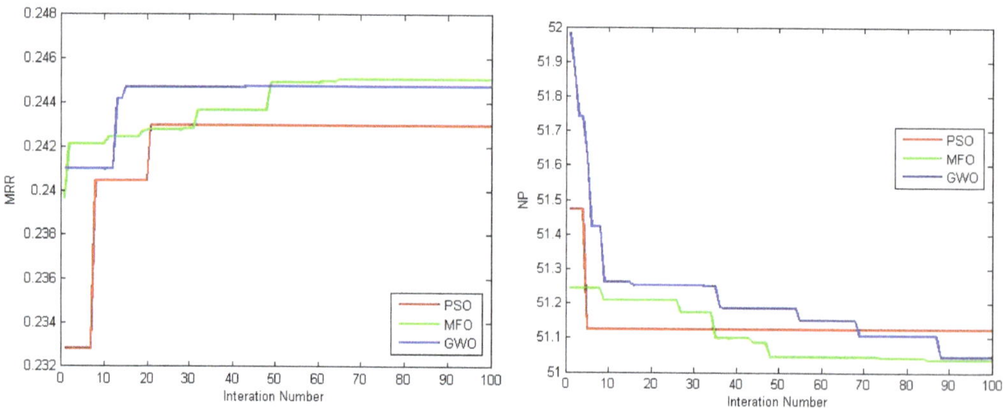

Figure 15. Comparison of convergence plots for various algorithms for MRR and NP.

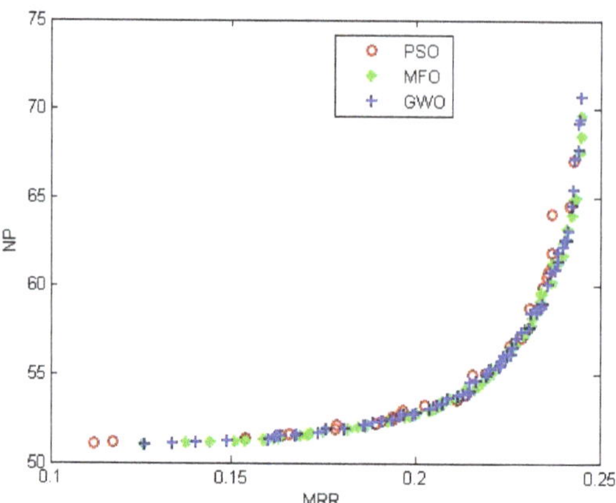

Figure 16. Pareto optimal solutions for various algorithms.

Table 14. The Pareto optimal solutions and their responses for various algorithms.

R.No.	GWO					MFO					PSO				
	EC	V	FR	MRR	NP	EC	V	FR	MRR	NP	EC	V	FR	MRR	NP
1	134.440	13.850	2.835	0.251	60.739	134.440	13.850	2.835	0.251	60.739	131.070	12.188	1.946	0.135	51.413
2	135.769	13.942	2.815	0.251	61.324	135.769	13.942	2.815	0.251	61.324	130.401	12.774	2.038	0.156	51.351
3	130.034	12.754	2.265	0.185	52.015	130.034	12.754	2.265	0.185	52.015	130.033	12.709	1.794	0.114	51.026
4	130.444	12.755	2.043	0.156	51.355	130.444	12.755	2.043	0.156	51.355	130.717	12.427	1.931	0.136	51.168
5	132.014	13.241	2.846	0.242	57.720	132.014	13.241	2.846	0.242	57.720	130.047	12.363	1.904	0.129	51.110
6	131.611	13.909	2.947	0.256	61.669	131.611	13.909	2.947	0.256	61.669	130.310	12.588	1.867	0.126	51.046
7	135.416	13.038	2.469	0.217	54.597	135.416	13.038	2.469	0.217	54.597	130.630	12.623	1.891	0.132	51.097
8	131.808	13.122	2.628	0.227	55.281	131.808	13.122	2.628	0.227	55.281	130.508	12.809	1.880	0.132	51.150
9	133.968	13.719	2.996	0.255	61.918	133.968	13.719	2.996	0.255	61.918	130.923	12.609	1.829	0.122	51.101
10	132.374	12.824	2.548	0.217	54.160	132.374	12.824	2.548	0.217	54.160	130.167	12.553	1.630	0.082	51.215
11	130.319	13.339	2.460	0.214	54.328	130.319	13.339	2.460	0.214	54.328	130.598	12.673	1.892	0.132	51.104
12	132.275	12.697	2.634	0.220	54.608	132.275	12.697	2.634	0.220	54.608	130.580	12.182	1.883	0.124	51.339

Both the MFO and GWO algorithms produced the better trade-off performances with the same solutions of EC = 132.014 g/L, V = 13. 2406 V, and FR = 2.8455 L/min for MRR = 0.242 g/min and NP = 57.7202 PPM.

7. Conclusions

In the present study, ECM operations were carried out on Monel 400 alloys by varying the applied voltage (11–15 V), flow rate (1–3 L/min), and electrolyte concentration (130–190 g/L). The meta-heuristic algorithms (i.e., moth-flame optimization (MFO) algorithm, grey wolf optimizer (GWO), and particle swarm optimization (PSO)) were implemented to find out the optimum process parameters for producing the best performance measures for the material removal rate (MRR) and nickel presence (NP) in the sludge. The regression model equations were developed to determine the optimum parametric combination. The high coefficient of determination values (R2 = 90.34% for MRR and R2 = 92.29% for NP) for both responses confirmed their adequacy. The algorithms were tuned to obtain the best possible feasible solution. This study proved that the MFO and GWO algorithms could be successfully utilized to find the best process parameter setting for the ECM process for Monel 400 alloys. The TOPSIS algorithm was implemented to convert multiple objectives into a single objective used in meta-heuristic algorithms. It was observed from the surface, main, and interaction plots of this ECM experimentation that all the process variables significantly affected the performances. The PSO algorithm converged very quickly in 10 iterations, while GWO and MFO took 14 and 64 iterations, respectively, for convergence of the MRR. In the case of the NP model, the PSO tool took only six iterations to converge, whereas MFO and GWO took 48 and 88 iterations, respectively. However, both MFO and GWO obtained the better trade-off performances with the same solutions of EC = 132.014 g/L, V = 13. 2406 V, and FR = 2.8455 L/min for MRR = 0.242 g/min and NP = 57.7202 PPM. The Pareto optimal set of solutions provided by the GWO, MFO, and PSO algorithms for the ECM process was beneficial to the industries, as it contained a wide range of optimal values. A particular solution from the Pareto optimal set can be chosen according to the specific needs of the objectives. Hence, it was confirmed that the metaheuristic algorithms of MFO and GWO were suitable for finding the optimum process parameters for machining Monel 400 alloys with ECM. Future studies may be explored by combining the artificial neural network (ANN) and fuzzy logic (FL) concepts with MFO and GWO to determine the optimum ECM process parameters.

Author Contributions: Conceptualization and methodology, V.N., A.S., L.N. and S.K.M.; writing—original draft preparation, V.N., A.S., L.N. and S.K.M.; writing—review and editing, S.S., E.A.N., R.S. and H.M.A.M.H. All authors have read and agreed to the published version of the manuscript.

Funding: The authors extend their appreciation to King Saud University for funding this work through Researchers Supporting Project number (RSP-2021/164), King Saud University, Riyadh, Saudi Arabia.

Institutional Review Board Statement: Not applicable.

Informed Consent Statement: Not applicable.

Data Availability Statement: Not applicable.

Conflicts of Interest: The authors declare no conflict of interest.

References

1. Skoczypiec, S.; Lipiec, P.; Bizoń, W.; Wyszyński, D. Selected Aspects of Electrochemical Micromachining Technology Development. *Materials* **2021**, *14*, 2248. [CrossRef] [PubMed]
2. Ge, Y.; Zhu, Z.; Wang, D. Electrochemical Dissolution Behavior of the Nickel-Based Cast Superalloy K423A in $NaNO_3$ Solution. *Electrochim. Acta* **2017**, *253*, 379–389. [CrossRef]
3. Zhang, Y.; Xu, Z.; Xing, J.; Zhu, D. Effect of tube-electrode inner diameter on electrochemical discharge machining of nickel-based super alloy. *Chin. J. Aeronaut.* **2016**, *29*, 1103–1110. [CrossRef]
4. Ge, Y.; Zhu, Z.; Zhu, Y. Electrochemical deep grinding of cast nickel-base superalloys. *J. Manuf. Process.* **2019**, *47*, 291–296. [CrossRef]
5. Bi, X.; Zeng, Y.; Qu, N. Wire electrochemical micromachining of high-quality pure-nickel microstructures focusing on different machining indicators. *Precis. Eng.* **2020**, *61*, 14–22. [CrossRef]
6. Tayal, A.; Kalsi, N.S.; Gupta, M.K.; Garcia-Collado, A.; Sarikaya, M. Reliability and economic analysis in sustainable machining of Monel 400 alloy. *Proc. Inst. Mech. Eng. Part C J. Mech. Eng. Sci.* **2021**, *235*, 5450–5466. [CrossRef]
7. Rajurkar, K.P.; Hadidi, H.; Pariti, J.; Reddy, G.C. Review of Sustainability Issues in Non-Traditional Machining Processes. *Procedia Manuf.* **2017**, *7*, 714–720. [CrossRef]
8. Chakraborty, S.; Bhattacharyya, B.; Diyaley, S. Applications of optimization techniques for parametric analysis of non-traditional machining processes: A Review. *Manag. Sci. Lett.* **2019**, *9*, 467–494. [CrossRef]
9. Mukherjee, R.; Chakraborty, S. Selection of the optimal electrochemical machining process parameters using biogeography-based optimization algorithm. *Int. J. Adv. Manuf. Technol.* **2013**, *64*, 781–791. [CrossRef]
10. Ayyappan, S.; Sivakumar, K. Investigation of electrochemical machining characteristics of 20MnCr5 alloy steel using potassium dichromate mixed aqueous NaCl electrolyte and optimization of process parameters. *Proc. Inst. Mech. Eng. Part B J. Eng. Manuf.* **2015**, *229*, 1984–1996. [CrossRef]
11. Ayyappan, S.; Sivakumar, K. Enhancing the performance of electrochemical machining of 20MnCr5 alloy steel and optimization of process parameters by PSO-DF optimizer. *Int. J. Adv. Manuf. Technol.* **2016**, *82*, 2053–2064. [CrossRef]
12. Sohrabpoor, H.; Khanghah, S.; Shahraki, S.; Teimouri, R. Multi-objective optimization of electrochemical machining process. *Int. J. Adv. Manuf. Technol.* **2016**, *82*, 1683–1692. [CrossRef]
13. Rao, R.V.; Rai, D.P.; Balic, J. A multi-objective algorithm for optimization of modern machining processes. *Eng. Appl. Artif. Intell.* **2017**, *61*, 103–125. [CrossRef]
14. Mehrvar, A.; Basti, A.; Jamali, A. Inverse modelling of electrochemical machining process using a novel combination of soft computing methods. *Proc. Inst. Mech. Eng. Part C J. Mech. Eng. Sci.* **2020**, *234*, 3436–3446. [CrossRef]
15. Singh, D.; Shukla, R. Parameter optimization of electrochemical machining process using black hole algorithm. *IOP Conf. Ser. Mater. Sci. Eng.* **2017**, *282*, 012006. [CrossRef]
16. Kalaimathi, M.; Venkatachalam, G.; Sivakumar, M.; Ayyappan, S. Multi-response optimisation of electrochemical machining process parameters by harmony search-desirability function optimiser. *Int. J. Manuf. Technol. Manag.* **2020**, *34*, 331–348. [CrossRef]
17. Diyaley, S.; Chakraborty, S. Optimization of electrochemical machining process parameters using teaching-learning-based algorithm. *AIP Conf. Proc.* **2020**, *2273*, 050010. [CrossRef]
18. Kasdekar, D.K.; Parashar, V.; Arya, C. Artificial neural network models for the prediction of MRR in Electro-chemical machining. *Mater. Today Proc.* **2018**, *5 Pt 1*, 772–779. [CrossRef]
19. Chakraborty, S.; Das, P.P.; Kumar, V. Application of grey-fuzzy logic technique for parametric optimization of non-traditional machining processes. *Grey Syst. Theory Appl.* **2018**, *8*, 46–68. [CrossRef]
20. Mirjalili, S.; Mirjalili, S.M.; Lewis, A. Grey Wolf Optimizer. *Adv. Eng. Softw.* **2014**, *69*, 46–61. [CrossRef]
21. Li, Y.; Lin, X.; Liu, J. An Improved Gray Wolf Optimization Algorithm to Solve Engineering Problems. *Sustainability* **2021**, *13*, 3208. [CrossRef]
22. Khalilpourazari, S.; Khalilpourazary, S. Optimization of production time in the multi-pass milling process via a robust grey wolf optimizer. *Neural Comput. Appl.* **2018**, *29*, 1321–1336. [CrossRef]
23. Kharwar, P.K.; Verma, R.K. Nature instigated Grey wolf algorithm for parametric optimization during machining (Milling) of polymer nanocomposites. *J. Thermoplast. Compos. Mater.* **2021**, 1–23. [CrossRef]
24. Kulkarni, O.; Kulkarn, S. Process Parameter Optimization in WEDM by Grey Wolf Optimizer. *Mater. Today Proc.* **2018**, *5 Pt 1*, 4402–4412. [CrossRef]
25. Chakraborty, S.; Mitra, A. Parametric optimization of abrasive water-jet machining processes using grey wolf optimizer. *Mater. Manuf. Process.* **2018**, *33*, 1471–1482. [CrossRef]
26. Sekulic, M.; Pejic, V.; Brezocnik, M.; Gostimirović, M.; Hadzistevic, M. Prediction of surface roughness in the ball-end milling process using response surface methodology, genetic algorithms, and grey wolf optimizer algorithm. *Adv. Prod. Eng. Manag.* **2018**, *13*, 18–30. [CrossRef]

27. Kharwar, P.K.; Verma, R.K. Exploration of nature inspired Grey wolf algorithm and Grey theory in machining of multiwall carbon nanotube/polymer nanocomposites. *Eng. Comput.* **2020**, 1–22. [CrossRef]

28. Mirjalili, S. Moth-Flame Optimization Algorithm: A Novel Nature-inspired Heuristic Paradigm. *Knowl.-Based Syst.* **2015**, *89*, 228–249. [CrossRef]

29. Li, Y.; Zhu, X.; Liu, J. An Improved Moth-Flame Optimization Algorithm for Engineering Problems. *Symmetry* **2020**, *12*, 1234. [CrossRef]

30. Yıldız, B.; Yildiz, A.R. Moth-flame optimization algorithm to determine optimal machining parameters in manufacturing processes. *Mater. Test.* **2017**, *59*, 425–429. [CrossRef]

31. Sivalingam, V.; Sun, J.; Mahalingam, S.K.; Nagarajan, L.; Natarajan, Y.; Salunkhe, S.; Nasr, E.A.; Davim, J.P.; Hussein, H.M.A.M. Optimization of Process Parameters for Turning Hastelloy X under Different Machining Environments Using Evolutionary Algorithms: A Comparative Study. *Appl. Sci.* **2021**, *11*, 9725. [CrossRef]

32. Kamaruzaman, A.F.; Mohd Zain, A.; Alwee, R.; Md Yusof, N.; Najarian, F. Optimization of Surface Roughness in Deep Hole Drilling using Moth-Flame Optimization. *ELEKTRIKA-J. Electr. Eng.* **2019**, *18*, 62–68. [CrossRef]

33. Karthick, M.; Anand, P.; Siva Kumar, M.; Meikandan, M. Exploration of MFOA in PAC parameters on machining Inconel 718. *Mater. Manuf. Process.* **2021**, 1–13. [CrossRef]

34. Lmalghan, R.; Rao, M.C.K.; Arunkumar, S.; Rao, S.S.; Herbert, M.A. Machining Parameters Optimization of AA6061 Using Response Surface Methodology and Particle Swarm Optimization. *Int. J. Precis. Eng. Manuf.* **2018**, *19*, 695–704. [CrossRef]

35. Abu-Mahfouz, I.; Banerjee, A.; Rahman, E. Evolutionary Optimization of Machining Parameters Based on Surface Roughness in End Milling of Hot Rolled Steel. *Materials* **2021**, *14*, 5494. [CrossRef]

36. Kalaimathi, M.; Venkatachalam, G.; Sivakumar, M.; Ayyappan, S. Experimental investigation on the suitability of ozonated electrolyte in travelling-wire electrochemical machining. *J. Braz. Soc. Mech. Sci. Eng.* **2017**, *39*, 4589–4599. [CrossRef]

37. Ayyappan, S.; Sivakumar, K. Experimental investigation on the performance improvement of electrochemical machining process using oxygen-enriched electrolyte. *Int. J. Adv. Manuf. Technol.* **2014**, *75*, 479–487. [CrossRef]

38. Qu, N.; Fang, X.; Li, W.; Zeng, Y.; Zhu, D. Wire electrochemical machining with axial electrolyte fushing for titanium alloy. *Chin. J. Aeronaut.* **2013**, *26*, 224–229. [CrossRef]

39. Qu, N.S.; Ji, H.J.; Zeng, Y.B. Wire electrochemical machining using reciprocated traveling wire. *Int. J. Adv. Manuf. Technol.* **2014**, *72*, 677–683. [CrossRef]

40. Leese, R.; Ivanov, A. Electrochemical micromachining: Review of factors affecting the process applicability in micro-manufacturing. *Proc. Inst. Mech. Eng. Part B J. Eng. Manuf.* **2018**, *232*, 195–207. [CrossRef]

41. Zhang, Y. Investigation into Current Efficiency for Pulse Electrochemical Machining of Nickel Alloy. University of Nebraska–Lincoln. 2010. Available online: http://digitalcommons.unl.edu/imsediss/2 (accessed on 10 November 2021).

42. Trimmer, A.L.; Hudson, J.L.; Kock, M.; Schuster, R. Single-step electrochemical machining of complex nanostructures with ultrashort voltage pulses. *Appl. Phys. Lett.* **2003**, *82*, 3327–3329. [CrossRef]

43. Rathod, V.; Doloi, B.; Bhattacharyya, B. Experimental investigations into machining accuracy and surface roughness of microgrooves fabricated by electrochemical micromachining. *Proc. Inst. Mech. Eng. Part B J. Eng. Manuf.* **2015**, *229*, 1781–1802. [CrossRef]

44. Shunmugesh, K.; Panneerselvam, K. Optimization of machining process parameters in drilling of CFRP using multi-objective Taguchi technique, TOPSIS and RSA techniques. *Polym. Polym. Compos.* **2017**, *25*, 185–192. [CrossRef]

45. Balaji, K.; Siva Kumar, M.; Yuvaraj, N. Multi objective taguchi–grey relational analysis and krill herd algorithm approaches to investigate the parametric optimization in abrasive water jet drilling of stainless steel. *Appl. Soft Comput. J.* **2021**, *102*, 107075. [CrossRef]

applied sciences

Article

Optimization of Process Parameters for Turning Hastelloy X under Different Machining Environments Using Evolutionary Algorithms: A Comparative Study

Vinothkumar Sivalingam [1], Jie Sun [1], Siva Kumar Mahalingam [2], Lenin Nagarajan [2,*], Yuvaraj Natarajan [2], Sachin Salunkhe [2], Emad Abouel Nasr [3], J. Paulo Davim [4] and Hussein Mohammed Abdel Moneam Hussein [5,6]

[1] Key Laboratory of High-Efficiency and Clean Mechanical Manufacture, National Demonstration Center for Experimental Mechanical Engineering Education, School of Mechanical Engineering, Shandong University, Jinan 250061, China; svkceg@gmail.com (V.S.); sunjie@sdu.edu.cn (J.S.)

[2] Department of Mechanical Engineering, Vel Tech Rangarajan Dr. Sagunthala R&D Institute of Science and Technology, Chennai 600 062, Tamilnadu, India; lawan.sisa@gmail.com (S.K.M.); drnyuvaraj@veltech.edu.in (Y.N.); drsalunkesachin@veltech.edu.in (S.S.)

[3] Industrial Engineering Department, College of Engineering, King Saud University, Riyadh 11421, Saudi Arabia; eabdelghany@ksu.edu.sa

[4] Department of Mechanical Engineering, Campus Universitário de Santiago, University of Aveiro, 3810-193 Aveiro, Portugal; pdavim@ua.pt

[5] Department of Mechanical Engineering, Faculty of Engineering, Helwan University, Cairo 11732, Egypt; hussein@h-eng.helwan.edu.eg

[6] Department of Mechanical Engineering, Faculty of Engineering, Ahram Canadian University, 6th of October City 19228, Egypt

* Correspondence: n.lenin@gmail.com; Tel.: +91-9976-9096-47

Citation: Sivalingam, V.; Sun, J.; Mahalingam, S.K.; Nagarajan, L.; Natarajan, Y.; Salunkhe, S.; Nasr, E.A.; Davim, J.P.; Hussein, H.M.A.M. Optimization of Process Parameters for Turning Hastelloy X under Different Machining Environments Using Evolutionary Algorithms: A Comparative Study. *Appl. Sci.* **2021**, *11*, 9725. https://doi.org/10.3390/app11209725

Academic Editor: Vladimir Modrak

Received: 6 September 2021
Accepted: 11 October 2021
Published: 18 October 2021

Publisher's Note: MDPI stays neutral with regard to jurisdictional claims in published maps and institutional affiliations.

Abstract: In this research work, the machinability of turning Hastelloy X with a PVD Ti-Al-N coated insert tool in dry, wet, and cryogenic machining environments is investigated. The machinability indices namely cutting force (CF), surface roughness (SR), and cutting temperature (CT) are studied for the different set of input process parameters such as cutting speed, feed rate, and machining environment, through the experiments conducted as per L_{27} orthogonal array. Minitab 17 is used to create quadratic Multiple Linear Regression Models (MLRM) based on the association between turning parameters and machineability indices. The Moth-Flame Optimization (MFO) algorithm is proposed in this work to identify the optimal set of turning parameters through the MLRM models, in view of minimizing the machinability indices. Three case studies by considering individual machinability indices, a combination of dual indices, and a combination of all three indices, are performed. The suggested MFO algorithm's effectiveness is evaluated in comparison to the findings of Genetic, Grass-Hooper, Grey-Wolf, and Particle Swarm Optimization algorithms. From the results, it is identified that the MFO algorithm outperformed the others. In addition, a confirmation experiment is conducted to verify the results of the MFO algorithm's optimal combination of turning parameters.

Keywords: Hastelloy X; turning; cutting force; surface roughness; liquid nitrogen; grass-hooper optimization algorithm; moth-flame optimization algorithm

1. Introduction

Nickel-based (Ni) alloys attract more researchers nowadays for their broader applications in the fields like aerospace, automobile, biomedical, and allied industries. Hastelloy is one of the Ni-based alloys, and it holds few unique characteristics like good strength-to-weight ratio, resistance to corrosion, higher melting temperature, good toughness, etc. [1]. Mainly, Hastelloy X is used to fabricate the combustion chamber of an aircraft engine because of its high heat-resisting property. However, the holding of all the above-said properties by Hastelloy X, resulting in very poor machinability. In this sense, the manufacturing industries face a difficult task in improving Hastelloy X machinability using traditional machining methods. [2]. Furthermore, the reduction of cutting forces (CF), surface roughness (SR), and cutting temperature (CT) during Hastelloy X machining adds to the difficulty of achieving good machinability. As a result, several researchers have worked on various research projects over time to increase the machinability of Hastelloy. Furthermore, they performed these tests under dry, wet, and cryogenic cooling conditions in order to demonstrate an increase in machinability. Therefore, these literatures are critically reviewed, and the extracted information is given here for ready reference to the readers.

Kadirgama et al. [3] studied the impact on cutting force by the parameters, namely axial depth, cutting speed, and feed rate while milling Hastelloy C-22HS. The models using Response Surface Methodology were developed using experimentation and Finite Element Analysis to predict the optimized cutting force. Kadirgama et al. [4] investigated the tool behavior such as tool wear and tool life during machining of Hastelloy C-22HS under wet conditions. PVD and CVD multilayer coated carbide tools were used for machining. The tool life was decreased in all the cases while increasing the cutting parameters, namely cutting speed (v_c), feed rate (f), and axial depth (a_p). Altin [2] studied the machinability of Ni-based (Hastelloy X) alloy under dry cutting conditions. The CF and SR were analyzed against the multilayer coated insert and various v_c. The experimentation results showed that the abrasiveness of the carbide particles on the tool and the mechanical loading had a growing influence on the CF. Sofuoğlu et al. [5] studied the impact of the v_c, tool extended length, and novel methods, namely Conventional Turning (CT), Ultrasonic Assisted Turning (UAT) and Hot-Ultrasonic Assisted Turning (HUAT) on the SR, a_p, and CT while machining Hastelloy X. The reduction in SR and increment in regular a_p and CT were attained in UAT and HUAT compared to CT. Dhananchezian [6] conducted the machinability study on Hastelloy C-276 under dry and cryogenic liquid nitrogen (LN_2) cooling conditions using turning operation. The output responses such as CT, CF, SR, chip morphology, and tool wear under dry turning were compared with LN_2 cooling-based turning. A considerable reduction in all the output responses was noted under liquid nitrogen cooling-based turning.

Kesavan et al. [7] conducted the CNC turning of Hastelloy C276 by varying v_c and the fixed values of f, a_p. The experimentation was executed under dry and LN_2 conditions. Further, Deform 3D analytical tool was used to create the simulation model based on the experimental design to identify the optimal cutting conditions. From the experimentation and simulated model results, it was evident that the cutting temperate and machining forces have been significantly reduced while machining under cryogenic cooling conditions rather than dry conditions. Dhananchezian and Rajkumar [8] examined the SR and Tool Wear characteristics of Nimonic 90 alloy and Hastelloy C-276 dry turning. During the turning process, various cutting inserts were used. In both cases, the roughness and tool wear metrics were observed to be larger as the turning length was increased. Dhananchezian and Rajkumar [8] made a comparative analysis on the tool wear rate and SR during the turning of Hastelloy C-22 underneath dry and LN_2 cooling conditions. A substantial drop in the SR was found in the turning of Hastelloy under LN_2 cooling rather than dry turning.

Oschelski et al. [9] used the Box-Behnken method to design the experiments by considering the ranges of parameters, namely v_c, a_p, lubricating conditions, constant f, and (wet, dry, and reduced quantity lubrication) for finish turning the Hastelloy X. The experimental results showed that the v_c, a_p, and interactions were the most significant factors affecting the SR. Next, Venkatesan et al. [10] reported the machinability study on Hastelloy X with PVD and CVD coated tools in comparison with dry and Minimum Quantity Lubrication (MQL) conditions. A mixture of coconut oil with Hexagonal Boron Nitride (HBN) nanoparticles was used as nanofluid for lubrication. Significant reductions in CF, SR, and tool wear were observed in MQL-PVD combination than MQL-CVD and dry-PVD. Finally, Sivalingam et al. [11] investigated the influence of whisker-reinforced ceramic tools on tool wear, SR, and tool chattering under dry and Atomization-based Cutting Fluid (ACF) cooling conditions when turning Inconel 718 material. Investigation results stated that the flank wear and SR of the tool were significantly reduced under ACF cooling conditions due to limited notching and fracture of the tool edge at the tool-chip interface.

Zhao et al. [12] investigated the characteristics of chip formation when machining NiTi shape memory alloys under different v_c with constant f, a_p. The shape of the chip and microstructure were examined to expose the chip flow behavior. The martensitic phase transformation seemed to have a noticeable effect on the material flow behavior and indeed on the chip formation. [11] investigated the possibility of improving the machinability of Inconel 718 alloy under a dry and atomized spray cutting fluid system. The turning of Inconel 718 alloy with ceramic inserts was carried out by varying the cutting parameters. The output responses such as tool wear, power consumption, surface topography, machine vibrations, chip morphology, and machining cost were analyzed against the experimental design of input parameters. It was observed that the atomized spray cutting fluid technique yielded better results than dry machining.

The effect of LN_2 cooling in improving the machinability of Hastelloy X is discussed in the following literature. Chetan et al. [13] investigated the turning of Nimonic 90 alloy using uncoated tungsten carbide inserts under the modes like dry, MQL, and cryogenic cutting. At lower v_c, the cutting performance of the cryogenically treated tool was good than the untreated tool. But, the performance of the tool under MQL and LN_2 was good in terms of minimum tool wear at a higher cutting speed. Further, a good SR was obtained under dry and MQL modes than LN_2 cooling mode at all levels of cutting speed. Iturbe et al. [14] compared the effects of liquid nitrogen and MQL based cryogenic cooling with conventional cooling. For short machining times, the cryogenic cum MQL cooling outperformed conventional cooling.

Sivaiah and Chakradhar [15] compared the results of LN_2 machining like tool wear, feed force, CF and CT, chip characteristics, and SR with the wet condition during machining of heat-treated 17-4 Precipitation Hardenable Stainless Steel. The LN_2 machining outperformed even at high f to reduce all the above-said parameters compared with wet machining. Tebaldo et al. [16] studied the machinability of Inconel 718 under different machining conditions and lubricating systems. The highest wear resistance was obtained while using the CVD-coated tools under conventional lubricated conditions. But, the MQL system provided good lubrication than cooling with lesser cost and low environmental impact. Shokrani et al. [17] investigated the impact of using different cooling systems, namely MQL, cryogenic and hybrid of cryogenic and MQL, during the CNC milling of Inconel 718 alloy material. Comparatively, the hybrid cooling system yielded better results in terms of good machinability, less SR, and greater tool life. Mehta et al. [18] studied the parameters such as SR, CF, and tool wear during machining of Inconel 718 material. During machining, various sustainable environments, namely dry state, MQL, LN_2 cooling, hybridization of cold air and MQL, and hybridization of MQL and LN_2, were used. The input parameters such as a_p, f, and v_c were kept constant during machining under all the above-said environments. Better surface finish and minimum cutting force were observed during the cold air and MQL environment. Alternatively, the very least tool wear was observed under MQL and LN_2 hybrid cutting environment than the dry environment.

Further, the researchers had used different optimization tools to identify the suitable process parameter values for minimizing the manufacturer's objectives. A few of them are discussed here. Khalilpourazari and Khalilpourazary [19] proposed an algorithm, namely Robust Grey Wolf Optimizer (RGWO), to minimize total production time by identifying the optimal input parameters multi-pass milling process. The parameter tuning during optimization was carried out using the Taguchi method. Further, an efficient constraint handling approach was implemented to handle the complex constraints of the problem. The results concluded that the RGWO outperformed the meta-heuristic algorithms such as the multi-verse optimizer and dragonfly algorithm and the other solution methods in the literature. Khalilpourazari and Khalilpourazary [20] developed the lexicographic weighted Tchebycheff method to obtain the optimal decision parameters of the grinding process for maximizing the quality of the surface and production rate and minimizing the machining time and cost. GAMS software was used for this purpose. Khalilpourazari and Khalilpourazary [21] used a novel strategy, namely Robust Stochastic Novel Search, to identify the optimal values of the grinding process parameters to minimize process cost and to maximize the rate of production and surface quality. The proposed method outperformed previously proposed methodologies and novel algorithms, including Multi-Population Ensemble Differential Evolution and Heterogeneous Comprehensive Learning Particle Swarm Optimization.

Rao et al. [22] optimized the abrasive water-jet machining parameters to minimize the kerf and surface roughness using the Jaya algorithm and the multi-objective Jaya algorithm. Better results were obtained through the used algorithms than the simulated annealing, particle swarm optimization, firefly algorithm, cuckoo search algorithm, blackhole algorithm, and biogeography-based optimization algorithms. Further, the PROMETHEE method was used to handpick a specific solution among the possible Pareto-optimal solutions obtained through the proposed algorithms based on the given requirements. Rao et al. [23] obtained the optimal set of process parameters of focused ion beam micro-milling, laser cutting, wire-electric discharge machining, and electrochemical machining processes. The maximization of material removal rate and minimization of SR were considered as the objectives in all the processes. The multi-objective Jaya algorithm was implemented to find the optimal solutions in all the cases. The test results showed that the implemented algorithms produced good results compared to other algorithms such as Genetic Algorithm, Non-dominated Sorting Genetic Algorithm, iterative search, and biogeography-based optimization algorithm, Khalilpourazari and Khalilpourazary [24] carried out the optimization of grinding process parameters to improve the SR and reduce production cost and time. A multi-objective dragonfly algorithm was employed for optimizing the process parameters. Results revealed that the proposed algorithm outperformed the Non-dominated Sorting Genetic Algorithm-II. Khalilpourazari and Khalilpourazary [25] proposed a novel hybrid algorithm, Sine–Cosine Whale Optimization Algorithm, to optimize the process parameters of the multi-pass milling process by minimizing the total production time. Almeida et al. [26] optimized the variable-angle composite cylinders via filament winding manufacturing process using GA. Similarly, Wang et al. [27] proposed a reliability-based design optimization technique to improve the buckling load of winding cylinders subjected to radial compression. The moving search windows in the Kriging metamodel are used to accelerate its convergence and reduce the number of training iterations. The results of this study demonstrated the advantages of adopting a variable stiffness design for achieving a maximum load capacity. Almeida et al. [28] proposed a genetic algorithm (GA) to enhance the strength of a cylindrical shell under internal pressure by optimizing the stacking sequence. The results offered asymmetric and non-conventional angles for internally pressured composite tubes, as opposed to the well-known $\pm 55°$ winding angle advice (for first ply failure approach).

In this research work, turning experiments are conducted on the Hastelloy X material using the PVD TiAlN carbide insert tool under dry, wet, and LN_2 environments. The v_c, f, a_p, and machining environment are considered input turning process parameters, and CF, SR, and CT are considered machinability indices. The evolutionary algorithms namely grasshopper optimization (GHO) [29–31], genetic algorithm (GA) [32–34], particle swarm optimization (PSO) [35,36], moth flame (MFO) [37,38], grey wolf optimization (GWO) [39–41] algorithms are used to identify the optimal set of turning process parameters (MATLAB R2020b version). A clear picture of the experimentation and subsequent processes are detailed in Figure 1.

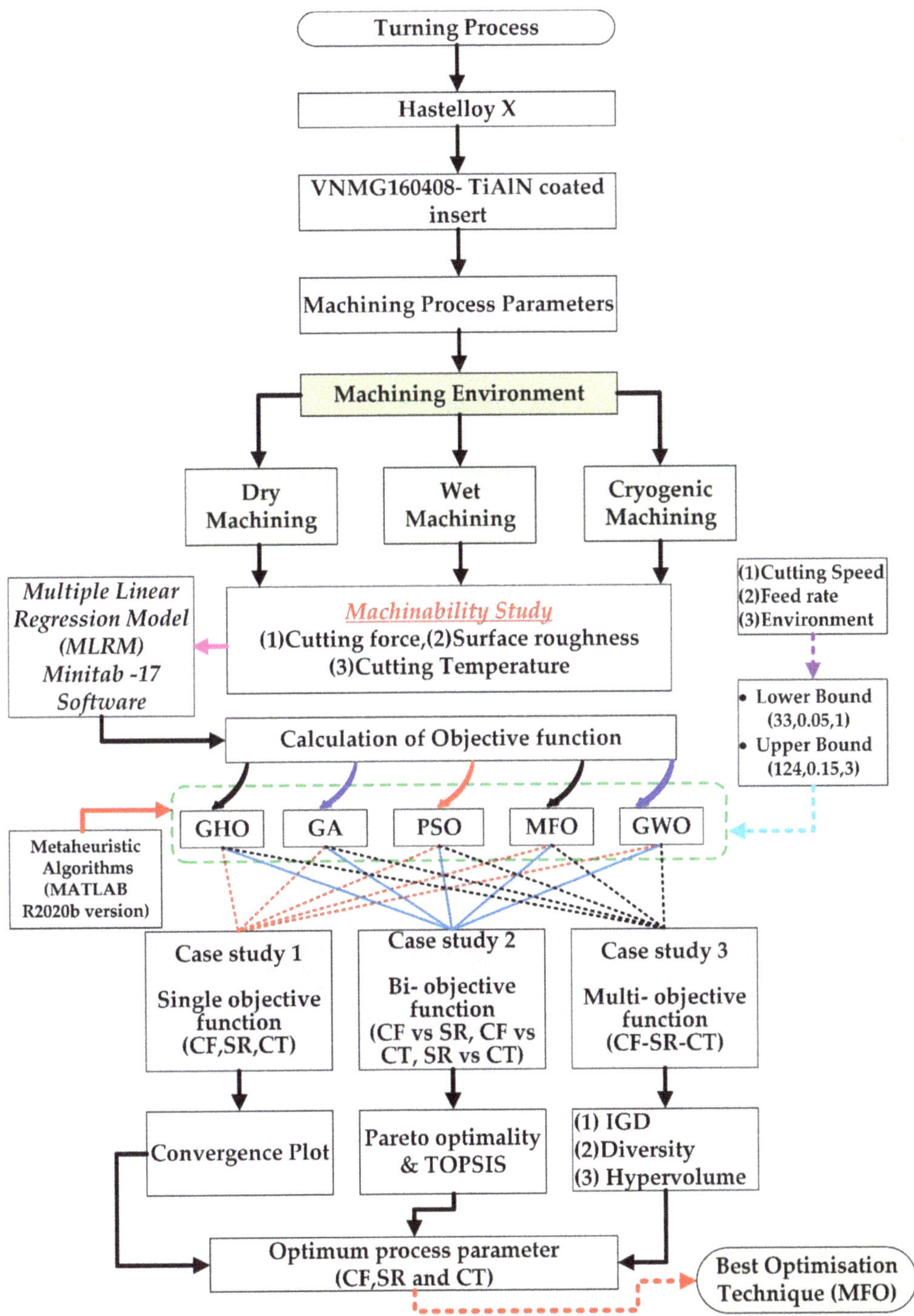

Figure 1. Flow chart of experimental part and metaheuristic algorithm.

2. Experimentation Details

The Hastelloy X bar having 20 mm diameter and 300 mm length is used for conducting the turning experiments on a C6140H turning machine. VNMG160408-SM1105 PVD TiAlN cutting tool inserts are used to do the turning operation. The impact of tool wear on the machinability indices is completely eliminated by using new inserts every time. The CF (Tangential Force, F_z) is calculated with a 9257B Kistler dynamometer, and the value is manipulated with dynoware software. The workpiece surface roughness (Ra) is measured using a contact-type surface roughness tester (TR200), a cutoff length of 0.8 mm, and a traverse length of 4 mm. For measuring the CT, a FORTIC 226 infrared imaging sensor is used. Cryogenic equipment consists of a self-pressurized pump, cryogenic dewar tank capacity of 50 L. At a pressure of 0.3 bar, LN_2 was sprayed onto the work-tool interface using a copper nozzle diameter of 3 mm. Figure 2 depicts a schematic representation of the experimental setup. The detailed experimental conditions are shown in Table 1.

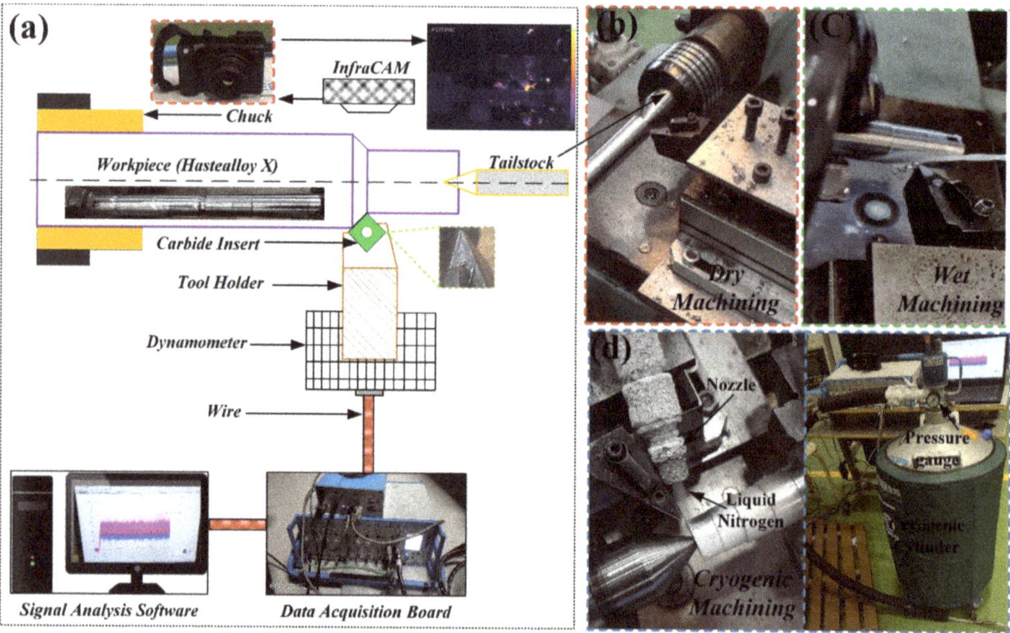

Figure 2. (**a**) Schematic View, Experimental Setup (**b**) Dry (**c**) Wet (**d**) Cryogenic Machining.

Table 1. Experimental Conditions.

Items	Descriptions
Workpiece	Hastealloy × (Ø20 × 300 mm)
Material Properties	**Chemical Composition (%):** Ni:50, Cr:21, Mo:17, Fe:2, Co:1,W:1, Mn:0.80, Al:0.05, Si:0.08, C:0.01, B:0.01 **Physical Properties:** Tensile strength:1370 MPa, Yield Strength: 1170 Mpa, Hardness:388 HB
Insert Specification	VNMG160408-SM1105, PVD TiAlN coated carbide insert, Sandvick
Nose radius	0.8 mm
Rake and relief angle	7°, 6°
Depth of cut (a_p)	0.1 mm

Table 1. *Cont.*

Items	Descriptions
Length of cut (Loc)	60 mm
Environment	Dry, Wet and Cryogenic machining
Cutting Fluid	Vegetable-based oil
Cutting force	Kistler 9257B dynamometer Cutting
Cutting temperature	FORTIC 226 infrared radiation imaging sensor
Surface Roughness	TR200 portable surface roughness tester Evaluation and sampling Lengths are 4 and 0.8 mm

In this work, turning experiments were performed using $L^3{}_{27}$ full factorial experimental design using Minitab 17. three factors are considered for this experiment: v_c, f, and environment; each factor has three different levels, as shown in Table 2.

Table 2. $L^3{}_{27}$ full factorial experimental design.

Factors	Unit	Symbol	Level 1	Level 2	Level 3
Cutting speed (v_c)	m/min	A	33	87	124
Feed rate (f)	mm/rev	B	0.05	0.1	0.15
Environment		C	1 (Dry)	2 (Wet)	3 (Cryogenic)

The experimental design and the corresponding measurement of machinability indices are presented in Table 3.

Table 3. Experimental design values.

S.no	Cutting Speed	Feed Rate	Environment	Cuting Force	Surface Roughness	Cutting Temperature
	m/min	mm/rev		Fz (N)	Ra (µm)	°C
1	33	0.05	Dry	256	3.42	380
2	87	0.05	Dry	192	3.01	416
3	124	0.05	Dry	165	2.98	472
4	33	0.1	Dry	339	2.96	435
5	87	0.1	Dry	281	2.83	477
6	124	0.1	Dry	220	2.72	515
7	33	0.15	Dry	430	2.75	510
8	87	0.15	Dry	385	2.62	550
9	124	0.15	Dry	322	2.53	596
10	33	0.05	Wet	245	3.25	250
11	87	0.05	Wet	186	2.86	313
12	124	0.05	Wet	156	2.80	347
13	33	0.1	Wet	302	2.87	386
14	87	0.1	Wet	276	2.74	414
15	124	0.1	Wet	208	2.68	472
16	33	0.15	Wet	412	2.69	491

Table 3. *Cont.*

S.no	Cutting Speed	Feed Rate	Environment	Cuting Force	Surface Roughness	Cutting Temperature
	m/min	mm/rev		Fz (N)	Ra (μm)	°C
17	87	0.15	Wet	368	2.53	515
18	124	0.15	Wet	308	2.43	565
19	33	0.05	Cryogeic	228	2.65	50
20	87	0.05	Cryogeic	168	2.48	95
21	124	0.05	Cryogeic	132	2.29	110
22	33	0.1	Cryogeic	275	2.2	90
23	87	0.1	Cryogeic	249	2.01	135
24	124	0.1	Cryogeic	168	1.96	140
25	33	0.15	Cryogeic	367	1.92	110
26	87	0.15	Cryogeic	320	1.84	130
27	124	0.15	Cryogeic	279	1.76	165

3. Results and Discussion

This section is divided into three case studies. Case study 1: Minimization of machinability indices individually; Case study 2: Simultaneous minimization of dual machinability indices by considering three combinations; Case study 3: Simultaneous minimization of all three indices. The quadratic Multiple Linear Regression Models (MLRM) are formulated for evaluating the minimum values of machinability indices in all the cases. The Moth-Flame Optimization (MFO) algorithm is proposed to identify the optimal set of turning process parameters in view of minimizing the objectives. The effectiveness of the proposed algorithm is tested against the results of other optimization algorithms such as Genetic Algorithm, Grass-Hooper Optimization (GHO), Grey-Wolf Optimization (GWO), and Particle Swarm Optimization (PSO). Pseudocode for optimization algorithms is shown in the Figure 3. The general parameters used in algorithms are the maximum population size: 50, and the maximum no. of iterations: 100 (MATLAB R2020b). Twenty-seven runs are executed for each algorithm in all the cases. The evaluated results from the case studies are discussed below.

3.1. Case Study 1

Cutting force analysis and its minimization plays a crucial role in machining operation and understanding the cutting phenomena of the work material in different environments (dry, wet, and LN_2 machining). Moreover, after machining, the work material must be superior in surface quality [33]. On the other hand, minimizing the machining zone's temperature is necessary to retain the cutting temperature as low as possible. In this case study, the MLRM for minimizing all the machinability indices is developed using Minitab 17. The developed MLRM are given in Equations (1)–(3).

Objective functions 1.

$$\begin{aligned} Minimize\ Cutting &= 212.7 - 0.244A + 703B + 20C \\ Force &\quad - 0.00571A^2 + 6289B^2 - 8.11C^2 \\ &\quad -0.60AB + 0.053AC - 143BC \\ R^2 &= 0.98, \quad Adj\ R^2 = 0.98 \end{aligned} \tag{1}$$

Objective functions 2.

$$\begin{aligned} Minimize\ Surface\ &= 3.56 - 0.008A - 9.65B + 0.73C \\ Roughness\quad &+ 0.000015A^2 + 19.3B^2 - 0.265C^2 \\ &+ 0.023AB + 0.00028AC - 0.65BC \\ R^2 = 0.98,\ &Adj\ R^2 = 0.97 \end{aligned} \tag{2}$$

Objective functions 3.

$$\begin{aligned} Minimize\ Cutting\ &= -0.083 + 0.076A + 2521B + 341.21C \\ Temperature\quad &+ 0.0033A^2 - 1400B^2 - 118C^2 \\ &- 1.42AB - 0.16AC - 396.67BC \\ R^2 = 0.97,\ &Adj\ R^2 = 0.96 \end{aligned} \tag{3}$$

Three responses were considered in this section R1, R2 and R3, CF, SR and CT, respectively, as shown in Table 4.

Table 4. Minimization of machinability indices using evolutionary algorithms for Case study 1.

Algorithms	Cutting Speed (m/min)	Feed Rate (mm/rev)	Environment	Machinability Index Value	Iteration No.
Cutting force					
MFO	124	0.05	3	127.10 N	2
GA	119.61	0.05	3	139.16 N	3
GHO	124	0.06	3	135.27 N	61
GWO	121.65	0.05	3	132.85 N	78
PSO	123.32	0.05	3	132.77 N	10
Surface roughness					
MFO	124	0.05	3	1.78 μm	1
GA	86.23	0.147	3	1.88 μm	3
GHO	124	0.129	3	1.85 μm	39
GWO	134.17	0.15	3	1.81 μm	79
PSO	126.32	0.052	3	2.33 μm	10
Cutting temperature					
MFO	34.04	0.05	3	33.19 °C	22
GA	80.77	0.05	3	78.98 °C	67
GHO	36	0.05	3	32.33 °C	65
GWO	39.57	0.06	3	48.44 °C	43
PSO	33.6	0.05	3	34.11 °C	26

Figure 4a present the convergence plot for cutting force using the evolutionary algorithms. It is observed that the response is converged in iteration no. 2 using the MFO algorithm. Provided the same response value is converged in iteration no. 3, 10, 61, and 78 based on GA, PSO, GHO, and GWO algorithms, respectively. Further, the response value is very minimum in the MFO algorithm and maximum in GA. The simplicity of the MFO algorithm along with the speed in searching is the prime reason for obtaining the best results in the present work. It is additionally inferred that the CF is very minimum in the LN_2 environment compared to the dry and wet machining environments. The LN_2 nozzle spray droplet acts as cushioning effect of the machining zone and, thus, minimizes the v_c [42].

(a) Moth-Flame Optimization Algorithm (MFO)

Initialize the parameters for Moth-flame
Initialize Moth position Mi randomly
For each $i=1{:}n$ do
Calculate the fitness function fi
End For
While $(iteration \leq max\_iteration)$ do
    Update the position of Mi
    Calculate the no. of flames
    Evaluate the fitness function fi
    If $(iteration==1)$ then
        $F = sort\ (M)$
        $OF = sort\ (OM)$
Else
        $F = sort\ (Mt\text{-}1, Mt)$
        $OF = sort\ (Mt\text{-}1, Mt)$
End if
    For each $i=1{:}n$ do
    For each $j=1{:}d$ do
    Update the values of r and t
    Calculate the value of D w.r.t. corresponding Moth
    Update $M(i,j)$ w.r.t. corresponding Moth
        End For
    End For
End While
Print the best solution

(b) Grass Hopper Optimization Algorithm (GHO)

Initialize the swarm Xi (i=1,2, 3,., n)
Initialize C_{max}, C_{min} and ite_{max}
Calculate the fitness of each agent
T = the best search agent
While $(I < ite_{max})$
    update C
    For each search agent
        Normalize the distance between the grasshoppers
        Update the position of current search agent
        Bring the current search agent back -
        -if it goes outside the boundaries
    End for
    Update T if there is a better solution
    $I = I + 1$
End While
Return

(c) Genetic Algorithm (GA)

{
Generation of initial population (Pi)
Evaluation of Pi
While (stopping criteria not satisfied) Repeat
    {
    For 1 to (no. of tournaments)
        {
        Select N chromosomes for tournament
        Find chromosome with lowest fitness
        Remove chromosome with lowest fitness
        Crossover (Creation of new chromosome)
        Evaluation of new chromosome
        }
        Mutation
        Evaluation of mutated chromosome
    }
}

(d) Grey Wolf Optimization Algorithm (GWO)

Initialize no. of grey wolf (X_{ij} – i =1,2,..nw and j=1,2,..nd)

While (it < nitr)

    Determine the fitness function F_{ik} (k=1,2..nf) of each wolf

    Calculate Pareto optimal distance f_i

    Sort f_i in descending order and set as sf_i and store the first wolf's data as X_{it} and F_{it}.

    Using the sorted data, assign $X_a = X_1$, $X_b = X_2$ and $X_d = X_3$.

    Compute

    For each wolf

        Update the position using

        $A1=2*a*rand()\text{-}a$ and $C1 = 2*rand()$ and
        $D_a=abs(C1*X_a\text{-}X_i)$ and $X_1 = X_a\text{-}A1*D_a$.

        $A2=2*a*rand()\text{-}a$ and $C2 = 2*rand()$ and
        $D_b=abs(C1*X_b\text{-}X_i)$ and $X_2 = X_b\text{-}A2*D_b$.

        $A3=2*a*rand()\text{-}a$ and $C3 = 2*rand()$ and
        $D_d=abs(C1*X_d\text{-}X_i)$ and $X_3 = X_d\text{-}A3*D_d$.

        $X_i = (X_1 + X_2 + X_3)/3$

        Check X_i within bounds

    End

End

Using TOPSIS method convert F_{it} into f_i

Sort f_i in descending order and display the first wolf's data which is optimum data

(e) Particle Swarm Optimization Algorithm (PSO)

P = Particle Initialization ();
For i=1 to itr_{max}
    For each particle p in P do
        fp = f(p);
        If fp is better than f(pBest);
        pBest = p;
            end
        end
    gBest = best p in P
    For each particle p in P do
    $v = v + c_1*rand*(pBest - p) + C_2*rand*(gBest\text{-}P)$;
        p = p+v;
    end
end

Figure 3. Pseudocode for optimization algorithms (**a**) MFO (**b**) GHO (**c**) GA (**d**) GWO (**e**) PSO.

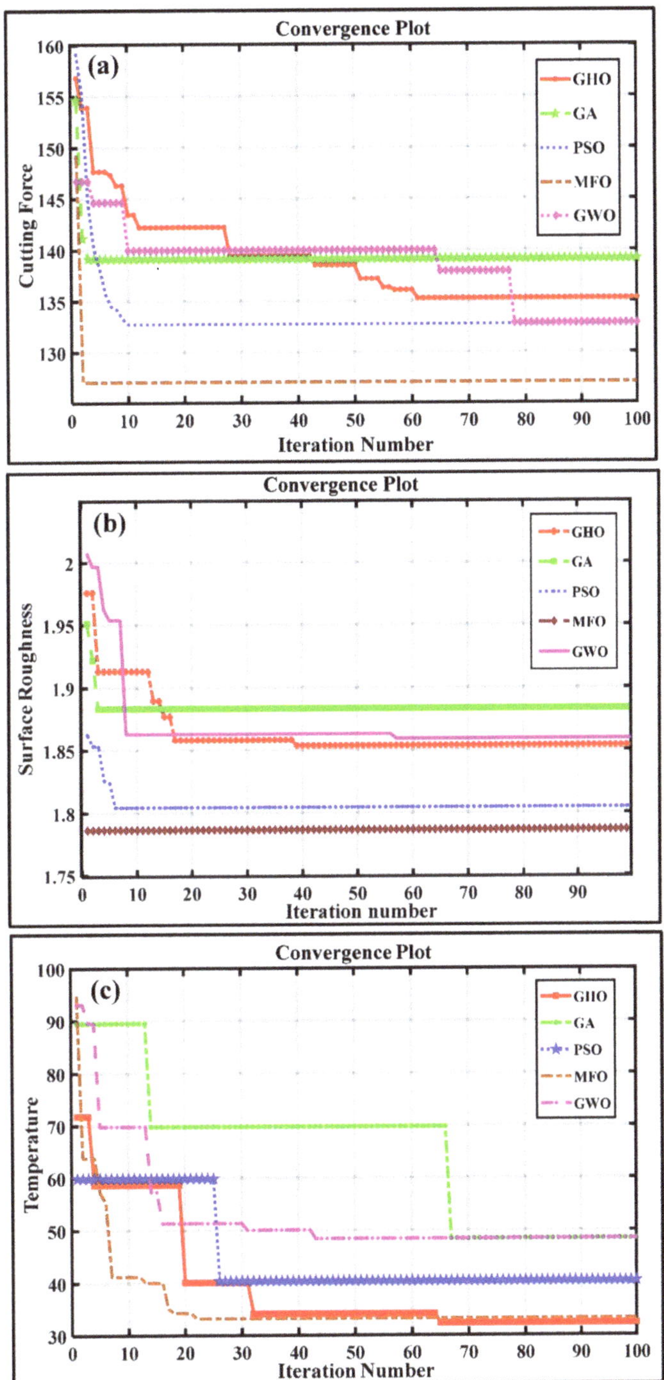

Figure 4. Convergence plot for case study 1 (**a**) Cutting Force (**b**) Surface Roughness (**c**) Cutting Temperature.

Similarly, the convergence plots for the other two machinability indices are given in Figure 4b,c, respectively. The order of preferences of algorithms based on their performance in obtaining the minimum response values is MFO-GWO-GHO-GA-PSO for SR and MFO-GHO-PSO-GWO-GA for CT. Chattered vibrations are generally existing in the machining process, and this is significantly affecting the SR. The SR values are lower in the cryogenic machining than the dry and wet machining from the experimental value and predicted data.

Further, the cutting temperature is directly related to CF and SR. CT increases with an increase in v_c. The flow of cryogenic LN_2 ($-196\,°C$) between tool and workpiece interface greatly reduces the CT in the machining zone compared with dry and wet machining [42].

3.2. Case Study 2

In this study, the simultaneous minimization of dual machinability indices with three different combinations using evolutionary algorithms is considered. The Pareto front analyses of all the algorithms with respect to CF vs. SR, CT vs. CF, and CT vs. SR are given in Figure 5a–c, respectively. Further, the TOPSIS method is used to convert the dual machinability indices into a single objective. Hence, the global minimum value of the machinability indices is obtained using TOPSIS results (Figure 5d–f) for all the evolutionary algorithms. In addition to that, the performance of evolutionary algorithms is validated using the hypervolume indicator. From these Figure 5, it is inferred that the MFO algorithm outperformed others in all three cases. The results are presented in Table 5.

Table 5. Minimization of machinability indices based on pareto front analysis and TOPSIS method for Case study 2.

Algorithms	Machinability Indices Considered	Cutting Speed (m/min)	Feed rate (mm/rev)	Environment	Machinability Index Value (MI$_1$)	Machinabilty Index Value (MI$_2$)	Hyper Volume (HV)
GHO		128.00	0.050	3	127.75 N	2.26 μm	0.301
GA	CF	126.00	0.062	3	127.12 N	2.27 μm	0.302
PSO	&	128.00	0.065	3	141.07 N	2.17 μm	0.319
MFO	SR	124.00	0.060	3	136.57 N	2.20 μm	0.324
GWO		129.00	0.060	3	136.57 N	2.20 μm	0.321
GHO		53.87	0.050	3	206.31 N	42.89 °C	0.652
GA	CF	34.06	0.060	3	218.52 N	32.79 °C	0.633
PSO	&	35.06	0.062	3	218.52 N	32.79 °C	0.624
MFO	TE	35.11	0.050	3	217.98 N	31.26 °C	0.697
GWO		46.90	0.050	3	211.12 N	39.03 °C	0.657
GHO		34.32	0.052	3	2.60 μm	36.16 °C	0.391
GA	SR	34.32	0.062	3	2.60 μm	36.16 °C	0.415
PSO	&	49.38	0.058	3	2.52 μm	43.72 °C	0.416
MFO	TE	50.00	0.053	3	2.52 μm	34.06 °C	0.443
GWO		52.00	0.056	3	2.52 μm	44.06 °C	0.441

The general observations are when CF increases, and SR worsens after machining; When CF increases, TE increases significantly and affects the tool life; and when TE increases, then SR decreases. Further, the LN_2 environment minimizes the CF, CT, and SR due to a fine droplet of LN_2 acting as a film barrier in the tool and workpiece interface, thus reducing chatter vibration and built-up edge formation. This is used to improve the tool life too.

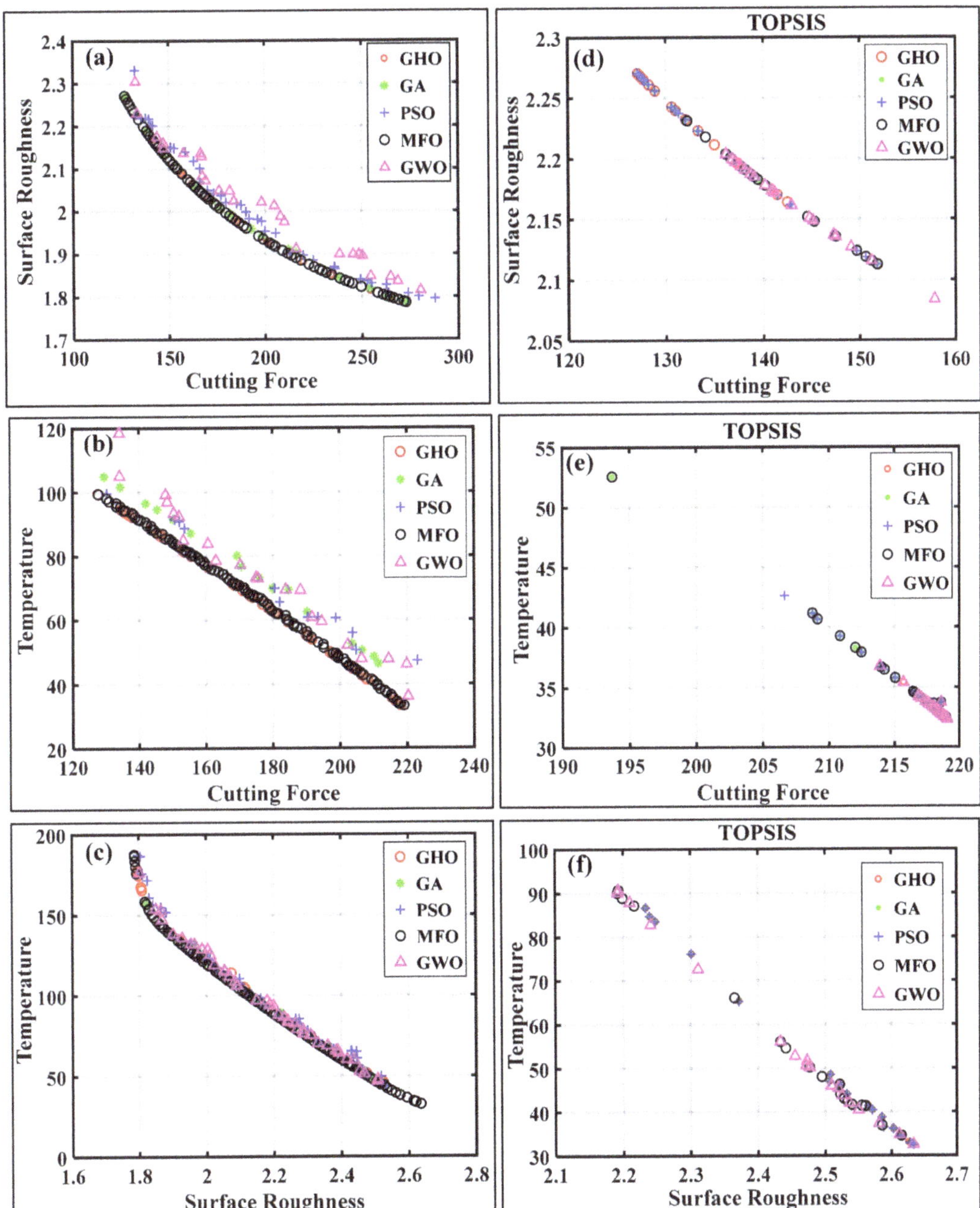

Figure 5. Pareto optimality (**a–c**) Multi-objective for single run (**d–f**) Multi-Objective for 27 runs.

3.3. Case Study 3

In this case study, the simultaneous minimization of all three machinability indices is carried out using evolutionary algorithms. As in case study 2, the TOPSIS method is used to convert all three machinability indices to a single objective. Further, the quality indicators, namely Diversity (DIV), Inverted Generational Distance (IGD), and Hyper Volume (HV), are used to opt out the best evolutionary algorithm which provides minimum values of machinability indices along with the corresponding set of turning process parameters. The algorithm, which has a higher DIV and HV value and a lower IGD value, is to be considered as the best algorithm among others. The DIV and IGD values are calculated using Equations (4) and (5), respectively. The hypervolume is calculated based on the Pareto analysis. The calculated values of quality indicators for all the algorithms are presented in Table 6.

$$DIV = \sqrt{\sum_{j=1}^{k} \left(f_j^{\max} - f_j^{\min}\right)^2} \tag{4}$$

$$IGD = \frac{\sqrt{\sum_{i=1}^{n} d_i^2}}{n} \tag{5}$$

$$d_i = \sqrt{\sum_{j=1}^{no} \left(o_{ij} - o_{bj}\right)^2} \tag{6}$$

where,

O_{ij}—ith run jth objective value;
O_{bj}—Best jth objective value;
d_i—Euclidean distance.

Table 6. Minimization of machinability indices based on IGD, DIV and HV for Case study 3.

Algorithms	IGD	DIV	HV	Cutting Speed (m/min)	Feed Rate (mm/rev)	Environment	Cutting Force (N)	Surface Roughness (μm)	Temperature (°C)
GHO	7.98	233.85	0.273	71.00	0.051	3	193.36	2.44	85.25
GA	9.81	279.61	0.267	72.00	0.051	3	194.36	2.48	86.25
PSO	5.79	264.68	0.265	94.62	0.052	3	173.13	2.45	73.28
MFO	5.10	286.72	0.286	92.62	0.052	3	171.13	2.35	72.28
GWO	6.70	267.48	0.269	96.62	0.052	3	174.13	2.55	74.28

Further the statistical analyses of the quality indicators are carried out using Minitab software to study the performance and consistency of the algorithms. The normal probability plots for the quality indicators IGD, DIV, and HV are given in Figure 6a–c, respectively, and the summary reports of the same are given in Figure 6d–f, respectively. From Figure 6d–f and as well as the findings (higher values of DIV and HV and Lower value of IGD) from Table 5, it is concluded that the MFO algorithm outperformed others.

The reasons for the better performance of MFO compared with other algorithms are explained here. The GA became gradually a dominant optimization technique compared to deterministic approaches mainly due to the higher probability of local solutions avoidance. However, the main drawback of GA was the stochastic nature of this algorithm which resulted in finding different solutions in every run. Despite the relatively high convergence rate of PSO, it has the drawback of premature convergence to local optimal and ineffectiveness in exploring the whole search space. Similarly, the original version of the GWO algorithm has the drawbacks of low solving accuracy, bad local searching ability, and slow convergence rate. Similarly, the disadvantage of GHO was being easy to fall into the local optimum which prevented the search process from finding a better solution. On the other hand, MFO is able to locate the local and global optimal solutions accurately with less computational time [43].

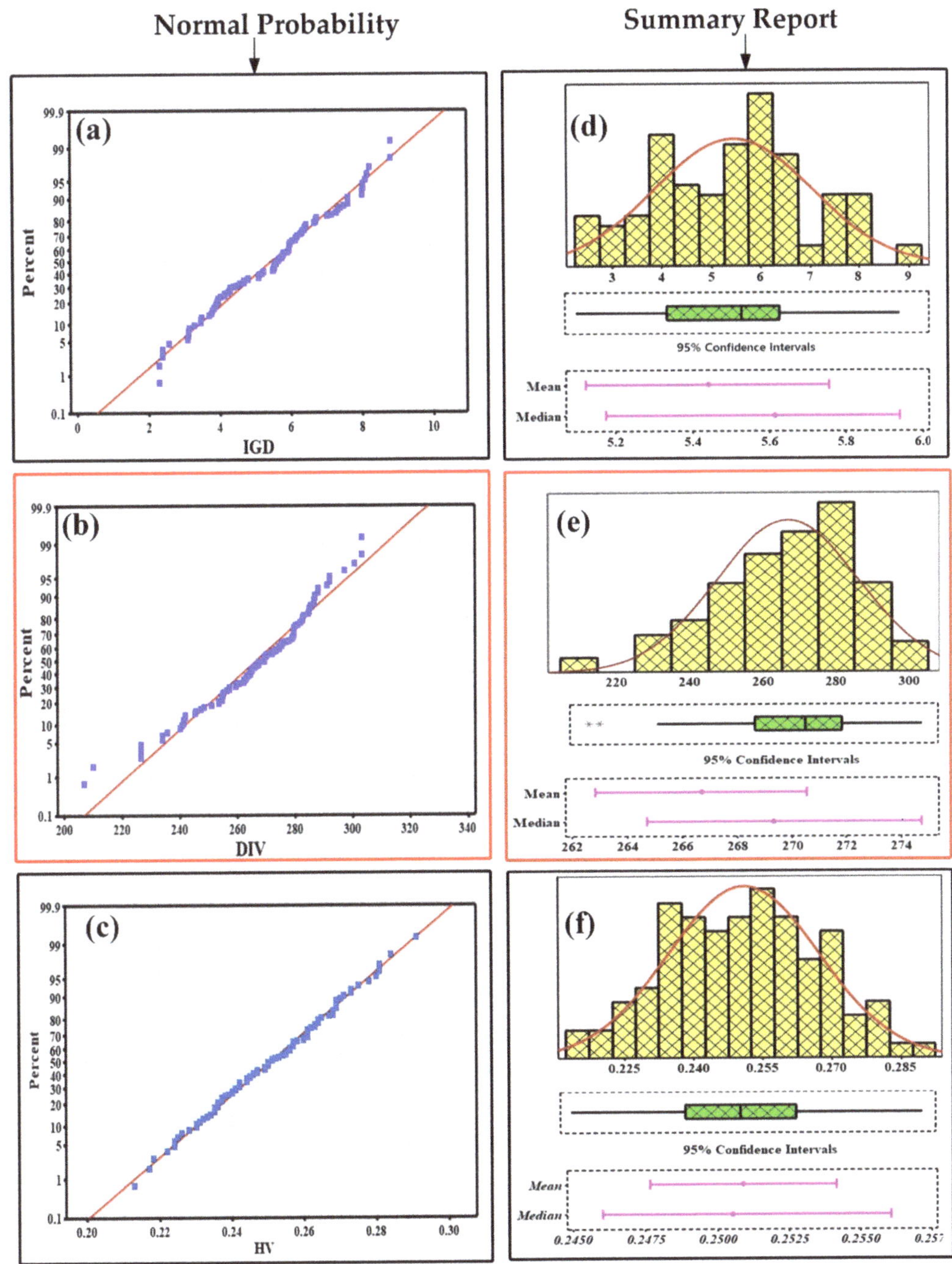

Figure 6. (**a**–**c**) Normal Probability plot (**d**–**f**) summary report.

4. Conclusions

The purpose of this study was to minimize machinability indices CF, SR, and CT while performing the turning of Hastelloy X. Three levels of turning process parameters namely cutting speed (v_c), feed rate f, and machining environment were considered for performing the experiments under L_{27} orthogonal array basis. Further, the MFO algorithm was used to identify the optimal set of turning process parameters to minimize the machinability indices individually and simultaneously. Three case studies were carried out for this purpose. The conclusion drawn from these case studies is given below.

1. From the case study 1 (minimization of machinability indices individually), as compared to other algorithms such as GHO, GA, PSO, and GWO, the MFO algorithm yielded the minimum values of CF = 127.1 N, SR = 1.78 μm, and CT = 33.19 °C for the optimal set of turning process parameters such as v_c = 124 m/min, f = 0.05 mm/rev, and cryogenic environment. The range of reduction in CF, SR, and CT values based on the MFO algorithm was 4–8 %, 1–23%, and 3–57%, respectively, compared with other algorithms.

2. The simultaneous minimization of dual machinability indices with three combinations were performed using the MFO algorithm in case study 2. The results were compared with the results obtained from other algorithms. Based on the hypervolume indicator identified from the Pareto analyses, again the MFO outperformed others, and the corresponding optimal set of input parameters were identified.

3. In case study 3, the simultaneous minimization of all three machinability indices was carried out using the MFO algorithm. The performance of MFO algorithm was compared with other algorithms using the quality indicators namely Diversity, Inverted Generational Distance, and Hyper Volume. From the analyses, the best results were obtained as CF = 171.13 N, SR = 2.35 μm and CT = 72.28 °C form the MFO algorithm for the inputs of v_c = 93 m/min, f = 0.05 mm/rev and cryogenic environment.

Based on the results of all three case studies, the MFO algorithm effectively predicted the optimal set of turning process parameters in view of minimizing the machinability indices individually and simultaneously when compared with other algorithms. Further, the other machinability indices such as tool life and machining cost will also be considered in addition to the existing indices as the future work.

Author Contributions: Conceptualization, V.S.; Methodology, V.S., J.S. and Y.N.; Experimental design, V.S. and J.S.; Experimental setup, V.S.; Measurements, V.S., S.K.M. and Y.N.; Investigation, L.N., S.K.M. and Y.N.; Resources, L.N., S.K.M. and Y.N.; Visualization, V.S., L.N., S.K.M. and S.S.; Writing—Original Draft Preparation, V.S., J.S., L.N., S.K.M., Y.N., S.S., E.A.N., J.P.D. and H.M.A.M.H.; Writing—Review & Editing, V.S., J.S., L.N., S.K.M., Y.N. and S.S.; Supervision, J.S. and V.S.; Project administration, V.S. and J.S.; Funding acquisition, J.S. and S.S. All authors have read and agreed to the published version of the manuscript.

Funding: This research has received funding from King Saud University through Researchers Supporting Project number (RSP-2021/164), King Saud University, Riyadh, Saudi Arabia. Moreover, this is part of the project was financially supported by Fundamental Research Fund of Shandong University under the funding code No. 2019HW040. the Future for Young Scholars of Shandong University, China under the funding code No. 31360082064026.

Institutional Review Board Statement: Not applicable.

Informed Consent Statement: Not applicable.

Data Availability Statement: Not applicable.

Acknowledgments: The authors extend their appreciation to King Saud University for funding this work through Researchers Supporting Project number (RSP-2021/164), King Saud University, Riyadh, Saudi Arabia.

Conflicts of Interest: The authors declare no conflict of interest.

References

1. Sivalingam, V.; Zhuoliang, Z.; Jie, S.; Baskaran, S.; Yuvaraj, N.; Gupta, M.K.; Aqib, M.K. Use of Atomized Spray Cutting Fluid Technique for the Turning of a Nickel Base Superalloy. *Mater. Manuf. Process.* **2021**, *36*, 373–380. [CrossRef]
2. METODO. Optimization of the turning parameters for the cutting forces in the Hastelloy X superalloy based on the Taguchi method. *Mater. Tehnol.* **2014**, *48*, 249–254.
3. Kadirgama, K.; Abou-El-Hossein, K.A.; Mohammad, B.; Al-Ani, H.; Noor, M. Cutting force prediction model by FEA and RSM when machining Hastelloy C-22HS with 90 holder. *J. Sci. Ind. Res.* **2008**, *67*, 421–427.
4. Kadirgama, K.; Abou-El-Hossein, K.; Noor, M.; Sharma, K.; Mohammad, B. Tool life and wear mechanism when machining Hastelloy C-22HS. *Wear* **2011**, *270*, 258–268. [CrossRef]
5. Sofuoğlu, M.A.; Çakır, F.H.; Gürgen, S.; Orak, S.; Kuşhan, M.C. Experimental investigation of machining characteristics and chatter stability for Hastelloy-X with ultrasonic and hot turning. *Int. J. Adv. Manuf. Technol.* **2018**, *95*, 83–97. [CrossRef]
6. Dhananchezian, M. Study the machinability characteristics of Nicked based Hastelloy C-276 under cryogenic cooling. *Measurement* **2019**, *136*, 694–702. [CrossRef]
7. Kesavan, J.; Senthilkumar, V.; Dinesh, S. Experimental and numerical investigations on machining of Hastelloy C276 under cryogenic condition. *Mater. Today Proc.* **2020**, *27*, 2441–2444. [CrossRef]
8. Dhananchezian, M.; Rajkumar, K. Comparative study of cutting insert wear and roughness parameter (Ra) while turning Nimonic 90 and hastelloy C-276 by coated carbide inserts. *Mater. Today Proc.* **2020**, *22*, 1409–1416. [CrossRef]
9. Oschelski, T.B.; Urasato, W.T.; Amorim, H.J.; Souza, A.J. Effect of cutting conditions on surface roughness in finish turning Hastelloy® X superalloy. *Mater. Today Proc.* **2021**, *44 Pt 1*, 532–537. [CrossRef]
10. Venkatesan, K.; Devendiran, S.; Nishanth Purusotham, K.; Praveen, V.S. Study of machinability performance of Hastelloy-X for nanofluids, dry with coated tools. *Mater. Manuf. Process.* **2020**, *35*, 751–761. [CrossRef]
11. Sivalingam, V.; Zan, Z.; Sun, J.; Selvam, B.; Gupta, M.K.; Jamil, M.; Mia, M. Wear behaviour of whisker-reinforced ceramic tools in the turning of Inconel 718 assisted by an atomized spray of solid lubricants. *Tribol. Int.* **2020**, *148*, 106235. [CrossRef]
12. Zhao, Y.; Li, J.; Guo, K.; Sivalingam, V.; Sun, J. Study on chip formation characteristics in turning NiTi shape memory alloys. *J. Manuf. Process.* **2020**, *58*, 787–795. [CrossRef]
13. Chetan; Gosh, S.; Rao, P.V. Environment friendly machining of Ni–Cr–Co based super alloy using different sustainable techniques. *Mater. Manuf. Process.* **2016**, *31*, 852–859. [CrossRef]
14. Iturbe, A.; Hormaetxe, E.; Garay, A.; Arrazola, P. Surface integrity analysis when machining Inconel 718 with conventional and cryogenic cooling. *Procedia CIRP* **2016**, *45*, 67–70. [CrossRef]
15. Sivaiah, P.; Chakradhar, D. Influence of cryogenic coolant on turning performance characteristics: A comparison with wet machining. *Mater. Manuf. Process.* **2017**, *32*, 1475–1485. [CrossRef]
16. Tebaldo, V.; di Confiengo, G.G.; Faga, M.G. Sustainability in machining: "Eco-friendly" turning of Inconel 718. Surface characterisation and economic analysis. *J. Clean. Prod.* **2017**, *140*, 1567–1577. [CrossRef]
17. Shokrani, A.; Al-Samarrai, I.; Newman, S.T. Hybrid cryogenic MQL for improving tool life in machining of Ti-6Al-4V titanium alloy. *J. Manuf. Process.* **2019**, *43*, 229–243. [CrossRef]
18. Mehta, A.; Hemakumar, S.; Patil, A.; Khandke, S.; Kuppan, P.; Oyyaravelu, R.; Balan, A. Influence of sustainable cutting environments on cutting forces, surface roughness and tool wear in turning of Inconel 718. *Mater. Today Proc.* **2018**, *5*, 6746–6754. [CrossRef]
19. Khalilpourazari, S.; Khalilpourazary, S. Optimization of production time in the multi-pass milling process via a Robust Grey Wolf Optimizer. *Neural Comput. Appl.* **2018**, *29*, 1321–1336. [CrossRef]
20. Khalilpourazari, S.; Khalilpourazary, S. A lexicographic weighted Tchebycheff approach for multi-constrained multi-objective optimization of the surface grinding process. *Eng. Optim.* **2017**, *49*, 878–895. [CrossRef]
21. Khalilpourazari, S.; Khalilpourazary, S. A Robust Stochastic Fractal Search approach for optimization of the surface grinding process. *Swarm Evol. Comput.* **2018**, *38*, 173–186. [CrossRef]
22. Rao, R.V.; Rai, D.P.; Balic, J. Multi-objective optimization of abrasive waterjet machining process using Jaya algorithm and PROMETHEE Method. *J. Intell. Manuf.* **2019**, *30*, 2101–2127. [CrossRef]
23. Rao, R.V.; Rai, D.P.; Balic, J. A multi-objective algorithm for optimization of modern machining processes. *Eng. Appl. Artif. Intell.* **2017**, *61*, 103–125. [CrossRef]
24. Khalilpourazari, S.; Khalilpourazary, S. Optimization of time, cost and surface roughness in grinding process using a robust multi-objective dragonfly algorithm. *Neural Comput. Appl.* **2020**, *32*, 3987–3998. [CrossRef]
25. Khalilpourazari, S.; Khalilpourazary, S. SCWOA: An efficient hybrid algorithm for parameter optimization of multi-pass milling process. *J. Ind. Prod. Eng.* **2018**, *35*, 135–147. [CrossRef]
26. Almeida, J.H.S., Jr.; St-Pierre, L.; Wang, Z.; Ribeiro, M.L.; Tita, V.; Amico, S.C.; Castro, S.G. Design, modeling, optimization, manufacturing and testing of variable-angle filament-wound cylinders. *Compos. Part B Eng.* **2021**, *225*, 109224. [CrossRef]
27. Wang, Z.; Almeida, J.H.S., Jr.; St-Pierre, L.; Wang, Z.; Castro, S.G. Reliability-based buckling optimization with an accelerated Kriging metamodel for filament-wound variable angle tow composite cylinders. *Compos. Struct.* **2020**, *254*, 112821. [CrossRef]
28. Almeida, J.H.S., Jr.; Ribeiro, M.L.; Tita, V.; Amico, S.C. Stacking sequence optimization in composite tubes under internal pressure based on genetic algorithm accounting for progressive damage. *Compos. Struct.* **2017**, *178*, 20–26. [CrossRef]

29. Sridhar, R.; Subramani, C.; Pathy, S. A grasshopper optimization algorithm aided maximum power point tracking for partially shaded photovoltaic systems. *Comput. Electr. Eng.* **2021**, *92*, 107124. [CrossRef]

30. Purushothaman, R.; Rajagopalan, S.; Dhandapani, G. Hybridizing Gray Wolf Optimization (GWO) with Grasshopper Optimization Algorithm (GOA) for text feature selection and clustering. *Appl. Soft Comput.* **2020**, *96*, 106651. [CrossRef]

31. Steczek, M.; Jefimowski, W.; Szeląg, A. Application of Grasshopper Optimization Algorithm for Selective Harmonics Elimination in Low-Frequency Voltage Source Inverter. *Energies* **2020**, *13*, 6426. [CrossRef]

32. Sangwan, K.S.; Kant, G. Optimization of machining parameters for improving energy efficiency using integrated response surface methodology and genetic algorithm approach. *Procedia CIRP* **2017**, *61*, 517–522. [CrossRef]

33. Hazir, E.; Ozcan, T. Response surface methodology integrated with desirability function and genetic algorithm approach for the optimization of CNC machining parameters. *Arab. J. Sci. Eng.* **2019**, *44*, 2795–2809. [CrossRef]

34. Almeida, J.H.S., Jr.; Bittrich, L.; Nomura, T.; Spickenheuer, A. Cross-section optimization of topologically-optimized variable-axial anisotropic composite structures. *Compos. Struct.* **2019**, *225*, 111150. [CrossRef]

35. Johari, N.F.; Zain, A.M.; Mustaffa, N.H.; Udin, A. Machining parameters optimization using hybrid firefly algorithm and particle swarm optimization. In *Journal of Physics: Conference Series*; IOP Publishing: Tokyo, Japan, 2017; Volume 892, p. 012005.

36. Lmalghan, R.; Rao, K.; ArunKumar, S.; Rao, S.S.; Herbert, M.A. Machining parameters optimization of AA6061 using response surface methodology and particle swarm optimization. *Int. J. Precis. Eng. Manuf.* **2018**, *19*, 695–704. [CrossRef]

37. Tamilarasan, A.; Rajmohan, T.; Ashwinkumar, K.; Dinesh, B.; Praveenkumar, M.; Reddy, R.D.; Kiran, K.S.; Elangumaran, R.; Krishnamoorthi, S. Hybrid WCMFO algorithm for the optimization of AWJ process parameters. In *IOP Conference Series: Materials Science and Engineering*; IOP Publishing: Tokyo, Japan, 2020; Volume 954, p. 012041.

38. Kamaruzaman, A.F.; Zain, A.M.; Alwee, R.; Yusof, N.M.; Najarian, F. Optimization of Surface Roughness in Deep Hole Drilling using Moth-Flame Optimization. *ELEKTRIKA J. Electr. Eng.* **2019**, *18*, 62–68. [CrossRef]

39. Fountas, N.; Koutsomichalis, A.; Kechagias, J.; Vaxevanidis, N. Multi-response optimization of CuZn39Pb3 brass alloy turning by implementing Grey Wolf algorithm. *Frat. Ed Integrità Strutt.* **2019**, *13*, 584–594. [CrossRef]

40. Sibalija, T.V.; Kumar, S.; Patel, G.M. A soft computing-based study on WEDM optimization in processing Inconel 625. *Neural Comput. Appl.* **2021**, *33*, 11985–12006. [CrossRef]

41. Sekulic, M.; Pejic, V.; Brezocnik, M.; Gostimirović, M.; Hadzistevic, M. Prediction of surface roughness in the ball-end milling process using response surface methodology, genetic algorithms, and grey wolf optimizer algorithm. *Adv. Prod. Eng. Manag.* **2018**, *13*, 18–30. [CrossRef]

42. Dhananchezian, M.; Rajkumar, K. Cryogenic turning of Hastelloy C-22. *Mater. Today Proc.* **2020**, *22*, 3075–3081. [CrossRef]

43. Yang, Z.; Shi, K.; Wu, A.; Qiu, M.; Hu, Y. A hybird method based on particle swarm optimization and moth-flame optimization. In Proceedings of the 11th International Conference on Intelligent Human-Machine Systems and Cybernetics (IHMSC), Hangzhou, China, 24–25 August 2019; Volume 2, pp. 207–210.

applied sciences

Article

Reflections on the Customer Decision-Making Process in the Digital Insurance Platforms: An Empirical Study of the Baltic Market

Gedas Baranauskas * and Agota Giedrė Raišienė *

Institute of Leadership and Strategic Management, Mykolas Romeris University, Ateities Str. 20, LT-08303 Vilnius, Lithuania
* Correspondence: gedasbaranauskas@mruni.eu (G.B.); agotar@mruni.eu (A.G.R.); Tel.: +370-62-151-887 (G.B.)

Abstract: Multifold effects of the COVID-19 global health crisis and economic lockdowns are reflected in the insurance industry, and are predicted to expand to the post-COVID-19 era. It is expected that, within a short period of time, the current worldwide situation, in regards to the coronavirus pandemic, will be reflected in new trends, regarding customer behavior, organizational management, and culture, as well as reveal improved business management models, legacy infrastructure, and service systems in insurance organizations. Here, a focus on end-user preferences, data, and their behavior modeling in digital platforms are major practical drivers within the modern insurance concept, but there is a paucity of researches within the theoretical synthesis of consumer decision-making (CDM) models, information system theories, and behavioral economics concerning modern insurance-specific value chains and digitalized decision-making processes. Therefore, the following research aims to expand upon the existing scientific knowledge of end-user behavioral patterns and process frameworks in the Baltic insurance market, by including and examining a factor group of technological enablers and a digital environment. Research results in digitalization, personalization, and customization levels within the Baltic non-life insurance market are homogenous with a leading position of Estonia and overall evaluations ranging between Satisfied and Rather Good. There are also three major factor groups and process stages identified, which influence insurance purchase decision-making in digital insurance platforms in the Baltic market.

Keywords: digital platforms; decision-making; insurance; Baltic; customization; personalization

Citation: Baranauskas, G.; Raišienė, A.G. Reflections on the Customer Decision-Making Process in the Digital Insurance Platforms: An Empirical Study of the Baltic Market. *Appl. Sci.* **2021**, *11*, 8524. https://doi.org/10.3390/app11188524

Academic Editors: Vladimir Modrak and Zuzana Soltysova

Received: 4 August 2021
Accepted: 6 September 2021
Published: 14 September 2021

Publisher's Note: MDPI stays neutral with regard to jurisdictional claims in published maps and institutional affiliations.

1. Introduction

Countries across the world underwent an unprecedented global health crisis and subsequent lockdowns in 2020–2021 due to the coronavirus disease (COVID-19) pandemic, moreover, social and economic outcomes are still hard to predict at a full scope. It is argued that insurance organizations should focus on the development of hybrid service-based and customer-driven business models, sustainable and innovative digital products, and consolidations, with InsurTech and technology companies, for better digitization of distribution channels [1,2]. The ongoing global pandemic has revealed a lack of flexibility, emotional connection, and data harmonization with the day-to-day needs of end-users, as well as personalized, situation-based, and easily customizable insurance services, as a part of practical improvements toward the modern insurance [3,4]. It is important to note intensive and dynamic changes in the market structure, which are mostly determined by a high competition among traditional peers and new virtual incumbents, non-traditional insurance service providers, such as "Big Tech", and product manufacturers [5,6].

From a theoretical point of view, the research refers to predecessor studies in the field of non-life insurance consumer decision-making, such as Hsee and Kunreuther (2000) [7], Kunreuther and Pauly (2006; 2015) [8,9], and a cross-border study report by the European Commission (2017) [10]. The present research also follows key findings and limitations of scientific studies on behavioral patterns in the Baltic insurance market, such as research by Kiyak and Pranckevičiūtė (2014) [11], and longitudinal studies conducted by Ulbinaitė et al. (2010, 2011, 2013) [12–15]. Additionally, novel studies of comparison websites [2], multi-sided platforms [1] in the insurance industry and the Service-Dominant Architecture (SDA) [16], as well as insurance literacy [16,17], were considered. However, a recent increase in practical demands for hyper-personalized support services and digitally customized insurance outcomes of on-demand and usage-based insurance reveal a research gap. There is a need for a multi-tier factor evaluation of the digital insurance environment and multi-agent-based model simulations oriented toward decision making in digital insurance platforms. Therefore, this research aims to extend the existing scientific knowledge on insurance end-user behavioral patterns and frameworks by including and examining factors of customization, personalization, and technological enablers. To investigate this situation, the following research questions have been raised:

1. How do digital environments and technological factors influence the decision-making process in insurance distribution platforms?
2. What are the predominant factors of insurance decision-making processes in digital insurance platforms?

From a methodological point of view, the research follows a triangulation logic, combining an online survey with Fuzzy and Likert logic, simplified art-based method outcomes for information presentation, and an embedded, descriptive case study with statistical data analysis methods, which summarize key findings of the online survey. The structure of this research paper is composed of five main sections. The first section introduces the theoretical background of the topic. The second section provides a detailed description of the research and survey methodology and applied methods. Results of the online survey are presented and analyzed in the third section. The fourth section presents a discussion on research limitations and possible future research directions. The fifth section concludes the main findings from theoretical and practical analyses.

2. The Theoretical Background of Digital Non-Life Insurance Concepts and Insurance Customer Decision-Making Processes

Although the importance of non-life insurance products for the financial wellbeing of individuals and society has been recognized for a long time, a practical spread is still considered vague, i.e., the lack of insurance literacy, financial decision-making skills, and a number of underinsured persons are observed [17]. On a theoretical level, it is argued that modern insurance-related decisions and customer behavior cannot be sufficiently explained by applying traditional neoclassical economic and financial theories and ignoring cognitive and emotional factors.

In general, fundamental studies on consumer decision-making in insurance services, insurers, and market behavior, were carried out by Hsee and Kunreuther in 2000 [7] and Kunreuther and Pauly in 2006 [8]. However, the phenomenon of insurance purchases under these studies has been analyzed from a holistic standpoint, resulting in a focus on high-level benchmark models of insurance demand and supply, legal regulation, and government insurance market principles. The interdisciplinary nature of the behavior of consumers in the insurance market was analyzed, mostly regarding monetary financial risks and positive theory perspectives, with limited attention to the influence of behavioral finance and economics theories, as well as heuristic and bias phenomena. This research gap was only partially covered in the last decade via a comprehensive analysis from the European Commission (2014) [10] study, and Kunreuther and Pauly in 2015 [9]. Here, assumptions on behavioral finance and economy and cognitive psychology domains have contributed to the foundation of the modern insurance concept by introducing the influence of individual risk perceptions, heuristic reflections in framing effects, and intuitive and deliberative thinking along general concepts of probability and risk, distribution channels, and social-demographic differences in Europe [9,10,17]. Nevertheless, afterwards, an increase in scientific attention to the phenomenon of insurance digitization and digitalization has been noticed, as the majority of research was mostly oriented to countries from Western and Central European regions, while analyses of the Baltic region (Lithuania, Latvia, and Estonia) resulted in low volumes, and were mostly limited to a one country-level (Lithuania or Latvia). A significant part of the existing scientific studies on digital insurance have been completed at a high-generality level, scattered to different parts of the insurance-specific value chain or technological solutions, without sufficiently considering the influence of modern mass customization and personalization concepts and marketing theories. A similar situation can be identified when analyzing consumer decision-making in insurance services, which, in the case of the Baltic region, are geographically limited, practically outdated, and, as in the case of Lithuania, the majority of studies were completed from 2008 to 2014, without considering customization, personalization, or digitalization factors. Therefore, the multidimensional and cross-border studies of digital insurance decision-making, insurance literacy, and platform business models in the Baltic region are required both theoretically and practically.

To conclude, the modern insurance combines derivatives from several concepts and theories:

- Risk management, regarding the relationships between insurance service counter-parties, the obligation of exchanging customer risks, risk handling methods, and the overall management of the asymmetric dominance effect legally and ethically [12,18];
- Customer behavior, regarding personal mental considerations and physical actions, behavioral and cognitive decision-making biases, purchase decision function, and interaction between the social system and the social reality [18,19];
- CDM and HCDM, regarding behavior and purchase decision-making of insurance customers. It can be defined as a continuous sequence of mental considerations and physical actions, which are divided into the two following main groups and two stages: a perception group of a need for insurance and a perception group of affordability; a personal evaluation stage of needs and affordability factors and objective evaluation stage of insurance content [12,13,18,19]. On the other hand, organizations, including the non-life insurance market, still focus on favorable customer experience management with limited attention and analysis on the multifaceted concept of customer experience and journey management [19]. This situation illustrates a need for conceptualization of insurance customer experience drivers and outcomes in the traditional model of three process stages and comprehensive evaluation categories. These categories combine customer experience drivers and customer integration via value co-creation efforts within different processes of insurance decision-making [20]. This type of holistic customer experience evaluation approach indicates that the customer value and experience are context-dependent, systematic, and interactive within all stages of the process [19,21];
- Information system theories and models of technology acceptance and self-service technologies (SSTs), in the form of constructs and determinant groups of behavioral intentions, usage behavior, and digital insurance platforms. The following constructs and determinants of behavioral intentions and usage behavior can be identified as slightly modified, but reflecting both traditional (face-to-face) and digital insurance distribution channels, such as performance expectancy, effort expectancy, social influence/boundaries, facilitating conditions (organizational and technical infrastructure) and attitude factors (cognitive processes and emotional reactions) [22,23].

Modern insurance specific value chain and co-creation activities depend on features of the core offering-purchase stage, but are also shaped by numerous internal and external factors: customer sourcing processes and an organization role within customer operations, high-level factors of sociodemographic and socioeconomic characteristics, insurance literacy, platform design and information layout, and social media and reality [12,13,17,19,24,25]. Figure 1 summarizes key elements and processes of insurance decision-making.

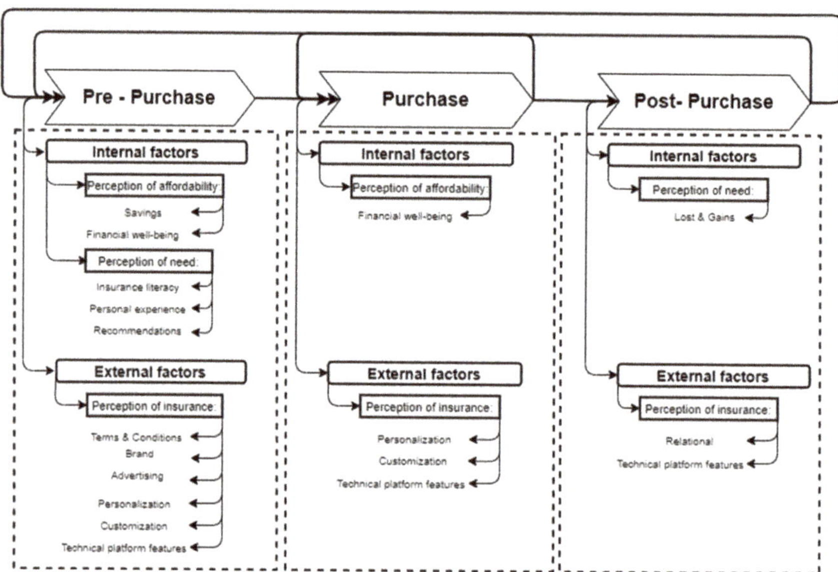

Figure 1. Features and background of the insurance decision-making process. Source: composed by the authors, by following [5,12,13,17].

Figure 1 illustrates the main process stages and influential driving factors, which affect the insurance decision-making and customer transition through the stages. In general, all influential driving factors can be divided into two groups. The first group is formed by internal factors, which are related to personal evaluation of an insurance need, financial affordability, and past and ongoing experience with insurance service providers. The second group compounds external factors, which define an objective and holistic evaluation of insurance decision-making from legal, marketing, and technical points of view. The financial and marketing-oriented factors have key role in the pre-purchase stage, where the final decision to purchase is made. It should also be outlined that technological factors and the digital environment have a significant influence on the modern insurance value chain and customer experience management. Contrary to the legal, informational, relational, or social-demographic driver factors, they are recognized equally important to the entire value co-creation process in all three stages of the insurance decision-making process. An intensive development of the internet technology, mobile devices, and digital platform business model have become critical enablers to balance the quality of service delivery, insurance personalization, and customization, with different customer expectations and experience levels:

- Operational activities and product-levels via the mediation of a customer's journey, by personalized easy-access communication tools and information exchange. It also allows ensuring operational capabilities for value co-creation processes and a spectrum of product customization options;
- Customer experience management activities and individual-levels via a well-designed purchase process in graphical interface solutions, situations involving flow interruptions or overwhelming information. It also promotes a continuous positive social interaction and an emotional brand connection after the purchase stage [26].

This theoretical setup reflects the modern non-life insurance meaning, which stands for a continuous (but not a simultaneous) sequence of the insurance-related processes and multiple internal and external influencing factors within the decision-making. The complex and non-linear non-life insurance market nature requires a decentralized, digitalized, and individual-centric approach to organizational management, product configuration, and customer experience management [12,13,17]. Otherwise, digital technologies, technological advancements, and global social networks increased the recognition of non-life insurance products and services, and have been rapidly embedded into daily operations and a value co-creation chain of insurance. These trends indicate the shift of traditional insurance concept boundaries and the non-life insurance market from a traditional, provider-centric management approach and service blueprint model, where a homogenous focus on the operational efficiency and cost-service level dichotomy is dominant [26,27]. The emerging prominence of the customer-centric philosophy and holistic design framework and availability of individual-level real-time data factors reflect in newly developed, innovative, and personalized touchpoints, on customer values and experience management [26,27]. Practically, it reflects on advanced online self-service platforms, which work in combination with techniques of personalization and customization and create a new cognitive framework. It embraces an intrinsic security need for customers and leads to a simpler creation, usage, and exchange of insurance knowledge and information [5,26,28].

It is also agreed that the last two decades of the globalization process, intensive digitization, and digitalization, and rapid socioeconomic changes, have brought about a remarkable, two-fold impact on the financial service market, organizations, and customers, including the insurance industry [29]. Therefore, it is recognized that existing integrated and sophisticated quantitative analysis techniques, financial and economic theories, and consumer decision-making (CDM) models are not fully sufficient to be applied in the modern, digitalized financial service, oriented to the platform business model [30]. The analysis and modeling of the optimal financial decision-making process should be supported by the application of behavioral reasoning theory (BRT), forecasting, and multi-criteria decision approach and methods with qualitative data, behavioral, and cognitive factor evaluations [29,31]. New theoretic models and conceptual frameworks should reflect both technical specifications and capabilities of digital platforms as well as recent behavioral patterns of fully digital end-users, which is no longer a linear progression through process stages, but more an iterative decision-making process [30]. The traditional CDM and hybrid decision-making models (HDCM) have been widely applied in practice since 1960s, but there is a paucity of research within the theoretical synthesis of CDM and HDCM models, concerning a modern insurance-specific value chain and digitalized decision-making process. A brief overview of the main CDM models and their influence points on digitalized insurance decision-making processes is presented in Table 1.

Table 1. The influence of the main CDM models on digitalized insurance decision-making processes. Source: created by the authors, following [30,32–36].

Model	Influence Points
F.M. Nicosia's model (1966) [37]	• A mutual relationship between the organization and the consumer. • Feedback area. • The interactive repurchase cycle.
J.F. Engel, D.T. Kollat, and R.D. Blackwell (EKB) model (1968) [38]	• Consideration of interactions between both internal and external inputs, information retrieval. • Presence of consumers' perspectives, feedback-search loop, and partial decision making. • The postponement of decision-making.
J.A. Howard and J.N. Sheth model (1969) [39]	• Inclusion of marketing, social influence, and exogenous variables.
S. Um and J.L. Crompton model (1990) [40]	• Stage of awareness set. • Passive and active information acquisition.
P. Kotler model (1997) [41]	• Identification of suspected target groups. • Differentiation and transformation of first time and repetitive customers.
J.E. McCarthy, W.E. Perreault and P.G. Quester model (1997) [42]	• Impact of social, situational, and mental factors.

Table 1. Cont.

Model	Influence Points
J. Walker and M. Ben-Akiva model (2002) [43]	• Flexible disturbances and irrational behavior. • Combination of latent psychological factors.
P. Kotler and K.L. Keller model (2006, 2012) [44,45]	• Post-purchase stage and behavior. • Skipping or reversing some stages.

For a long period of time, main CDM models defined in Table 1, together with the expected utility theory have been accepted as a dominant paradigm by economic and marketing researchers, and used practically by organizations to determine and model customer responses to products or services, purchase offerings, and motivational appeals [33,36]. On the other hand, modeling of the financial attitude and decision-making process in digital platforms has specific circumstances to consider. The stage of the initial problem, problem recognition, or need arousal, together with a non-standard combination of high psychological and monetary risks, internal and external biases associated with certain consequences of financial decision-making, have a strong influence in this field [30,34]. These particular characteristics of financial decision-making require reframing processes and variables of existing CDM and HCDM models. It is also essential to identify and apply the fundamentals of customer behavior models to the comprehensive examination of the insurance service customer's engagement and characteristics. The following essential points of traditional consumer decision-making models can remarkably contribute to a conceptual framework of the digital insurance decision-making process. Fundamental outcomes of Nicosia's (1966) [37] model, as a separation of the buying process to the multiple stages, iterative and constant connection between the organization and the customer (in the form of a feedback area and repurchase cycle) can be identified in the insurance. Relevant factors, considering various endogenous and exogenous variables, marketing stimuli components, process options of partiality, and postponement of decision-making, were presented in the EKB (1968) [38] and J. A. Howard and J. N. Sheth (1969) [39] models. These elements support the content and the context of the modern insurance and decision-making process, but are also limited to apply fully, due to the logic of the linear process and the unmeasured relationships between variables [36,46]. Later revisions and elaborations of these limitations in the choice set model of S. Um and J. L. Crompton (1990) [40] and the hybrid choice model of J. Walker and M. Ben-Akiva (2002) [43] introduce relevant factors of possible disturbances, irrational behavior, and an unreliable memory of customers, as well as the initial stage of the awareness set [32,33,36]. Furthermore, simplified models by P. Kotler (1997) [43], P. Kotler and K. L. Keller (2006, 2012) [44,45], and J. E. McCarthy, W. E. Perreault, and P. G. Quester (1997) [42] reflect early complex models of buyer behavior, and in this way, support the holistic approach to the decision-making process. Overall, these models outline the pre-purchase and post-purchase stages and customer-orientated activities of identification differentiation and transformation of target groups, which are also important parts of the modern insurance concept.

3. Materials and Methods

The present research is a continuation and a supplementary portion of the authors' multi-step research (2020, 2021) within a practical status, regarding insurance digitalization, customization and personalization level, and digital insurance platforms in the Baltic non-life insurance market. The research also contributes to the research field, providing empirical validation of a conceptual framework of the digital insurance customer decision-making process. Therefore, data collection and analyses were carried out by using a convergent parallel research design and the following triangulation of scientific methods:

1. A descriptive thematic analysis and synthesis of scientific findings within digital non-life insurance and decision-making models.

2. Modeling a conceptual process flow and framework of digital insurance purchase-decision making. A simplified logic of the Robinson's [47–49] conceptual modeling framework in combination with a logical data flow diagram was applied.
3. The online survey of 157 insurance-related specialists from three Baltic countries, using a structured questionnaire of 24 questions, a full-blown Likert evaluation scale, and visualized prototypes of online customization frameworks. Visual outcomes were illustrated by using the design and prototyping software Axure RP Pro (version 8).
4. Descriptive statistics, multiple factors, and correlation analyses of online survey results. The exploratory factor analysis (EFA), confirmatory factors analyses (CFA), and Pearson correlation analysis were applied by using the statistical analysis software IBM SPSS Statistics 26 (Armonk, NY, USA: IBM Corp.).

It is argued that a convergent parallel research design and triangulation of scientific methods allow simultaneous collection, analysis, and interpretation of quantitative data and qualitative evaluations in a single research study. Moreover, complementary data can be used and transformed to a more holistic understanding of the research phenomenon [50,51]. The details of the questionnaire are presented in Table 2.

Table 2. Presentation of the questionnaire structure, content, and methodological foundation. Source: composed by the authors.

Element	Content
Logic of structure	Twenty-four close-ended questions and statements: Three demographic-type questions about the respondent's geographic location, age group, and gender; Four questions about the "state-of-the-art", regarding country specifics, within levels of insurance digitalization, insurance service provider preparation for digitalization, customization, and personalization in digital insurance platforms; Three comparative statements about the visualized prototypes of three online customization frameworks; Fourteen questions oriented to the conceptual framework about the multi-criteria influencing customer decision making in digital insurance platforms.
Methodological foundation	A full-blown Likert scale of 9 points and linguistic variables in the following parts of the questionnaire: - In questions 4–7, point 5 has a value of "Neutral"; 1—"Very poor"; 9—"Excellent"; - In questions 10–24, point 5 has a value of "Neutral"; 1—"Extremely low"; 9—"Extremely Strong".

The questionnaire was formulated by using a combined full-blown Likert scale and a Fuzzy set of 9 points with numerical and linguistic values due to several methodological reasons:

- The research object of the conceptual framework regarding the customer decision-making processes in digital insurance platforms follows the multi-criteria decision making (MCDM) approach. It concerns the design of the criteria evaluation scale and validity of results, using the traditional ranking scale [52,53]. It is also recommended to reduce possible risks of uncertainty, subjective interpretation, and bias within responses, by applying specific wording techniques, reversed forms, as well as an improved scale for evaluations [52];
- The research aims to evaluate the influence of the digital environment and technological factors as well as identify the predominant factors, as it is not aligned to the classical set theory, binary terms, and bivalent conditions. It requires a comparison concerning a certain criterion (a degree of influence to purchase), gradual membership, and a combined linguistic and visual analog scale for easier interpretation and understanding of questions [54];

- Provision of responses grounded with a well-balanced and gradual assessment for further interpretation within methods of descriptive statistics, and comparative and correlation analysis [54].

The selection of the fundamental scale of nine points (from 1 to 9) follows the multi-criteria decision making (MCDM) approach and the logic of the analytic hierarchy process (AHP). It determines a comparison of criteria and their weight importance within the suggested conceptual framework [55]. It is also argued that a combined full-blown Likert and fundamental AHP 1 to 9 scale is relevant under the cross of the threshold at n = 10 criteria [56].

The online survey was conducted through multiple forms, distribution channels, and periods. The detailed summary of the surveying process is as follows:

- The introductory section and the questionnaire were translated to English and three local languages of the Baltic countries, Lithuanian, Latvian, and Estonian respectively. Translation was handled in collaboration with native speakers of local languages and a qualified English linguist;
- The main survey distribution channel involved a direct contact with insurance service distributors and institutions via publicly available and/or personal contact emails. Supplementary channels and forms involved messaging and posting on Facebook and via the authors' personal professional network;
- The online survey process was held between 16 February and 22 May 2021.

The research sample was formulated by following a logic of probability sampling and a simple sample type. The probability sampling selection was related to the specific research and target population, a methodological set of selection criteria to participate in the survey, and the overall possibility of reducing the sample bias and gathering higher, unbiased data quality. Authors accept key limitations and risks of using simple random sampling—a high level of the standard error of estimate [57]. The target population was defined by using two critical selection criteria: (a) working positions and experiences in the non-life insurance field, and (b) the workplace as an insurance service provider, physically located and operating in the Baltic region (Lithuania, Latvia, or Estonia).

4. Results

A combination of statistical techniques, involving descriptive statistics and factor analyses, are recommended within the applied research to simplify an interpretation of quantitative variables and, overall, to ensure a comprehensive examination of the empirical dataset [58]. This type of multi-level qualitative and quantitative analysis allows researchers to examine the validity of theoretical constructs and the consistency of research instruments, sampling adequacy, and underlining structures and relationships among latent variables (factors) [58,59]. The selected forms of the descriptive statistics involved a manual calculation of a simple and grouped frequency distribution and measures of central tendency. The factor analysis was conducted with the SPPS statistical software (version 26) by completing exploratory factor analysis (EFA), confirmatory factor analysis (CFA), and Pearson's correlation analysis. The logical sequence of statistical analysis procedures is presented in Figure 2.

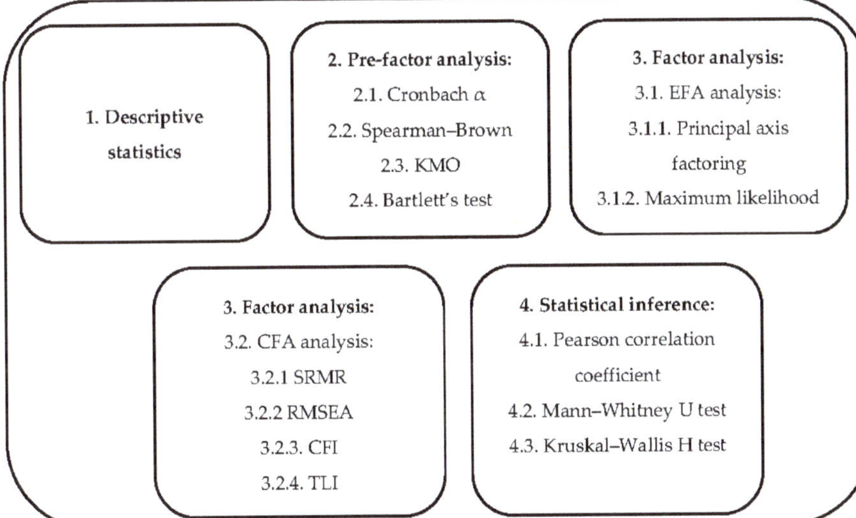

Figure 2. Logical sequence and content of statistical analysis. Source: created by the authors.

4.1. Descriptive Statistics

Application of the descriptive statistics provides a valuable summary of the dataset characteristics and research sample in a manageable and organized format [60]. It also contributes to the research conclusions and validation of the conceptual framework by revealing the influence of sociodemographic factors. The main characteristics of the research sample are presented in Table 3.

Table 3. Characteristics of the research sample. Source: composed by the authors.

Variables	Data Values	Absolute Number	%
Gender	Female	123	78
	Male	34	32
Age group	18–25	36	23
	26–35	54	34
	36–45	47	30
	46–55	17	11
	56–65	3	2
	+65	0	0
Country	Estonia	32	20
	Latvia	31	20
	Lithuania	94	60

In total, 157 insurance-related specialists from three Baltic countries agreed to participate in the online survey; the majority (64% of all respondents) belong to the 26–45 year-old age group, are female (78%), and from Lithuania (60%). One important reliability feature of the research sample is the variety of respondent age groups. Accordingly, a representation of five different age groups was identified, while the majority was composed of three age groups within the age range of 18–55. Nevertheless, the analysis of results shows that sociodemographic factors do not have any significant influence on digitalization, personalization, and customization evaluation or factor groups of the insurance decision-making process in digital platforms.

4.2. Factor and Correlation Analysis

4.2.1. Pre-Factor Analysis

The pre-factor analysis of four indicators was conducted using the SPPS statistical software (version 26). It was oriented to investigate an internal consistency of questionnaire items, a level of questionnaire reliability, sampling adequacy, and usefulness of factor analysis. Results of the pre-factor analysis are presented in Table 4.

Table 4. Results of indexes in the pre-factor analysis.

Indices	Result
Cronbach α	0.875
Spearman–Brown	0.701
Kaiser–Meyer–Olkin (KMO)	0.839
Bartlett's test of sphericity χ^2	0.000

As per Table 4, the Cronbach α indicator has a value of 0.875, which shows that the internal consistency of questionnaire items is relatively high and acceptable. Test reliability and acceptance to perform data reduction using EFA and CFA techniques were confirmed by the Spearman–Brown and Kaiser–Meyer–Olkin coefficients. The value of KMO was 0.839, exceeding the suggested cut-off value of 0.6. Moreover, it indicates a meritorious selection of variables. These results indicate that the sampling was adequate and supports the application of factor analysis [61]. Finally, the Bartlett's test of sphericity (a test of at least one significant correlation between two of the items studied) was significant: χ^2 (153) = 1140.42, $p < 0.05$. Here, the p-value is smaller than the significance level ($\alpha = 0.05$); therefore, it confirms that the dataset is suitable to continue investigation within procedures of EFA and CFA.

4.2.2. EFA, CFA, and Pearson's Correlation Analysis

EFA and CFA are generally accepted and widely used as a combination of statistical techniques, allowing researchers to reach high effectiveness in both statistical data content analysis and testing of construct and criteria [58]. Firstly, EFA was applied to determine a structure of latent dimensions among the observed variables reflected in the items of an instrument. Construct validity was determined by using the principal component analysis (PCA) extraction method and varimax rotation. In addition to the indicators of EFA, which were presented in the pre-factor analysis, as per Table 2, it is important to "ground" the size of the sample. Authors followed the position offered by Guadagnoli and Velicer (1988), who claim that 150 participants could be interpreted as adequate if factor loads of several items exceed 0.80 [58,62].

The CFA principal procedure was used for a better interpretation of factor structures, to test the validity of the structure obtained after EFA. Here, three groups of model fit indices were used to examine the goodness-of-fit of the model with a given dataset: videlicet the absolute fit indices (coefficients of standardized root mean square residual, SRMR), parsimonious indices (a coefficient of root means square error of approximation, RMSEA), and comparative indices (coefficients of comparative fit index, CFI; non-normed fit index, also known as the Tucker–Lewis index, TLI-NNFI) [58]. Results of these indices are provided in Table 5.

Table 5. Results of CFA. Source: composed by the authors, using IBM SPSS Statistics 26 (Armonk, NY: IBM Corp.).

Indices	Result
SRMR	0.063
RMSEA	0.065
CFI	0.921
TLI-NNFI	0.903

It is important to note that the chi-square/df ratio was 1.66, indicating that the model fitted the data well. Overall, the results of fit indices in Table 5 are under the recommended cut-off points for a good model-data fit. The value of SRMR was less than 0.08, the value of RMSEA did not exceed the acceptable limit of 0.08, and values of CFI and TLI-NNFI indices were above the recommended value of 0.90. Finally, the following five groups of related factors were identified after completing EFA and CFA:

1. The first-factor group (F1) consists of six internal factors, which together represent a factor group of personal evaluation and considerations. These six factors represent the evaluation of the influence level within insurance purchases in digital insurance platforms: perception of a need of insurance, financial well-being, potential financial savings, consideration of loss and gains probability, recommendations, and insurance literacy.
2. The second-factor group (F2) was formed of four external factors of technological features and the marketing domain, which influence the insurance purchase process in digital insurance platforms: advertising, the brand of insurance service provider, key technical platform features, and graphical user interface features.
3. The third-factor group (F3) has two factors inside, representing a status regarding evaluation of the digitalization level and level of preparation of insurance service providers to apply digital solutions.
4. The fourth-factor group (F4) composes four combined factors of insurance knowledge and operational features level, which influence the insurance purchase process in digital insurance platforms: insurance literacy, product terms, and conditions acceptability, customization of insurance products, personalization of insurance processes, and services.
5. The fifth-factor group (F5) has two factors, representing an "as-is" evaluation status of the service personalization level and product customization level in existing digital insurance platforms.

Results of the statistical factor analysis show the existence of five-factor groups, but only three of them are directly related to the digital behavior and purchase decision-making process of Baltic insurance customers: the first-factor group, the second-factor group, and the fourth-factor group. Additionally, calculation of the total rank-sum was conducted to identify the most influential factor groups; the first-factor group with 6648 points. The second-factor and the fourth-factor groups have rank-sums of 4482 and 4352 points, respectively. The Pearson correlations between the five-factor groups were calculated in a parallel way. Details of the Pearson's correlation calculation are provided in Table 6.

It should be outlined that three factors groups are identified as having a strong positive Pearson correlation: the first-factor group, the second-factor group, and the fourth-factor group. This result has a two-fold outcome. First, it supports the previous finding of these three factor groups being predominant factors within insurance purchase–decision-making processes at Baltics online insurance platforms. Second, it grounds the theoretical assumption about the digital environment and technological factor influences to end-users and extends the foundation of the theoretical insurance decision-making framework, with new components in the pre-purchase and purchase stages.

Table 6. The calculation of Pearson's correlation. Source: composed by authors by using using IBM SPSS Statistics 26 (Armonk, NY: IBM Corp.).

		Factor Group 1	Factor Group 2	Factor Group 3	Factor Group 4	Factor Group 5
Factor Group 1	Pearson Correlation	1	0.654 **	0.327 **	0.726 **	0.528 **
	Sig. (2-tailed)		0.000	0.000	0.000	0.000
	N	157	157	157	157	157
Factor Group 2	Pearson Correlation	0.654 **	1	0.173 *	0.758 **	0.279 **
	Sig. (2-tailed)	0.000		0.030	0.000	0.000
	N	157	157	157	157	157
Factor Group 3	Pearson Correlation	0.327 **	0.173 *	1	0.444 **	0.763 **
	Sig. (2-tailed)	0.000	0.030		0.000	0.000
	N	157	157	157	157	157
Factor Group 4	Pearson Correlation	0.726 **	0.758 **	0.444 *	1	0.663 **
	Sig. (2-tailed)	0.000	0.000	0.000		0.000
	N	157	157	157	157	157
Factor Group 5	Pearson Correlation	0.528 **	0.279 **	0.763 **	0.663 **	1
	Sig. (2-tailed)	0.000	0.000	0.000	0.000	
	N	157	157	157	157	157

** Correlation is significant at the 0.01 level (2-tailed). * Correlation is significant at the 0.05 level (2-tailed).

Finally, the aim was to analyze the views of insurance-related specialists regarding their sociodemographic characteristics, i.e. age, country, and gender. As the normal distribution assumptions were not met, Mann–Whitney U and Kruskal–Wallis tests were applied to identify whether there were any statistically significant differences between different groups of respondents in the three most influential factor groups. Results of the Kruskal–Wallis test are presented in Table 7.

Table 7. Results of Kruskal–Wallis H test. Source: composed by the authors by using IBM SPSS Statistics 26 (Armonk, NY: IBM Corp.).

Variable	Factor Group	Results of Kruskal–Wallis H Tests		
		Kruskal–Wallis H	Df	Asymptotic Significance
Age group	F1	3.453	4	0.485
	F2	15.209	4	0.004
	F4	2.736	4	0.603

Table 7. *Cont.*

Variable	Factor Group	Results of Kruskal–Wallis H Tests		
		Kruskal–Wallis H	Df	Asymptotic Significance
Country	F1	0.986	2	0.611
	F2	2.317	2	0.314
	F4	1.732	2	0.421

Results of the Mann–Whitney U test are presented in Table 8.

Table 8. Results of Mann–Whitney U test. Source: composed by the authors by using IBM SPSS Statistics 26 (Armonk, NY: IBM Corp.).

Variable	Factor Group	Results of Mann–Whitney U Test Statistics Table			
		Mann–Whitney U	Wilcoxon W	Z	Asymptotic Significance. (2-Tailed)
Gender	F1	1980.000	2575.000	−0.473	0.636
	F2	2086.000	9712.000	−0.021	0.986
	F4	2011.00	2606.000	−0.341	0.733

The Kruskal–Wallis identified that the difference between the mean values of the second-factor group for age groups was statistically significant (p-value = 0.004 is less than the significance level 0.05). It was identified that the highest mean rank of 103.53 was in the age group 46–55, compared to the lowest mean rank of 59.22 in the age group 18–25. Moreover, it was noticed that the value of a mean rank increases constantly within age groups. Several conclusions can be formulated in regards to the above-defined observations:

1. External factors of technological platform features and marketing domains do not have any significant influence on the digital insurance decision-making process for the customers in the age group 18–25, because members of this very age group are considered tech-savvy-type customers without any strong brand recognition in insurance.
2. External factors of technological platform features and marketing domains have a significant influence on the digital insurance decision-making process for customers in the age group of 46–55. Members of this age group tend to have a higher need for a technologically friendly platform, are already experienced with a specific brand of insurance service providers, and possess an overall emotional connection to the brand.

The Mann–Whitney U test was applied to test for gender differences in evaluations of factors groups. Results, presented in Table 8, do not show any statistically significant differences in the evaluation of factor groups according to gender.

5. Discussion and Limitations

First, the above-presented factor and correlation analysis validate theoretical assumptions of internal and external main driving factors and the framework of the digital insurance decision-making process, and disclose the actual Baltic insurer's perception towards this process reflection in digital distribution platforms. The main finding of a strong, positive Pearson correlation among three factor groups both support and extend the findings of Milner and Rosenstreich (2013) [30] and Rocha and Botelho (2018) [63] studies. The personal insurance experience factor has the highest rank (7.4) among all influential factors, supporting the findings by the structural model of attitude towards insurance (ATI) by Rocha and Botelho (2018) [63], where factors of perception of risk in relation to the good/asset and personal concern with finances are indicated as having the highest positive influence towards consumers' willingness to pay for insurance. Additionally, the high rank (7.2) and influence of the insurance service provider brand factor confirm marketing domain-oriented findings of Milner and Rosenstreich (2013) [30], where marketing mixes are recognized as an outlying component of the final financial decision-making frameworks, and the structural ATI model of Rocha and Botelho (2018) [63], where the factor of trust in the industry is also outlined among three of the most influential factors. From the perspective of the Baltic insurance market research, the empirical investigation and results of marketing-oriented factors extend the finding by Kiyak and Pranckevičiūtė (2014) [11], where no significant statistical relationship has been identified between the insurance purchase intention and the discount factor. Moreover, this contradiction of evaluation results in insurance research on marketing-oriented factors will reveal a possible scope of the marketing domain inclusion in further studies on digital insurance decision-making. Second, the findings on the insurance literacy factor have a two-fold meaning and scientific contribution. The lower average evaluation rank (6.9) on the insurance literacy factor partially supports the findings by Weedige and Ouyang [18], Weedige et al. [64], and Allodi et al. [17], where insurance literacy was identified as a potential mitigation action and an influential factor to reduce the problem of underinsurance and low-level insurance knowledge. Otherwise, recognition of the insurance literacy factor in non-standard combination, with the factors product terms and conditions acceptability, customization, and personalization foster an additional scientific discussion and contribute, as a standpoint, to further conceptual modeling and empirical analysis.

The research also has methodological and empirical limitations. First, in methodological terms, the validity and reliability of CFA outcomes within the selected sample of 157 respondents are questionable. It contradicts the Hoelter's critical N (C.N.) recommendation, where the acceptable size of the sample (N) should be above 200 to accept a model at the 0.05 level [65,66]. Nevertheless, authors follow the methodological position by Guadagnoli and Velicer (1988) [62], claiming that a sample's size, as a function of the number of variables, is not an important factor to determine the sample pattern as being a stable and approximate population pattern. It is outlined that a research sample size between 100 and 200 observations is valid, and under this sample, "a correlation coefficient becomes an adequate estimator of the population correlation coefficient" [62]. Additionally, the value of the KMO of 0.839 confirms the adequacy of sampling. Additionally, the scope of the research instrument and assessment scale, by including more than 10 items, is considered to bring some negative effects on the reliability of measurement results and affect the perception of respondents [67]. Otherwise, it is confirmed that if the index of the Cronbach's alpha is equal to 0.7 or above, the research instrument of 10 or more items is treated as sufficient [61]. Previous research in this field also confirmed that 9-point scale scores correlated better than the 5-point scale, and, overall, an increase in the reliability level was noticed due to the transition to a higher number scale [67]. The main empirical limitation of the research can be interpreted as a future research discourse within this topic. Application of the traditional fuzzy analytic hierarchy process (FAHP), the modified version of extended fuzzy analytic hierarchy process (E-FAHP), or a combination of different multi-criteria decision-making (MCDM) techniques can reduce the vagueness inherent in a stand-alone linguistic assessment and an uncertain comparison of opinions [68]. Accordingly, future scientific discussions and empirical investigations should be focused on using this set of methodological techniques in continuous scientific investigations. A more extensive and overall continuous validation of the authors' suggested influential driving factors within a higher sample of external customers in the Baltic insurance market is required. Additionally, a simultaneous theoretical analysis within the presented topic, by adding the foundation of organization buyer behavior (OBB) and/or technology acceptance models (TAM), should be completed by the authors.

6. Conclusions

The theoretical analysis of scientific literature and recent practical trends have revealed that the global insurance market and insurers are in a multidimensional transition period, including an intensive digital acceleration (mostly due to COVID-19), dynamic changes in the market structure, and a spread of the new hybrid operational business model. The digital environment, big data analytics, and technological and cognitive-emotional factors have become major drivers of the modern insurance concept. Otherwise, it has also been recognized that insurance customer behavior and decision-making processes in digital platforms reflect on the fundamental outcomes of the traditional CDM and hybrid choice models. Parallelly, the following features of the complex and simplified customer behavioral models, expected utility theory and behavioral reasoning theory, could be identified: interactive multi-step logic, the complexity of perceived risks, personal bias, and contextual variables in decision-making, influence of the marketing domain, and feedback-repurchase loop. The conceptual modeling confirmed that, within digital platforms, the behavior and purchase decision making of insurance customers is a continuous, but not a simultaneous, sequence of three-stage processes. Additionally, the statistical factor analysis confirmed that the combination of three internal and external factor groups influence the insurance purchase–decision-making in Baltics digital insurance platforms. Here, a strong, positive Pearson correlation was identified among six internal factors of personal evaluation and considerations, four external factors of technological features and marketing domains, and four combined factors of insurance knowledge, levels of product customization, and service personalization. Otherwise, the analysis of sociodemographic factors, such as age group, gender, and country, showed that there are no statistically significant relationships among these types of factors and evaluations of digitalization, personalization, and customization levels within the Baltic non-life insurance market and digital insurance platforms.

Author Contributions: Conceptualization, G.B.; methodology, G.B. and A.G.R.; software, G.B. and A.G.R.; validation, G.B. and A.G.R.; formal analysis, G.B.; investigation, G.B.; resources, G.B.; data curation, G.B.; writing—original draft preparation, G.B.; writing—review and editing, G.B. and A.G.R.; visualization, G.B.; supervision, A.G.R. Both authors have read and agreed to the published version of the manuscript.

Funding: This research received no external funding.

Institutional Review Board Statement: Not applicable.

Informed Consent Statement: Informed consent was obtained from all subjects involved in the study.

Data Availability Statement: The data used in this research work are available from the corresponding authors upon request.

Acknowledgments: Technical software IBM SPSS Statistics 26 (Armonk, NY: IBM Corp.) support was provided by a colleague Rūta Juozaitienė.

Conflicts of Interest: The authors declare no conflict of interest.

References

1. Pousttchi, K.; Gleiss, A. Surrounded by middlemen—How multi-sided platforms change the insurance industry. *Electron. Mark.* **2019**, *29*, 609–629. [CrossRef]
2. Porrini, D. The effects of innovation on market competition: The case of the insurance comparison websites. *Mark. Manag. Innov.* **2018**, *3*, 324–332. [CrossRef]
3. Wiesböck, F.; Matt, C.; Hess, T.; Li, L.; Richter, A. How management in the German insurance industry can handle digital transformation. *Manag. Rep.* **2017**, *1*, 1–26.
4. Warg, M.; Zolnowski, A.; Frosch, M.; Weiß, P. From product organization to platform organization—Observations of organizational development in the insurance industry. In Proceedings of the Conference of the 10 Years Naples Forum on Service, Ischia, Italy, 4–7 June 2019; pp. 2–16.
5. Łyskawa, K.; Kędra, A.; Klapkiv, L.; Klapkiv, J. Digitalization in insurance companies. In Proceedings of the Conference Contemporary Issues in Business, Management and Economics Engineering, Vilnius, Lithuania, 9–10 May 2019; pp. 842–852.
6. Zarina, I.; Voronova, I.; Pettere, G. Internal model for insurers: Possibilities and issues. In Proceedings of the Conference Contemporary Issues in Business, Management and Economics Engineering, Vilnius, Lithuania, 9–10 May 2019; pp. 255–265.
7. Hsee, C.K.; Kunreuther, H.C. The affection effect in insurance decisions. *J. Risk Uncertain.* **2000**, *20*, 141–159. [CrossRef]
8. Kunreuther, H.; Pauly, M.V. Insurance decision-making and market behavior. *Found. Trends Microecon.* **2006**, *1*, 63–127. [CrossRef]

9. Kunreuther, H.; Pauly, M. Behavioral economics and insurance: Principles and solutions. In *Research Handbook on the Economics of Insurance Law*; Schwarcz, D., Siegelman, P., Eds.; Edward Elgar Publishing Ltd.: Cheltenham, UK, 2015.

10. Suter, J.; Duke, C.; Harms, A.; Joshi, A.; Rzepecka, J.; Lechardoy, L.; Hausemer, P.; Wilhelm, C.; Dekeulenaer, F.; Lucica, E. Study on consumers' decision making in insurance services: A behavioural economics perspective. In *Final Report Prepared by London Economics, Ipsos and VVA Europe*; Publications Office of the European Union: Luxembourg, 2017.

11. Kiyak, D.; Pranckevičiūtė, L. Causal survey of purchase of non-life insurance products for Lithuanian consumers. *Reg. Form. Dev. Stud.* **2014**, *3*, 112–122. [CrossRef]

12. Ulbinaitė, A.; Kučinskienė, M.; Moullec, Y.L. Conceptualising and simulating insurance consumer behaviour: An agent-based-model approach. *Int. J. Modeling Optim.* **2011**, *1*, 250–257. [CrossRef]

13. Ulbinaitė, A.; Moullec, Y.L. Towards an ABM-based framework for investigating consumer behaviour in the insurance industry. *Ekonomika* **2010**, *89*, 250–257. [CrossRef]

14. Ulbinaitė, A.; Kučinskienė, M. Insurance service purchase decision-making rationale: Expert-based evidence from Lithuania. *Ekonomika* **2013**, *92*, 137–155. [CrossRef]

15. Ulbinaitė, A.; Moullec, Y.L. Determinants of insurance purchase decision making in Lithuania. *Eng. Econ.* **2013**, *24*, 144–159. [CrossRef]

16. Zolnowski, A.; Warg, M. Let's get digital: Digitizing the insurance business with service platforms. *Cut. Bus. Technol. J.* **2017**, *30*, 3–8.

17. Allodi, E.; Cervellati, E.M.; Stella, G.P. A new proposal to define insurance literacy: Paving the path ahead. *Risk Gov. Control Financ. Mark. Inst.* **2020**, *10*, 22–32.

18. Weedige, S.S.; Ouyang, H. Consumers' insurance literacy: Literature review, conceptual definition, and approach for a measurement instrument. *Eur. J. Bus. Manag.* **2019**, *11*, 49–65.

19. Åkesson, M.; Edvardsson, B.; Tronvoll, B. Customer experience from a self-service system perspective. *J. Serv. Manag.* **2014**, *25*, 677–698. [CrossRef]

20. Rosebaum, M.S.; Otalora, M.L.; Ramírez, G.C. How to create a realistic customer journey map. *Bus. Horiz.* **2017**, *60*, 143–150. [CrossRef]

21. Ulaga, W.; Eggert, A. Value-based differentiation in business relationships: Gaining and sustaining key supplier status. *J. Mark.* **2006**, *70*, 119–136. [CrossRef]

22. Taherdoost, H. A review of technology acceptance and adoption models and theories. *Procedia Manuf.* **2018**, *22*, 960–967. [CrossRef]

23. Monami, A.M. The unified theory of acceptance and use of technology: A new approach in technology acceptance. *Int. J. Sociotechnol. Knowl. Dev.* **2020**, *12*, 79–98.

24. Klauss, P.; Edvardsson, B.; Maklan, S. Developing a typology of customer experience management practice—From preservers to vanguards. In Proceedings of the 12th International Research Conference in Service Management, La Londe les Maures, France, 29 May–1 June 2012.

25. Baek, J.; Lee, J.A. Conceptual framework on reconceptualizing customer share of wallet (SOW): As a perspective of dynamic process in the hospitality consumption context. *Sustainability* **2021**, *13*, 1423. [CrossRef]

26. Tueanrat, Y.; Papagiannidis, S.; Alamanos, E. Going on a journey: A review of the customer journey literature. *J. Bus. Res.* **2021**, *125*, 336–353. [CrossRef]

27. Ponsignon, F.; Smart, A.P.; Phillips, L. A customer journey perspective on service delivery system design: Insights from healthcare. *Int. J. Qual. Reliab. Manag.* **2018**, *35*, 1–16. [CrossRef]

28. Germanakos, P.; Tsianos, N.; Lekkas, Z.; Mourlas, C.; Belk, M.; Samaras, G. intelligent authoring tools for enhancing mass customization of e-services—The smarTag framework. In Proceedings of the 11th International Conference on Enterprise Information Systems—Human-Computer Interaction, Milano, Italy, 6–10 May 2009; pp. 91–96.

29. Zopounidis, C.; Doumpos, M. Multi-criteria decision aid in financial decision making: Methodologies and literature review. *J. Multi-Criteria Decis. Anal.* **2002**, *11*, 167–186. [CrossRef]

30. Milner, T.; Rosenstreich, D. A review of consumer decision-making models and development of a new model for financial services. *J. Financ. Serv. Mark.* **2013**, *18*, 106–120. [CrossRef]

31. Sahu, A.K.; Padhy, P.; Dhir, A. Envisioning the future of behavioral decision-making: A systematic literature review of behavioral reasoning theory. *Australas. Mark. J.* **2020**, *28*, 145–159. [CrossRef]

32. Vij, A.; Walker, J.L. Hybrid choice models: The identification problem. In *Handbook of Choice Modelling*; Hess, S., Daly, A., Eds.; Edward Elgar Pub: Cheltenham, UK; Northamptom, MA, USA, 2014; pp. 519–564.

33. Goodhope, O.O. Major classic consumer buying behaviour models: Implications for marketing decision-making. *J. Econ. Sustain. Dev.* **2013**, *4*, 164–173.

34. Gómez-Díaz, J.A. Reviewing a consumer decision making model in online purchasing: An ex-post-fact study with a Colombian sample. *Av. En Psicol. Latinoam.* **2016**, *34*, 273–292. [CrossRef]

35. Ragothaman, N.; Shanmugam, V. Impact of market drivers on consumers purchase decision with reference to steel products in tamilnadu. *Int. J. Appl. Bus. Econ. Res.* **2017**, *15*, 35–40.

36. Holland, J. Navigating Uncertainty: Tourists' Perceptions of Risk in Ocean Cruising. Ph.D. Thesis, University of Brighton, Brighton, UK, 2019.

37. Nicosia, F.M. *Consumer Decision Processes: Marketing and Advertising Implications*; Prentice-Hall: Englewood Cliffs, NJ, USA, 1966.
38. Engel, J.F.; Kollat, D.T.; Blackwell, R. *Consumer Behavior*; Holt Rinehart and Winston: New York, NY, USA, 1968.
39. Howard, J.A.; Sheth, J.N. *The Theory of Buyer Behavior*; John, Wiley and Sons: Hoboken, NJ, USA, 1969.
40. Um, S.; Crompton, J.L. Attitude determinants in tourism destination choice. *Ann. Tour. Res.* **1990**, *17*, 432–448. [CrossRef]
41. Kotler, P. *Marketing Management: Analysis, Planning, Implementation, and Control*, 9th ed.; Prentice Hall: Upper Saddle River, NJ, USA, 1997.
42. McCarthy, J.E.; Perreault, W.D.; Quester, P.G. *Basic Marketing: A Managerial Approach*, 1st Australasian ed.; Irwin: Sydney, Australia, 1993.
43. Walker, J.; Ben-Akiva, M. Generalized random utility model. *Math. Soc. Sci.* **2002**, *43*, 303–343. [CrossRef]
44. Kotler, P.; Keller, K.L. *Marketing Management*, 12th ed.; Prentice Hall: Upper Saddle River, NJ, USA, 2006.
45. Kotler, P.; Keller, K.L. *Marketing Management*, 14th ed.; Prentice Hall: Upper Saddle River, NJ, USA, 2012.
46. Muzondo, N. Modelling consumer behaviour conceptually through the seven Ps of marketing: A revised theoretical generic consumer stimulus-response mode. *Univ. Zimb. Bus. Rev.* **2016**, *4*, 89–107.
47. Robinson, S. Conceptual modelling for simulation part II: A framework for conceptual modelling. *J. Oper. Res. Soc.* **2008**, *59*, 291–304. [CrossRef]
48. Robinson, S. Conceptual modelling for simulation part I: Definition and requirements. *J. Oper. Res. Soc.* **2008**, *59*, 278–290. [CrossRef]
49. Robinson, S. A tutorial on conceptual modeling for simulation. In Proceedings of the 2015 Winter Simulation Conference, Huntington Beach, CA, USA, 6–9 December 2015; pp. 1820–1834.
50. Edmonds, W.A.; Kennedy, T.D. Convergent-parallel approach. In *An Applied Guide to Research Designs: Quantitative, Qualitative, and Mixed Methods*, 2nd ed.; Edmonds, W.A., Kennedy, T.D., Eds.; SAGE Publications, Inc.: Los Angeles, CA, USA, 2017; pp. 181–188.
51. Razali, F.M.; Aziz, N.A.A.; Rasli, M.R.; Zulkefly, N.F.; Salim, S.A. Using convergent parallel design mixed method to assess the usage of multi-touch hand gestures towards fine motor skills among pre-school children. *Int. J. Acad. Res. Bus. Soc. Sci.* **2019**, *9*, 153–166.
52. Suárez-Álvarez, J.; Pedrosa, I.; Lozano, L.; García-Cueto, E.; Cuesta, M.; Muñiz, J. Using reversed items in Likert scales: A questionable practice. *Psicothema* **2018**, *30*, 149–158. [PubMed]
53. Musani, S.; Jemain, A. A fuzzy MCDM approach for evaluating school performance based on linguistic information. *AIP Conf. Proc.* **2013**, *1571*, 1006–1012.
54. Quirós, P.; Alonso, J.M.; Pancho, D.P. Descriptive and comparative analysis of human perceptions expressed through Fuzzy rating scale-based questionnaires. *Int. J. Comput. Intell. Syst.* **2016**, *9*, 450–467. [CrossRef]
55. Peculis, R.; Rogers, D.; Campbell, P. Task model of software intensive acquisitions: Integrated tactical avionics system for a maritime helicopter case study. In Proceedings of the Twelfth Australian Aeronautical Conference, Melbourne, Australia, 17–19 October 2007; pp. 510–522.
56. Goepel, K. Comparison of judgment scales of the analytical hierarchy process—A new approach. *Int. J. Inf. Technol. Dec. Making* **2019**, *18*, 445–463. [CrossRef]
57. Taherdoost, H. Sampling methods in research methodology: How to choose a sampling technique for research. *Int. J. Acad. Res. Manag.* **2016**, *5*, 18–27. [CrossRef]
58. Koyuncu, I.; Kılıç, A.F. The use of exploratory and confirmatory factor analyses: A document analysis. *Eğitim Bilim* **2019**, *44*, 361–388. [CrossRef]
59. Dhillon, H.K.; Zain, M.; Zain, A.Z.; Quek, K.F.; Singh, H.; Kaur, G.; Nordin, R. Exploratory and confirmatory factor analyses for testing validity and reliability of the Malay language Questionnaire for Urinary Incontinence Diagnosis (QUID). *Open J. Prev. Med.* **2014**, *4*, 844–851. [CrossRef]
60. Mathur, B.; Kaushik, M. Data analysis of students marks with descriptive statistics. *Int. J. Recent Innov. Trends Comput. Commun.* **2014**, *2*, 1188–1191.
61. Chan, L.L.; Idris, N. Validity and reliability of the instrument using exploratory factor analysis and Cronbach's alpha. *Int. J. Acad. Res. Bus. Soc. Sci.* **2017**, *7*, 400–410.
62. Guadagnoli, E.; Velicer, W. Relation of sample size to the stability of component patterns. *Psychol. Bull.* **1988**, *103*, 265–275. [CrossRef] [PubMed]
63. Rocha, A.Q.; Botelho, D. Attitudes towards insurance for personal assets: Antecedents and consequents. *Eur. J. Bus. Soc. Sci.* **2018**, *6*, 62–80.
64. Weedige, S.S.; Ouyang, H.; Gao, Y.; Liu, Y. Decision making in personal insurance: Impact of insurance literacy. *Sustainability* **2019**, *11*, 6795. [CrossRef]
65. Shadfar, S.; Malekmohammadi, I. Application of Structural Equation Modeling (SEM) in restructuring state intervention strategies toward paddy production development. *Int. J. Acad. Res. Bus. Soc. Sci.* **2013**, *3*, 576–618. [CrossRef]
66. Bollen, K.; Liang, J. Some properties of Hoelter's CN. *Sociol. Methods Res.* **1988**, *16*, 492–503. [CrossRef]
67. Tarka, P. Likert scale and change in range of response categories vs. the factors extraction in EFA model. *Acta Univ. Lodz. Folia Oeconomica* **2015**, *1*, 27–35. [CrossRef]
68. Reig-Mullor, J.; Pla-Santamaria, D.; Garcia-Bernabeu, A. Extended fuzzy analytic hierarchy process (E-FAHP): A general approach. *Mathematics* **2020**, *8*, 2014. [CrossRef]

MDPI

St. Alban-Anlage 66

4052 Basel

Switzerland

Tel. +41 61 683 77 34

Fax +41 61 302 89 18

www.mdpi.com

Applied Sciences Editorial Office

E-mail: applsci@mdpi.com

www.mdpi.com/journal/applsci